Progress in Optical Science and Photonics

Volume 39

Indexed by Scopus

The purpose of the series Progress in Optical Science and Photonics is to provide a forum to disseminate the latest research findings in various areas of Optics and its applications. The intended audience are physicists, electrical and electronic engineers, applied mathematicians, biomedical engineers, and advanced graduate students.

Vikas Dubey · Neha Dubey ·
Marta Michalska Domańska · Jagjeet Kaur Saluja
Editors

Upconversion Luminescence of Phosphors

Springer

Editors
Vikas Dubey
Department of Physics
North-Eastern Hill University (NEHU)
Shillong, Meghalaya, India

Neha Dubey
Department of Physics
Government V.Y.T. PG Autonomous College
Durg, Chhattisgarh, India

Marta Michalska Domańska
Military University of Technology
Warsaw, Poland

Jagjeet Kaur Saluja
Department of Physics
Government V.Y.T. PG Autonomous College
Durg, Chhattisgarh, India

ISSN 2363-5096 ISSN 2363-510X (electronic)
Progress in Optical Science and Photonics
ISBN 978-981-95-1761-9 ISBN 978-981-95-1762-6 (eBook)
https://doi.org/10.1007/978-981-95-1762-6

This Springer imprint is published by the registered company Springer Nature Singapore Pte Ltd.
The registered company address is: 152 Beach Road, #21-01/04 Gateway East, Singapore 189721, Singapore

Preface

This book highlights various applications of up conversion phosphors mainly for solar cell upconverters, biomedical application of upconverter phosphors. The eventuality of tuning chemical and physical properties in upconversion phosphors has a strong influence on a number of assorted applications including solid-state lasers, photovoltaic, and bioimaging. To have a fundamental understanding of upconversion luminescence, it is essential to focus on the theoretical elucidation of spectroscopic properties and luminescence dynamics of photon upconversion in rare earth-activated phosphors. The process of upconversion generally originates from specific properties of rare earth ions. Upconversion phenomenon has been validated as a reliable process for producing highly excited species that emits spontaneous or stimulated emission at significantly shorter wavelengths than those of excitation. Auzel and several other researchers have put forward several mechanisms to provide a better understanding of the upconversion phenomenon. According to them, upconversion luminescence consists of intermediate electronic states and the complete process satisfies the energy conversion principle. Upconversion is an appealing process whose optical properties have been sagaciously applied in rationally designed materials. For this purpose, various hosts doped with transition metal ions and actinide doped hosts have been investigated. However, due to poor optical properties, they are still under the arena of basic research. Fortunately, rare earth-doped hosts have managed to exhibit high upconversion efficiencies even at room temperature. It is expected that with upcoming advances in the study of upconversion, nanostructure synthesis, and their modification technology along with a boost in information regarding optimization and enhancement of intricate optical processes in the upconversion phenomenon will provide a potential background and many more applications will be explored.

Chapter "Introduction to up Conversion Phosphor" describes the mechanism of up conversion is well known since 1960's but in past two decades a wide variety of up conversion materials has been investigated by tuning shape and size of materials. Indeed, many researchers paid a huge attention specifically on rare earth-based phosphor materials due to their interesting properties to utilize for real time applications. This chapter is opted to provide a general introduction of up conversion phosphor materials dealing with different up conversion mechanisms such as excited-state absorption, energy transfer, ground state absorption, and photo avalanche. Moreover, a brief overview also discussed fruitfully on charming applications of up conversion phosphors such as bioimaging, photodynamic therapy, latent fingerprint detection, temperature sensor, and security links.

Chapter "Basic Principle and Mechanism of Upconversion Luminescence in Phosphor Materials" represents investigated with the rapid development of nanotechnology. The upconversion materials can efficiently convert near-infrared light into visible or ultra-visible luminescence using a stepwise multiphoton process which make these materials ideal candidates for various applications such as lasers, solar cells, sensors, security inks, and optical imaging.

Upconversion can take place in both organic and inorganic materials through a number of different processes. It usually takes place in polycyclic aromatic hydrocarbons (PAHs) in organic matter and in d or f-block elements in inorganic materials. Nowadays lanthanide-doped upconversion materials are proved to be more promising owing to wide range applications in solid-state lighting and solar cells. Upconversion materials can efficiently convert sub-band gap photons into above band gap-light in solar cells has led to vast research in upconversion materials and its applications in the energy field. The efficiency of upconversion materials can be elevated through the optimization of the following factors: added impurity, absorption, energy transfer and emission. The appropriate knowledge of basic principles and mechanism involved in upconversion can lead to novel approaches in enhancing the upconversion luminescence. The proposed chapter thoroughly discusses and focused on the mechanism of upconversion and its differentiation from analogous processes such as simple harmonic generation and two-photon absorption. Various processes possible for upconversion in materials such as energy transfer upconversion (ETU), excited-state absorption (ESA), and photon avalanche (PA) are explained with necessary conditions in detail. This chapter provide the readers an insight to realize how emission energy is high than excitation energy while following the law of conservation of energy. In addition, the chapter elaborates the structure and function of upconversion materials. The efficiency of the upconversion process, fundamental physics involved, and atomic level details behind the upconversion process is also discussed. The reported research works in the field over the years is reviewed while simultaneously bringing to the forefront the future scope of the field. The chapter finally summarizes with the limitations, challenges in improving the upconversion luminescence in materials and possible ways to overcome.

Chapter "A Review Report on Rare Earth Activated Phosphors for Bioimaging Applications" presents an overview of various synthesis and characterization techniques used for rare earth-activated phosphors intended for bioimaging applications. Here, various researchers developed the rare earth-activated phosphors using a variety of synthesis techniques, including solid state, sol-gel, hydrothermal, combustion, etc. As reported in their studies, the resulting phosphors have been compared by using various characterization techniques for its structural analysis, morphological analysis, optical analysis, to assess their potential in bioimaging applications. Previously some authors have highlighted the importance of bioimaging phosphor to improve the quality of medical optical imaging and its low-cost effect. We can stimulate the biological window for tissue imaging experiments using rare earth-activated phosphors. Herein, some important wavelength ranges relevant for tissue imaging fall within the biological optical window of 650–950 nm, where reduced absorption and scattering enable deeper light penetration into biological tissues. Current phosphors have a significant drawback that they are only ideal for one time use because their persistent brightness fades after a few hours. To address this problem, persistent phosphors with photo-stimulated properties are used. In this article, an in-depth assessment has been provided that surely will be a good resource for researchers in exploring various fields related to biomedical imaging.

Chapter "Synthesis of Upconversion Phosphors" is to present a thorough discussion of the synthesis methods used for the development of upconversion phosphors. Upconversion phosphors are extensively utilized in a variety of optoelectronic devices, displays, bio-sensors, scintillators, and cathode-ray tubes. Upconversion phosphors can be developed into micro-sized particles, nanoparticles, colloidal nanoparticles, and thin films. Different types of physical and chemical methods are followed to synthesize the nanomaterials and thin films. These phosphors can be synthesized through numerous methods, including solid-state, solvothermal, combustion, co-precipitation, sol-gel, electrospinning, and laser ablation. The chosen synthesis route significantly impacts the luminescent properties of the materials. Therefore, understanding the synthesis methods and their effects is crucial for developing phosphors for practical use. This chapter is structured into three main sections. The first section discusses the impact of different factors on the synthesis processes in the top-down approach. The second section introduces various synthesis methods of the bottom-up approach and emphasizes the precursors, solvents, temperatures, and other synthesis conditions. The final section focuses on the various techniques used to develop thin films of upconversion materials and explores future advancements in synthesis techniques.

Chapter "Spectroscopy Techniques for Upconversion Phosphors" shows upconversion phosphors, typically doped with rare earth ions are capable of converting multiple lower energy photons into a single higher-energy photon. Owing to their photostability, sharp emission bands, anti-Stokes emission, and long luminescence lifetime, makes them promising candidates for applications in bioimaging, optoelectronics, solid-state lighting, and anticounterfeiting measures. However, in order to optimize these materials for such applications requires a detailed understanding of structure-property relationships. This chapter provides a comprehensive review on the spectroscopy techniques aimed at characterizing distinct properties of the upconversion phosphors. The discussion is structured around six characterization goals which includes (a) analysis of electronic structure using PL and UV-Visible, (b) investigation of surface ligands bound to the upconversion phosphors using IR and NMR, (c) examination of vibrational modes using FTIR and Raman, (d) determination of elemental composition using XPS, EDXS, and EELS (e) evaluation of local atomic environment around dopant ions using XAS, and (f) assessment of luminescence lifetime dynamics using TRS. By correlating spectroscopic data with optical properties, this chapter highlights the role of spectroscopy as a vital role in the rational design and performance enhancement of upconversion phosphors.

Chapter "Upconversion Phosphors for Forensic Applications" shows the public's knowledge and interest in this field of criminal investigation have grown in recent years thanks to the forensic science-focused television shows, online series, films and dramas as well as the extensive coverage of several criminal trials. Although not brand-new, the field of using upconversion (UC) for forensic applications is still developing. UC luminescence is an optical phenomenon that arises from anti-Stokes luminescence and involves the absorption of two or more low-energy photons to produce high-energy photons. It has attracted a lot of attention. Compared to organic dyes and quantum dots, UC nanomaterials are preferable due to their unique properties. They are promising luminophores for forensic-based applications. This chapter provides a thorough introduction to lanthanide-doped UC phosphors, covering the basic idea, the synthesis and design of desired properties, a variety of applications in forensic-based applications, applied techniques and applied phosphors, reported research works, the benefits and drawbacks of current methods and anticipated future results.

Chapter "A Comprehensive Study on the Thermoluminescence Dosimetry of Phosphors" shows detailed exploration of recent advancements in thermoluminescence dosimetry (TLD), with a particular focus on inorganic phosphor materials. Thermoluminescence, the light emitted by a material when heated after exposure to ionizing radiation, plays a vital role in measuring radiation doses. In this study, a wide array of phosphors—spanning chemical families such as aluminates, borates, fluorides, phosphates, silicates, sulfides, and titanates—are examined in terms of their synthesis techniques, structural characteristics, and suitability for radiation detection. Special attention is given to essential dosimetric attributes, including glow curve profiles, dose-response behavior, sensitivity, fading resistance, and biological tissue equivalence. Additionally, the article sheds light on the applications of these materials across various domains, such as medical, environmental, space, and personal radiation monitoring. By compiling insights from numerous studies, this review serves as a valuable resource for researchers and professionals seeking to optimize the performance of TLD materials for specific applications.

Chapter "Thermoluminescence in Upconversion Materials: An Application to TL Dosimetry" represents defect levels in upconversion phosphors play a significant role in its applicability as a suitable luminescent material. Thermoluminescence (TL) is a defect related phenomenon and considered as an inevitable tool to understand the nature and distribution of traps/defects in the crystal lattice. The present chapter reviews the physics of the thermoluminescence related to dosimetric applications in upconversion materials. Enormous research has been carried out to comprehend the fundamental mechanism and improve the materials characteristics as well as to evolve advance and novel thermoluminescence materials. Nowadays, thermoluminescence dosimetry (TLD) is a well-known dosimetric technique with wide range applications in personnel, environmental and clinical dosimetry. TLD is based on the materials which are previously exposed to ionizing radiations, emit visible radiations, when they are heated. The impurities in the upconversion TL materials encourage localized energy levels within the forbidden energy band gap which plays a vital role in TL process. By detecting the presence and nature of these defect levels, the sensitivity of

TL is unrivalled. This chapter focuses on the investigation into the relation between the luminescence and defects involved in the process. Moreover, various models to understand the TL properties in the upconversion materials through energy band theory are discussed in detail. Different type of radiation exposure employed in TL studies display a remarkable impact on the trap distribution resulting in shallow and deep traps. Various techniques to determine different trapping parameters such as activation energy, order of kinetics and frequency factor are discussed in detail which provides a comparative framework to interpret many observations. The knowledge of kinetic parameters namely order of kinetics equips one to understand recombination process and trapping profile, comprehensively. The present chapters also discuss the behavior of TL glow peaks on the basis of order of kinetics which indicates the trapping/retrapping of charge carriers resulting to TL emissions in the phosphors.

Chapter "Upconversion Phosphors for Targeted Drug Delivery" focuses on lanthanide-based upconversion phosphors that have drawn significant interest in the field of material science and biomedicine for targeted drug delivery. The unique optical and chemical properties of lanthanide ions for having at least one electron in the 4f orbital make it suitable for upconversion nanoparticles (UCNPs) possess distinctive attributes, including autofluorescence, a high signal-to-noise ratio, minimal photodamage to living organisms, and deep penetration in biological samples, which render them highly suitable for drug delivery applications. UCNPs appear to be a suitable drug-making technique that can be used to treat cancer with nearly minimal side effects, among other applications. Upconversion materials play a critical part in sensing applications. Utilizing these nanoparticles allows for the conversion of longer-wavelength photons into shorter-wavelength fluorescence, enabling the detection of temperature, pH, ions, and trace biomolecules. Delivering the drug to the tumor site stands out as a primary issue, as the traditional approach of drug distribution in the biological context leads to various side effects in individuals. Researchers from many fields have been interested in photodynamic treatment because of its capacity to activate the manufactured medicine during the upconversion process using light. Hence, there is a growing prominence of Ln3+-based upconversion materials, indicating their potential as groundbreaking options for new biomedical applications. This article review specifically examines the chemical makeup and development of upconversion nanoparticles, as well as the methods for creating targeted drug delivery systems. Additionally, the article delves into the application of these particles in different cancer treatment approaches, including therapies such as photodynamic therapy (PDT).

Chapter "Future Prospects of Up-Conversion Materials: Advancements, Applications, and Challenges" shows, over the last few decades, a significantly important category of nanoparticles made of 'upconversion' materials has emerged as a promising candidate for various scientific applications. Upconversion phosphors (UCPs) are well known for their ability to convert multiple longer-wavelength photons to shorter-wavelength fluorescence. Because of their relatively bright and narrow-band fluorescence, they have been useful and utilized in a variety of applications. The current book chapter explores the potential applications of upconversion materials in commercial energy-efficient lighting, safe bioimaging, high-efficiency solar cells, and high-density data storage applications. Recent advances highlight the capacity of UCPs to revolutionize technologies and contribute to sustainable energy solutions. UCPs provide a distinct advantage in lighting applications by converting low-energy light into high-energy emissions. This process can be used to improve the overall efficiency and light yield of lighting systems, thereby making them more energy-efficient and environmentally friendly. Furthermore, the versatility of UCPs enables the generation of specific colors by controlling the material composition. This capability expands the possibilities for dynamic and customizable lighting solutions. Furthermore, the chapter investigates the use of UCPs in emerging technologies like quantum dots and nanophotonics, which broadens the possibilities for advanced lighting systems. It investigates how combining upconversion processes with these cutting-edge technologies can result in novel solutions for next-generation displays and illumination. These phosphors have distinct properties in bioimaging that make them useful tools for improving imaging resolution and sensitivity. The chapter explores recent developments, including how UCPs can be tailored for specific biological markers, allowing for more precise and efficient imaging techniques in medical diagnostics and research. Furthermore, the incorporation of UCPs into solar cells represents an exciting opportunity for advancements in renewable energy. These materials' significant ability to convert low-energy sunlight into higher-energy photons demonstrates their potential to improve solar energy conversion efficiency. The chapter investigates strategies for optimizing upconversion processes within solar cells, addresses challenges, and proposes future research directions to maximize the impact of this technology on sustainable energy solutions. Data storage is another area where UCPs have exciting applications. Their optical properties could transform data encoding and retrieval methods. The chapter discusses recent experiments and theoretical frameworks that investigate the viability of using UCPs for high-density data storage and pave the way for novel approaches to information technology. Despite the optimistic outlook, challenges remain, and the chapter critically examines current limitations and potential barriers. Addressing efficiency, cost-effectiveness, and scalability issues, the discussion seeks to guide future research efforts and foster collaboration within the scientific community.

Shillong, India — Vikas Dubey
Durg, India — Neha Dubey
Warsaw, Poland — Marta Michalska Domańska
Durg, India — Jagjeet Kaur Saluja

Contents

Introduction to up Conversion Phosphor

Shalu Atri, Vinod Bala Taxak, and Sachin Pal

Abstract The mechanism of upconversion is well known since 1960s but in past two decades, a wide variety of upconversion materials has been investigated by tuning shape and size of materials. Indeed, many researchers paid a huge attention specifically on rare earth-based phosphor materials due to their interesting properties to utilize for real time applications. This chapter is opted to provide a general introduction of up conversion phosphor materials dealing with different upconversion mechanisms such as excited state absorption, energy transfer, ground state absorption, and photo avalanche. Moreover, a brief overview also discussed fruitfully on charming applications of up conversion phosphors such as bio-imaging, photodynamic therapy, latent fingerprint detection, temperature sensor and security links.

Keywords Down and Up Conversion · Electronic Transitions · Energy Transfer · Laser · Diode

1 Introduction

In 1966, the topic gained more attention when François Auzel reported the first photon upconversion in Yb^{3+}-Er^{3+} doped glasses for lasers that produced green emission under infrared excitation [1]. After some time, blue emission was produced by infrared excitation with coupled ions Yb^{3+}-Tm^{3+} [1]. This phenomenon was understood to be a sequential summation of two or three photons by energy transfer. Similar experiments were carried out independently by Ovsyankin and Feofilov to utilize the

S. Atri
Department of Inorganic Chemistry, Faculty of Natural Sciences, Comenius University Bratislava, lkovicova 6, Mlynska Dolina, SK-842 15 Bratislava, Slovakia

V. B. Taxak
Department of Chemistry, Maharshi Dayanand University, Rohtak 124001, Haryana, India

S. Pal (✉)
Department of Chemistry, Netaji Subhas University of Technology, Dwarka 110078, Delhi, India
e-mail: sachinranga555@gmail.com; sachin.pal@nsut.ac.in

V. Dubey et al. (eds.), *Upconversion Luminescence of Phosphors*, Progress in Optical Science and Photonics 39, https://doi.org/10.1007/978-981-95-1762-6_1

ion coupling Yb^{3+}-Tm^{3+}, Yb^{3+}-Er^{3+}, and Yb^{3+}-Ho^{3+}, but they suggested an alternative interpretation involving a two-ion cooperative process [2]. These investigations gained a boost in 1969 when the General Electric Co. used a GaAs: Si light-emitting diode as a pump source to produce a visible light source for practical applications [3, 4].

1.1 Energy Absorption and Emission

The fluorescence and phosphorescence process begins with photon absorption ($E_1 = h\nu_1$), exciting the activator from its ground state to an excited state (Fig. 1a). The activator then returns to the ground state by emitting a lower-energy photon ($E_2 = h\nu_2$) (Fig. 1b), with energy loss ($\Delta E = \Delta E_1 + \Delta E_2$) occurring through non-radiative transitions as heat or phonons (Fig. 1c) [5].

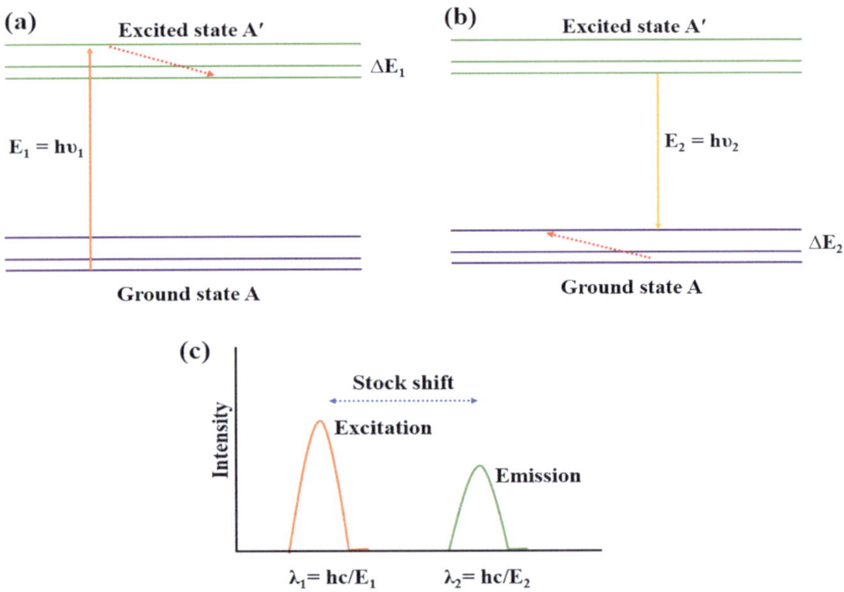

Fig. 1 Schematic representation of energy transfer in phosphors: **a** absorption by an activator, **b** luminescent photon emission and **c** the stokes shift between wavelength of the excitation and emission pulses

1.2 Luminescence and Phosphors

Luminescence, a term introduced by Eilhardt Wiedemann, refers to light emission at low temperatures [6]. It is classified into fluorescence (emission ceases when excitation stops) and phosphorescence (emission persists after excitation). Photoluminescence, a type triggered by photon absorption, typically follows Stokes' law and involves spontaneous emission due to electronic excitation or internal vibrations. Stokes luminescence materials include most organic dyes [7, 8], fluorescent proteins [9], organically modified silica [10, 11], semiconductor quantum dots [12], and metal complexes [13, 14]. The photoluminescence mechanism in these materials can be explained using the Jablonski energy diagram [15]. The luminescence is now classified into a variety of categories based on the nature of the exciting source (Table 1).

The term "phosphor," derived from phosphorus, refers to materials that exhibit fluorescence or phosphorescence. The recent research has focused on developing phosphors using various host materials doped with lanthanide and/or transition metal ions [16].

Table 1 Types of luminescence depending upon the energetic sources

Type	Definition	Source of energy
Fluorescence	Electronic decay between allowed transitions	UV and visible photons
Phosphorescence	Electronic decay between forbidden transitions	Ultraviolet and visible photons
Bioluminescence	Luminescence in a living organism	Gibbs energy of chemical reactions
Cathodoluminescence	Luminescence caused electron bombardment (cathode 'rays')	Electron kinetic energy
Chemiluminescence	Luminescence during a chemical reaction	Gibbs energy of chemical reaction
Electroluminescence	Luminescence resulting from the application of an electric field	Electrical potential energy
Photoluminescence	Luminescence after irradiation by visible or ultraviolet light	Ultraviolet and visible photons
Radioluminescence	Luminescence because of radioactivity	Energetic particles and γ-rays
Thermoluminescence	Luminescence following an increase in temperature	Thermal energy

1.3 Stoke Luminescence and Anti-stoke Luminescence

Stoke Luminescence

The emission spectrum of most luminescent materials always lies at a longer wavelength than the excitation spectrum. Stokes-type emission includes the down-conversion and quantum cutting of emission. The mechanisms of down-conversion and quantum cutting for the conversion of high-energy photons into low-energy photons will be discussed in Chapter "Basic Principle and Mechanism of Upconversion Luminescence in Phosphor Materials". Sommerdjik et al. were the first to detect the Quantum cutting in YF_3: Pr^{3+}, where one UV photon was transformed into two lower-energy visible photons [17, 18]. A 140% quantum yield for 185 nm excitation was obtained at room temperature. This phenomenon has also been investigated for other lanthanide ions, such as Tm^{3+} doped LaF_3 [19]. Trupke et al. proposed a detailed technique utilizing quantum-cutting materials as a down-converter to improve the photoelectric conversion efficiency of solar cells [20].

Anti-stoke Luminescence

Upconversion, a type of anti-Stokes emission, involves converting low-energy infrared photons into higher energy visible light, with applications in infrared detection and white LED lighting. Unlike nonlinear processes like SHG or THG, upconversion relies on luminescence (Fig. 2). The majority of the research on upconversion has focused on the behaviour of lanthanoid ions, particularly Er^{3+}, Tm^{3+}, Ho^{3+}, and Yb^{3+}. These ions have been utilized in a variety of host structures, including binary oxides Y_2O_3, Gd_2O_3, and ZrO_2, perovskite structures ABO_3 (A = Sr, Ba, Pb; B = Ti), fluorides (e.g. $NaYF_4$), and oxyfluoride glasses, which are discussed in Sect. 2.1.

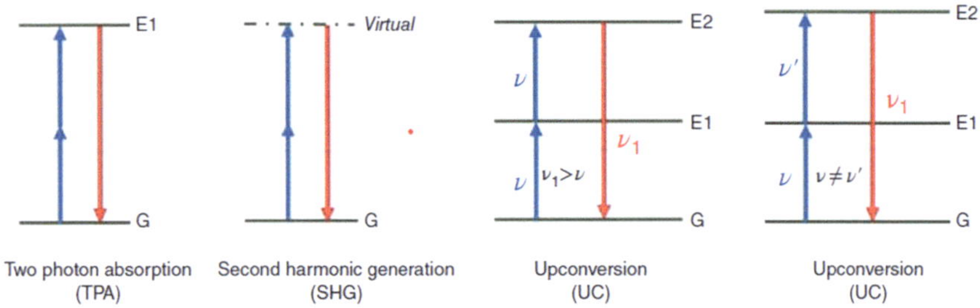

Fig. 2 Basic energy level diagrams depicting typical anti-stokes lines

2 Types of Upconversion Phosphors

A typical up conversion material is made up of two components: an inorganic host matrix and dopant ions. The synthesis of new upconversion materials is dependent on the nature of the host lattice and dopant ions in the matrix. The changing host lattice may substantially alter the radiative and non-radiative (multi-phonon relaxation and energy transfer) properties, resulting in entirely changing the upconversion luminescence behaviour. For instance, shifting from a lattice with high phonon energies (fluoride and oxide) to one with low phonon energies (chloride, bromide, and iodide) could reduce the efficiency of multi-phonon relaxation processes. However, many of these compounds belong to a hydrophilic nature. Selecting a host that has specific optical or magnetic features can also affect the dopant ion's upconversion emission properties by exchange interaction perturbation or sensitization. A modification in the concentration of metal dopant ions has a significant impact on the emission characteristics of an ion. The most obvious impact is an alteration in the colour of the emitted light. The flexibility of choice for the host lattice, dopant, and co-dopant covers a very large spectrum of possibilities in the production of compounds with new and unprecedented upconversion capabilities. The rare earth-containing upconversion phosphors, transition metal-containing upconversion phosphors, and mixed rare earth/transition metal phosphors will be discussed in this section [21].

2.1 *Rare Earth-Containing Upconversion Phosphors*

The active components of well-known upconverting phosphors are trivalent rare earth ions, which are commonly Pr^{3+}, Nd^{3+}, Ho^{3+}, Er^{3+}, Tm^{3+}, or Yb^{3+} [22]. The outermost electronic configurations for the trivalent lanthanide ions are $[Xe]4f^{0-14}$. Their inner 4f shell contains the spectroscopically relevant electronic states. The 4f states are relatively insensitive to the exact nature of the surrounding host lattice (i.e. the specific crystal field and the local site symmetry) because their outer 5s and 6p shells are involved in the bonding. In general, there are two main reasons for lanthanide elements which are especially suitable for utilization in upconversion processes. First, they possess distinct spectroscopically active atomic-like levels that are perfect for energy transfer between discrete levels of the same and different ions, mixing of distinct colour output, and selective colour output. Second, all lanthanides, except La^{3+}, Ce^{3+}, Yb^{3+}, and Lu^{3+}, have lifetimes in the microsecond (μs) to millisecond (ms) range and are metastable, because they are parity-forbidden but partially allowed in optical transitions. These long-lived levels may form transient populations because of their delayed decay, which allows them to function as reservoirs and storage for the energy of the excitation photons. As a result, many of the excited 'f' states have long enough lifetimes for upconversion processes to play a role. In most cases, host matrixes do not exhibit upconversion phosphors at room temperature. But the luminescent centre is produced by the dopants in it.

Lanthanide ions are typically used as a dopant, and a very small amount of it is required. The efficiency of upconversion depends on the interaction between the host and dopant ion. Thus, two crucial components of an effective lanthanide-based upconverter are (1) a highly stable host material acting as a protector, and (2) an efficient dopant ion acting as the luminescent centre. When a dopant ion is excited to a higher energetic state due to the non-radiative transfer of energy from another dopant ion, this is known as sensitized luminescence. The radiation-emitting ion is known as an activator, whereas the energy donor is known as a sensitizer. The excited state absorption (ESA), energy transfer upconversion (ETU), and photon avalanche (PA) are the three primary categories of lanthanide upconversion mechanisms (will be discussed in Chapter "Basic Principle and Mechanism of Upconversion Luminescence in Phosphor Materials") [23, 24]. Upconversion luminescence can be detected in various lanthanide activator-sensitizer combination systems, and most of the crystalline materials may be investigated as host materials.

2.1.1 Activator

Important optical features of Ln^{3+}, like its sharp and narrow f-f transition bands, result from weak electron–phonon coupling, which is caused by the filled $5s^2$ and $5p^6$ subshells shielding the 4f electrons of Ln^{3+}. Furthermore, the f-f transitions are Laporte-forbidden, leading to low transition probabilities and long excited states. Most lanthanide ions exhibit more than one excited 4f energy level except La^{3+}, Ce^{3+}, Yb^{3+}, and Lu^{3+} [25, 26]. However, to produce practically significant upconversion emission, the energy difference between each excited level and its lower-lying intermediate level (ground level) must be small enough to allow for photon absorption and energy transfer stages in upconversion processes. As a result, many lanthanide ions are not appropriate for this kind of excitation due to a lack of sufficient energy levels. Because of their energy levels arranged in a ladder-like fashion, Pr^{3+}, Nd^{3+}, Ho^{3+}, Er^{3+}, and Tm^{3+} are now the most popular activators in upconversion phosphors (Fig. 3a–d) (Table 1) [27]. Er^{3+}, Ho^{3+}, and Tm^{3+}-doped materials generate the most efficient green, red, and blue UC emissions to date. Other Ln^{3+} ions, including Tb^{3+}, Eu^{3+}, and Pr^{3+} have also been shown to be potential activators of UC luminescence. Table 2 provides the most likely electronic transitions and their corresponding emission wavelengths.

Pr^{3+}

Two-photon frequency upconversion has been examined in a fluoride glass matrix with Pr^{3+} as a dopant. One pump wavelength of 1014 nm irradiation promotes GSA up to the metastable $^1G_4$ energy level. This pump cannot be used to excite any further, however, irradiation at the same time with a second suitable pump with 850 nm wavelength can excite the GSA centres via ESA to the $^3P_3$ level. The non-radiative decay of the doubly excited ions to the $^3P_0$ level releases their energy. These eventually decrease to the $^3H_6$ level and generate a red light ($^3P_0 \rightarrow {}^3H_6$; 645 nm),

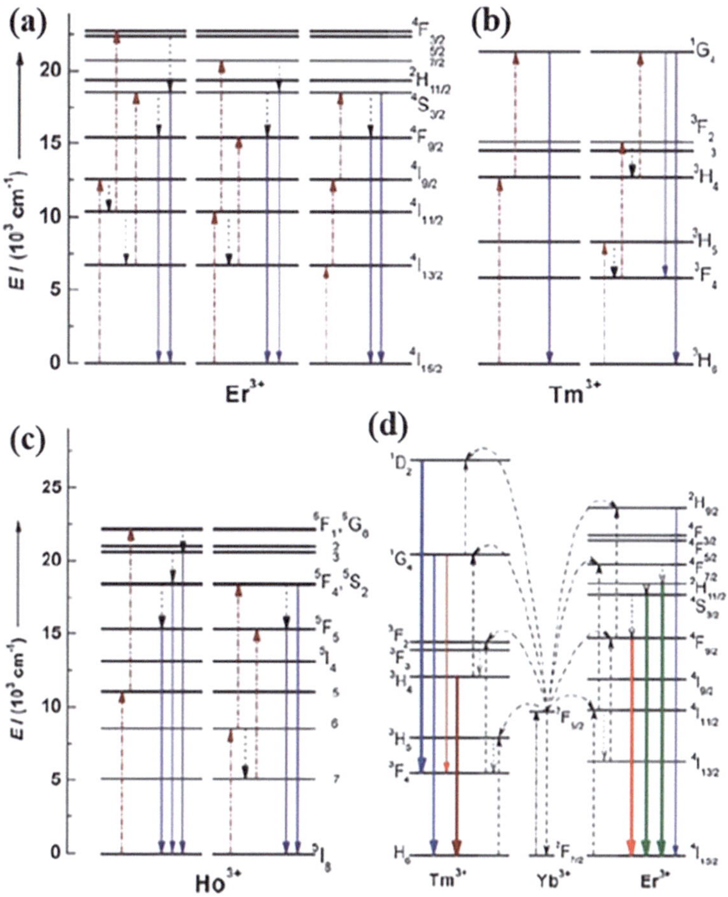

Fig. 3 Energy level diagram for typical upconversion process of **a** Er^{3+}, **b** Tm^{3+}, and **c** Ho^{3+} and **d** proposed energy transfer mechanisms showing the upconversion processes in Er^{3+}, Tm^{3+}, and Yb^{3+}-doped crystals under 980-nm diode laser excitation. Reprinted with permission from Ref. [27]

(Fig. 4a) [28]. The upconversion and visible emission occur only at the intersection of the two beams.

Nd^{3+}

The visible upconversion emissions of Nd^{3+} ions doped in oxyfluoride glass ceramics were recorded at 414 nm ($^2P_{1/2} \rightarrow {}^4I_{9/2}$), 450 nm ($^2P_{1/2} \rightarrow {}^4I_{11/2}$), 523 nm ($^2P_{1/2} \rightarrow {}^4I_{13/2}$), 587 nm ($^2P_{1/2} \rightarrow {}^4I_{15/2}$), and 655 nm ($^4G_{7/2} \rightarrow {}^4I_{13/2}$) under 796 nm laser excitation [29]. Ground state absorption (GSA) excites Nd^{3+} ions from its initial ground state ($^4I_{9/2}$) to the $^4F_{5/2}$, $^2H_{9/2}$ states, and then fast non-radiative relaxation populates the $^4F_{3/2}$ state. Subsequently, another infrared photon is absorbed from the excited state $^4F_{3/2}$ to the $^2P_{1/2}$ state, where it can undergo non-radiative decay to

Table 2 Lanthanides with their typical UC emissions and transitions

Ln^{3+}	Electronic transitions and their emission wavelengths
Pr^{3+}	$^3P_0 \rightarrow {}^3H_4$ (486 nm), $^3P_1 \rightarrow {}^3H_5$ (520 nm), $^3P_0 \rightarrow {}^3H_5$ (540 nm), $^1D_2 \rightarrow {}^3H_4$ (605 nm), $^3P_0 \rightarrow {}^3F_2$ (635 nm), $^3P_0 \rightarrow {}^3H_6$ (645 nm), $^3P_1 \rightarrow {}^3F_3$ (670 nm), $^3P_0 \rightarrow {}^3F_3$ (690 nm), $^3P_0 \rightarrow {}^3F_4$ (720 nm)
Nd^{3+}	$^2P_{1/2} \rightarrow {}^4I_{9/2}$ (430 nm), $^2P_{1/2} \rightarrow {}^4I_{11/2}$ (482 nm), $^2P_{1/2} \rightarrow {}^4I_{13/2}$ (525 nm), $^4G_{7/2} \rightarrow {}^4I_{9/2}$ (535 nm), $^2P_{1/2} \rightarrow {}^4I_{15/2}$ (586 nm), $^4G_{7/2} \rightarrow {}^4I_{11/2}$ (600 nm), $^4G_{7/2} \rightarrow {}^4I_{13/2}$ (665 nm), $^4G_{7/2} \rightarrow {}^4I_{15/2}$ (766 nm)
Sm^{3+}	$^4G_{7/2} \rightarrow {}^6H_{5/2}$ (520 nm), $^4F_{3/2} \rightarrow {}^6H_{5/2}$ (541 nm), $^4G_{5/2} \rightarrow {}^6H_{5/2}$ (555 nm), $^4G_{5/2} \rightarrow {}^6H_{7/2}$ (590 nm), $^4G_{7/2} \rightarrow {}^6H_{9/2}$ (646 nm), $^4G_{5/2} \rightarrow {}^6H_{9/2}$ (657 nm), $^4G_{5/2} \rightarrow {}^6H_{11/2}$ (700 nm), $^6F_{11/2} \rightarrow {}^6H_{5/2}$ (800–875 nm)
Eu^{3+}	$^5D_3 \rightarrow {}^7F_1$ (415 nm), $^5D_3 \rightarrow {}^7F_2$ (429 nm), $^5D_2 \rightarrow {}^7F_2$ (490 nm), $^5D_2 \rightarrow {}^7F_3$ (510 nm), $^5D_1 \rightarrow {}^7F_1$ (535 nm), $^5D_1 \rightarrow {}^7F_2$ (554 nm), $5D_0 \rightarrow {}^7F_1$ (590 nm), $^5D_0 \rightarrow {}^7F_2$ (618 nm)
Tb^{3+}	$^5D_3 \rightarrow {}^7F_6$ (380 nm), $^5D_3 \rightarrow {}^7F_5$ (415 nm), $^5D_3 \rightarrow {}^7F_4$ (438 nm), $^5D_4 \rightarrow {}^7F_6$ (490 nm), $^5D_4 \rightarrow {}^7F_5$ (541 nm), $^5D_4 \rightarrow {}^7F_4$ (585 nm), $^5D_4 \rightarrow {}^7F_3$ (619 nm)
Ho^{3+}	$^5F_4$, $^5S_2 \rightarrow {}^5I_8$ (541 nm), $^5F_5 \rightarrow {}^5I_8$ (660 nm)
Er^{3+}	$^2H_{9/2} \rightarrow {}^4I_{15/2}$ (415 nm), $^2H_{11/2} \rightarrow {}^4I_{15/2}$ (525 nm), $^4S_{3/2} \rightarrow {}^4I_{15/2}$ (541 nm), $^4F_{9/2} \rightarrow {}^4I_{15/2}$ (652 nm)
Tm^{3+}	$^3P_0 \rightarrow {}^3H_6$ (295 nm), $^1I_6 \rightarrow {}^3F_4$ (345 nm), $^1D_2 \rightarrow {}^3H_6$ (362 nm), $^1D_2 \rightarrow {}^3F_4$ (455 nm), $^1G_4 \rightarrow {}^3H_6$ (475 nm), $^1G_4 \rightarrow {}^3F_4$ (644 nm), $^3F_3 \rightarrow {}^3H_6$ (694 nm), $^3H_4 \rightarrow {}^3H_6$ (800 nm)
Yb^{3+}	$2 \times {}^2F_{5/2} \rightarrow {}^2F_{7/2}$ (CET; 470–500 nm)

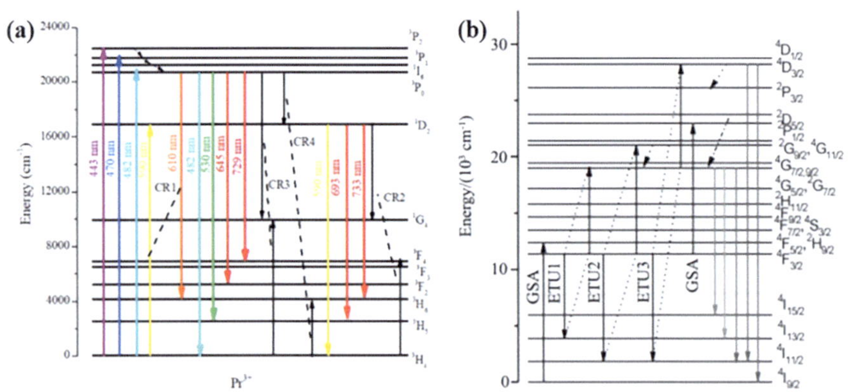

Fig. 4 Energy level diagram of **a** Pr^{3+} ions for absorption and emission at visible wavelengths and **b** Nd^{3+} ions and possible upconversion mechanisms under excitation at 796 nm. Reprinted with permission from Refs. [28, 29]

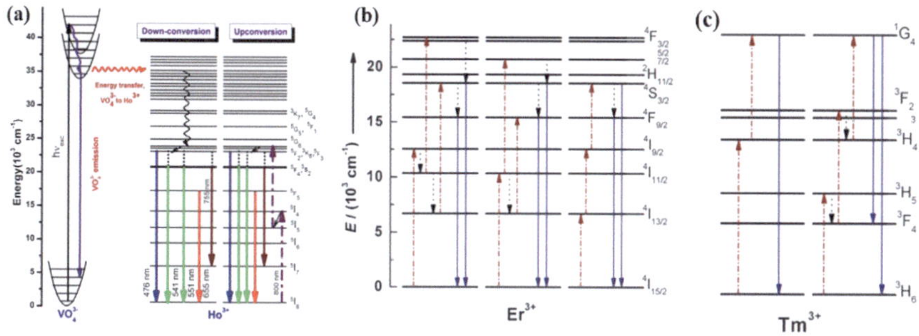

Fig. 5 Energy level diagrams showing typical UC processes for **a** Ho^{3+}, **b** Er^{3+} and **c** Tm^{3+}. Reprinted with the permission from Refs. [26, 30]

$^4G_{7/2,9/2}$. The excited $^2P_{1/2}$ and $^4G_{7/2,9/2}$ states are showing the all-possible visible transitions (Fig. 4b). In addition, the two UV intense upconversion emissions of Nd^{3+} ions were also showed at 354 nm ($^4D_{3/2} \rightarrow {}^4I_{9/2}$), 382 nm ($^4D_{3/2} \rightarrow {}^4I_{11/2}$) under 796 nm laser excitation through energy transfer upconversion (ETU).

Ho^{3+}

Ho^{3+} doped YVO_4 is known well as a green, $^5F_4$, $^5S_2 \rightarrow {}^5I_8$ (541 and 551 nm) and red, $^5F_5 \rightarrow {}^5I_8$ (655 nm) phosphor [30]. The ground state absorption (GSA) process promotes the electron of Ho^{3+} ions to $^5I_4$ under 800 nm laser excitation. Non-radiative relaxation of electrons from the $^5I_4$ level to the $^5I_5$ level takes place. The second photon is absorbed by the electrons at the $^5I_5$ level, causing it to migrate to the $^5F_2$ level. Now, non-radiatively de-excited electrons are being excited to $^5F_3$, $^5F_4$, $^5S_2$, and $^5F_5$ levels through single phonon or multi-phonon processes, resulting in emission bands at 476, 541, 551, 655, and 755 nm because of transitions between $^5F_3 \rightarrow {}^5I_8$, $^5F_4 \rightarrow {}^5I_8$, $^4S_2 \rightarrow {}^5I_8$, $^5F_5 \rightarrow {}^5I_8$, and $^5F_4$, $^5S_2 \rightarrow {}^5I_7$, respectively (Fig. 5a). Thus, two-photon absorption processes elevating the Ho^{3+} ion are responsible for all the emission bands.

Er^{3+}

Er^{3+} ion exhibits an extremely high upconversion efficiency because its energy gap is similar across various energy levels. The energy difference (10,350 cm^{-1}) in Er^{3+} between the $^4I_{11/2}$ and $^4I_{15/2}$ levels, for instance, is comparable to that (10,370 cm^{-1}) between the $^4F_{7/2}$ and $^4I_{11/2}$ levels. Thus, upconversion emission can be produced with a 980 nm excitation at the energy levels of $^4I_{15/2}$, $^4I_{11/2}$, and $^4F_{7/2}$. Furthermore, as there is an energy difference in the same region between the $^4F_{9/2}$ and $^4I_{13/2}$ states, at least three distinct transitions in Er^{3+} ions are triggered by 980 nm excitation, resulting in the sequential absorption of two photons followed by the emission of green and red light [26].

$$^2H_{11/2} \rightarrow ^4 I_{15/2} + h\nu(525\,\text{nm, green})$$
$$^4S_{3/2} \rightarrow ^4 I_{15/2} + h\nu(541\,\text{nm, green})$$
$$^4F_{9/2} \rightarrow ^4 I_{15/2} + h\nu(652\,\text{nm, red})$$

Three major peaks made up the upconversion spectrum (Fig. 5b).

All upconversion spectra from Er^{3+} are similar. However, the relative intensities and positions of the three major peaks vary depending on the concentration of activator ions, sensitizer ions, and the composition of the host matrix. In general, absorption can be enhanced by increasing the concentration of lanthanide ions in the material

Tm^{3+}

The Tm^{3+} doped low-silica calcium aluminosilicate (LSCAS) glass displays the transitions peaks at 295 (UV; $^3P_0 \rightarrow {}^3H_6$), 360 (UV; $^1D_2 \rightarrow {}^3H_6$), 456 (blue; $^1D_2 \rightarrow {}^3F_4$), 480 (blue; $^1G_4 \rightarrow {}^3H_6$), 650 (red; $^1G_4 \rightarrow {}^3F_4$), 667 (red; $^3F_2 \rightarrow {}^3H_6$), 694 (NIR; $^3F_3 \rightarrow {}^3H_6$), and 800 nm (NIR; $^3H_4 \rightarrow {}^3H_6$) under 290 mW excitation power [16, 31]. It is important to note that the NIR emission is 15 times stronger than the UV and visible emissions (Fig. 5c).

2.1.2 Sensitizer

The concentration of the activators has several limitations due to cross-relaxation. The excitation energy may be quenched at high concentrations of dopant. Therefore, in order to prevent the quenching effect, the concentration of activator ions needs to be carefully controlled and kept low. However, low amounts of doping may result in low pump light absorption and low emission phosphor capacity. The host material is frequently further doped with highly absorbing ions known as sensitizers to enhance the absorption of lanthanide-doped phosphors. This should also effective energy transfer to the activator [26]. A sensitizer with a sufficient absorption cross-relaxation in the near-infrared (NIR) spectrum can be co-doped to greatly increase the efficiency of upconversion luminescence. In order to benefit from the successful ETU interaction between the sensitizer and activator, the sensitizer and activator are typically co-coped in the host phosphors [26]. The Yb^{3+} ion is the most widely utilized sensitizer for Pr^{3+}, Nd^{3+}, Ho^{3+}, Er^{3+}, and Tm^{3+} as an activator. Yb^{3+} has a simple energy level framework that consists of only one excited 4f level, particularly the $^2F_{5/2}$ level. The energy difference between the ground state ($^2F_{7/2}$) and excited state ($^2F_{5/2}$) of Yb^{3+} corresponds well to the energy difference between the $^4I_{11/2}$ and $^4I_{15/2}$ and the $^4F_{7/2}$ and $^4I_{11/2}$ states of Er^{3+}, permitting for effective cooperative energy transfer (CET) between the two ions. A similar methodology can be applied to other activators ions, Pr^{3+}, Nd^{3+}, Ho^{3+}, and Tm^{3+}. These optical properties make Yb^{3+} an excellent upconversion sensitizer.

Yb^{3+}/Pr^{3+}

In Yb^{3+}/Pr^{3+} doped $NaYF_4$, the excitation of Yb^{3+} ion from its ground state to its excited state under 980 nm laser diode wavelength correspond to the $^2F_{7/2} \rightarrow {}^2F_{5/2}$ transition of Yb^{3+} ion [32]. Exciting Pr^{3+} ions from the $^3H_4$ ground state to the $^1G_4$ excited state can be achieved through an energy transfer process from Yb^{3+} ions to an adjacent Pr^{3+} ion. Energy transfer from another excited Yb^{3+} ion and absorption of the pumping photon's energy can both accelerate the population in the $^1G_4$ level to the $^3P_0$ levels. The excited electron may lose its energy by producing visible emissions from populated $^3P_0$ level. The radiative decay from the $^3P_0$ state to the ground state results in blue emission at 486 nm. In general, pumping energy at 980 nm is difficult to populate in $^3P_1$ and higher energy levels of Pr^{3+} ions. Cross-relaxation between $^3P_0 \rightarrow {}^3P_2$ and $^3H_5 \rightarrow {}^3H_4$ in Pr^{3+} ions boosts occupation in the $^3P_2$ state, which populates $^1I_6$ and $^3P_1$ states via non-radiative relaxation (Fig. 6a). Excited Pr^{3+} ions in the $^1I_6$ and $^3P_1$ levels decay to the $^3H_5$ state, causing green emission cantered at 523 and 540 nm. The red emissions at 605 and 639 nm arise from the transitions from the $^3P_0$ state to the $^3H_6$ and $^3F_2$ states, respectively.

Yb^{3+}/Nd^{3+}

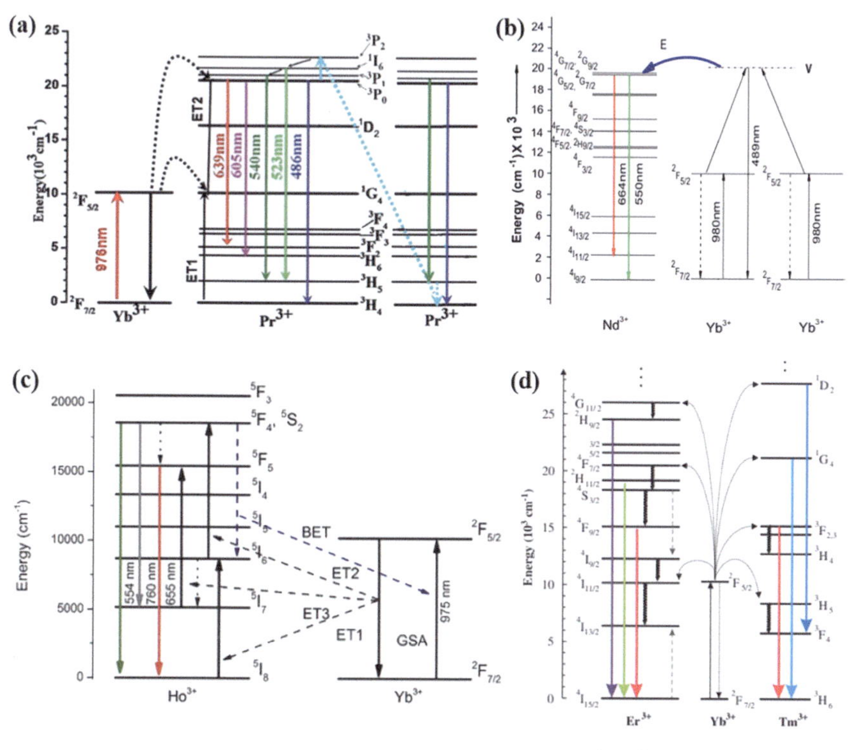

Fig. 6 Energy level diagrams showing typical UC processes for **a** Yb^{3+}/Pr^{3+}, **b** Yb^{3+}/Nd^{3+}, **c** Yb^{3+}/Ho^{3+}, **d** Yb^{3+}/Er^{3+} and Yb^{3+}/Tm^{3+}. Reprinted with the permission from Refs. [32–34, 36]

The Yb^{3+}/Nd^{3+} doped Y_2O_3 phosphor exhibits green, and red upconversion emissions that peak at 550 nm, and 664 nm when excited by a 980 nm diode laser [33]. The ground state Nd^{3+} ions are partially excited by the excited Yb^{3+} ion and are pumped directly to the ($^4G_{7/2}$, $^2G_{9/2}$) level. Therefore, energy transfer from the excited Yb^{3+} ions and cooperative sensitization between the Yb^{3+} ions in the $^2F_{5/2}$ level populated the $^4G_{7/2}$ level. The departed Nd^{3+} ions from the $^4G_{7/2}$ level relax radiatively into low-energy levels, producing radiation in the green and red regions, which correspond to the $^4G_{7/2}$ $^4I_{9/2}$ and $^4G_{7/2}$ $^4I_{11/2}$ transitions, respectively. The energy level diagram displays these transitions in Fig. 6b.

Yb^{3+}/Ho^{3+}

The Yb^{3+}/Ho^{3+} doped $BaTiO_3$ reveals strong upconversion in the green (538 nm; $^5F_4 \rightarrow {}^5I_8$ and 548 nm; $^5S_2 \rightarrow {}^5I_8$) and red (655 nm; $^5F_5 \rightarrow {}^5I_8$) phosphors under 980 nm laser excitation (Fig. 6c) [34, 35]. In addition to these, the appearance of low-intensity emission bands at 394, 424, and 464 nm can be assigned to the Ho^{3+} transitions at $^5G_4$, $^3K_7 \rightarrow {}^5I_8$, $^5G_5 \rightarrow {}^5I_8$, and $^5F_1$, $^5G_6 \rightarrow {}^5I_8$, respectively (Fig. 6c). More details of mechanisms and photophysical pathways on the upconversion processes will be discussed in Chapter "Basic Principle and Mechanism of Upconversion Luminescence in Phosphor Materials".

Yb^{3+}/Er^{3+}

The $NaYF_4$ nanoparticles co-doped with Yb^{3+}/Er^{3+} pair involves an upward transition from $^2F_{7/2}$ to $^2F_{5/2}$ of the Yb^{3+} ion under 980 nm excitation laser diode wavelength, which then excited the neighbour Er^{3+} ion from its ground state ($^4I_{15/2}$) to the excited state ($^4I_{11/2}$) [36]. The earlier excited Er^{3+} ion is then promoted to the higher energy $^4F_{7/2}$ level by another Yb^{3+} ion that was excited into the $^2F_{5/2}$ level. The small energy gap lying between high-energy level, $^4F_{7/2}$ and low-energy level, $^4S_{3/2}$ state, phonon-induced relaxation takes place. Finally, upconversion luminescence is produced via red emission at 656 nm via the transition $^4F_{9/2} \rightarrow {}^4I_{15/2}$ and green emission at 540 nm via the transition $^4S_{3/2} \rightarrow {}^4I_{15/2}$ (Fig. 6d). A relatively high (10% or more) concentration of Yb^{3+} is needed for the efficient resonant energy transfer of the stimulation among Yb^{3+} ions to increase the upconversion emission.

Yb^{3+}/Tm^{3+}

In the case of Yb^{3+}/Tm^{3+} pairs that showed the analogous behaviour to the Yb^{3+}/Er^{3+} doped $NaYF_4$ under excitation at 980 nm, resulting UV emissions at 290 nm ($^3P_0 \rightarrow {}^3H_6$), 345 nm ($^1I_6 \rightarrow {}^3F_4$) and 362 nm ($^1D_2 \rightarrow {}^3H_6$), visible emissions at 450 nm ($^1D_2 \rightarrow {}^3F_4$), 475 nm ($^1G_4 \rightarrow {}^3H_6$), 644 nm ($^1G_4 \rightarrow {}^3F_4$) and 694 nm ($^3F_3 \rightarrow {}^3H_6$), and NIR emission at 800 nm ($^3H_4 \rightarrow {}^3H_6$) (Fig. 6d) [36].

2.1.3 Comparison of Various Host Matrices for Upconversion

The host matrix and its interaction with the dopant ion determine the upconversion behaviour of phosphors. The interionic distance, coordination number, and relative structural position of the dopant ion significantly affect the phosphor upconversion performance. Thus, picking an appropriate host material is also crucial for designing upconversion phosphors with attractive optical properties like high efficiency and controllable emission colour. Host lattices must satisfy a few essential requirements to promote effective up conversion luminescence: (1) perfect lattice matching with dopant ions to manage lattice distortion, (2) low phonon energies to increase emission efficiency and minimize non-radiative losses, (3) clearness to both excitation light and upconversion emission, (4) asymmetrical crystal field resulting in mixed parity levels [37]. Here's a comparison of various host matrices commonly used for upconversion:

Inorganic Host Matrices: $NaYF_4$ is one of the most widely used host matrices for upconversion due to its high chemical stability, low phonon energy, and efficient energy transfer properties. Doping lanthanide ions (e.g. Erbium, Ytterbium) into $NaYF_4$ can result in efficient up conversion [38]. β-$NaYF_4$ is like $NaYF_4$ but with a slightly different crystal structure and it can also be utilized for efficient upconversion processes.

Oxide Host Matrices: Yttrium oxide (Y_2O_3) is another popular host matrix for upconversion materials. It offers good chemical stability and can support various lanthanide dopants. Additionally, Gd_2O_3 is also used as a host matrix for upconversion materials, offering similar advantages to Y_2O_3.

Fluoride Host Matrices: Calcium fluoride (CaF_2) is known for its optical transparency and low refractive index, making it suitable for upconversion applications. On the other hand, SrF_2 shares similar properties with CaF_2 and can be used as an alternative host matrix.

Glass Host Matrices: Silicate glasses doped with lanthanide ions can exhibit upconversion properties. These glasses offer flexibility in terms of composition and can be tailored for specific applications. Phosphate glasses have been explored as host matrices for upconversion materials due to their good chemical durability and thermal stability.

Perovskite Host Matrices: Perovskite structures based on lanthanum halides have shown promise for upconversion applications due to their unique electronic and optical properties.

Lead Halide Perovskites: Lead halide perovskites have also been investigated as potential host matrices for up conversion materials, although concerns about lead toxicity exist.

Polymer Host Matrices: Certain conjugated polymers can serve as host matrices for upconversion materials, offering advantages such as flexibility, processability, and compatibility with solution-based processing techniques.

Among the existing host matrices, fluoride materials ($NaLnF_4$; Ln = Y, La, Gd, and Lu) series is widely used [39]. The $NaLnF_4$ hexagonal phases are more efficient than their cubic phases among this series [40]. A hexagonal shape of Ln^{3+} with a less symmetric crystal field increases the possibility of Laporte-prohibited intra-4f transitions. For example, when the site symmetry of Er^{3+} dopants is reduced, from the D_{4h} symmetry of tetragonal ZrO_2 to the C_{2h} symmetry of monoclinic ZrO_2, and from the O_h symmetry of cubic $NaYF_4$ to the C_{3h} symmetry of hexagonal $NaYF_4$ [41], the upconversion luminescence of Er^{3+} is 10 times more luminescent due to changed transition probability. The efficiency of phosphors has increased the upconversion intensity through both internal and external approaches. It was discovered that co-doping into a host matrix with non-Ln^{3+} (Li^+, Na^+, K^+, Mg^{2+}, Zn^{2+}, and Bi^{3+}) was a helpful technique to significantly increase the upconversion luminescence intensity by modifying the local symmetry surrounding the Ln^{3+} ions [42-44].

The upconversion efficiency and brightness of oxides are lower than that of fluorides due to non-radiative losses, however, oxides have greater mechanical strength, thermal stability, and chemical stability. Another crucial concern with oxides and halides used as host materials for upconverters is stability. For oxides, air stability coexists with weak upconversion luminescence because of high phonon energies ($> 500\ cm^{-1}$). On the other hand, halides have small phonon energies ($< 350\ cm^{-1}$), which is favourable for high upconversion efficiency, however, these hosts are unstable in air and moisture-sensitive [45]. Moreover, the heavy halides (I, Br, and Cl) are the perfect luminescent hosts. However, due to their hygroscopic nature, they are not very useful. For example, Yb^{3+}/Er^{3+} doped $NaYF_4$ has an upconversion efficiency that is 20 times more than Yb^{3+}/Er^{3+} doped La_2O_3 and 6 times higher than Yb^{3+}/Er^{3+} doped $La_2(MoO_4)_3$ [46].

The reported study suggested each of these host matrices has its own set of advantages and disadvantages, and the choice depends on factors such as desired upconversion efficiency, spectral properties, environmental stability, and compatibility with fabrication processes. Researchers continue to explore new host matrices and optimize existing ones to enhance the performance of up conversion materials for various applications.

3 Conclusions

The chapter discusses the discovery and technological significance of photon upconversion in rare earth-doped materials, particularly those using Yb^{3+}-Er^{3+} and Yb^{3+}-Tm^{3+} ions. It highlights the key roles of luminescence, specifically Stokes and anti-Stokes processes, in the development of upconversion phosphors. These materials, enabled by the unique properties of trivalent rare earth ions, have become essential in applications like infrared detection, solar cell enhancement, and advanced optical devices, with ongoing research focused on optimizing their efficiency and expanding their technological applications. Various host matrices, including fluoride, oxide, and glass, offer distinct advantages and challenges for upconversion applications. Fluoride materials like $NaYF_4$, particularly in hexagonal phases, demonstrate superior upconversion efficiency due to favourable crystal symmetry. Oxides provide better mechanical and thermal stability but suffer from higher non-radiative losses.

The choice of host matrix depends on the specific application, balancing upconversion efficiency, stability, and compatibility. Ongoing research aims to optimize these materials to enhance upconversion performance across different applications.

Acknowledgements This work has been carried out in the frame of the PhotoMXene project No. 3305/03/02 co-funded by the EU Horizon 2020 Research and Innovation Programme under the Programme SASPRO2 COFUND Marie Sklodowska-Curie grant agreement No. 945478. This work forms part of the "synthesis and structural analysis of energy efficient luminescent nanomaterials for multifunctional applications", CSIR No. 21(1114)/20/EMR-II, funded by CSIR-Research Scheme.

References

1. F.E.C.R. Auzel, Compteur quantique par transfert d'energie entre deux ions de terres rares dans un tungstate mixte et dans un verre. CR Acad. Sci. Paris **262**, 1016–1019 (1966)
2. V.V. Ovsyakin, P.P. Feofilov, Cooperative sensitization of luminescence in crystals activated with rare earth ions. J. Exp. Theor. Phys. **4**, 317 (1966)
3. S.V. Galginaitis, G.E. Fenner, in *Proceedings of 2nd International Conference on Gallium Arsenide*. A Visible Light Source Utilizing a GaAs Electroluminescent Diode and a Stepwise Excitable Phosphor (1968), p. 131
4. R.A. Hewes, J.F. Sarver, Infrared excitation processes for the visible luminescence of Er^{3+}, Ho^{3+}, and Tm^{3+} in Yb^{3+}-sensitized rare earth trifluorides. Phys. Rev. **182**(2), 427 (1969)
5. A. Kitai (ed), *Luminescent Materials and Applications* (Wiley, 2008)
6. H.S. Virk, in *Defect and Diffusion Forum*. History of Luminescence from Ancient to Modern Times, vol 361. (Trans Tech Publications Ltd., 2015), pp. 1–13
7. M.V. Sednev, V.N. Belov, S.W. Hell, Fluorescent dyes with large Stokes shifts for super-resolution optical microscopy of biological objects: a review. Methods Appl. Fluoresc. **3**(4), 042004 (2015)
8. S. Gao, B. Balan, K. Yoosaf, F. Monti, E. Bandini, A. Barbieri, N. Armaroli, Highly efficient luminescent solar concentrators based on benzoheterodiazole dyes with large stokes shifts. Chemi. A Eur. J. **26**(48), 11013–11023 (2020)
9. Y. Chen, S. Wang, F. Zhang, Near-infrared luminescence high-contrast in vivo biomedical imaging. Nat. Rev. Bioeng. **1**(1), 60–78 (2023)
10. E. Herz, A. Burns, D. Bonner, U. Wiesner, Large stokes-shift fluorescent silica nanoparticles with enhanced emission over free dye for single excitation multiplexing. Macromol. Rapid Commun. **30**(22), 1907–1910 (2009)
11. S. Dash, S. Mishra, S. Patel, B.K. Mishra, Organically modified silica: synthesis and applications due to its surface interaction with organic molecules. Adv. Colloid Int. Sci. **140**(2), 77–94 (2008)
12. C. Li, W. Chen, D. Wu, D. Quan, Z. Zhou, J. Hao, J. Qin, Y. Li, Z. He, K. Wang, Large Stokes shift and high efficiency luminescent solar concentrator incorporated with $CuInS_2$/ZnS quantum dots. Sci. Rep. **5**(1), 17777 (2015)
13. K.K.W. Lo, L.J. Charbonnière, Luminescent materials: metal complexes, clusters, and nanomaterials. Eur. J. Inorg. Chem. **2017**(44), 5055–5057 (2017)
14. V.W.W. Yam, K.M.C. Wong, Luminescent metal complexes of d^6, d^8 and d^{10} transition metal centres. Chem. Commun. **47**(42), 11579–11592 (2011)
15. S. Sinha, M.K. Mahata, in *Upconversion Nanophosphor.*What are Upconversion Nanophosphors: Basic Concepts and Mechanisms (Elsevier, 2022), pp. 19–48
16. C.C. Klick, J.H. Schulman, in *Solid State Physics*. Luminescence in Solids, vol 5. (Academic Press, 1957), pp. 97–172

17. W.W. Piper, J.A. DeLuca, F.S. Ham, Cascade fluorescent decay in Pr^{3+}-doped fluorides: achievement of a quantum yield greater than unity for emission of visible light. J. Lumin. **8**(4), 344–348 (1974)
18. J. Zhou, Y. Teng, S. Zhou, J. Qiu, Quantum cutting in luminescent glasses and glass ceramics. Int. J. Appl. Glass Sci. **3**(4), 299–308 (2012)
19. R. Pappalardo, Calculated quantum yields for photon-cascade emission (PCE) for Pr^{3+} and Tm^{3+} in fluoride hosts. J. Lumin. **14**(3), 159–193 (1976)
20. T. Trupke, M.A. Green, P. Wurfel, Improving solar cell efficiencies by down-conversion of high-energy photons. J. Appl. Phys. **92**(3), 1668–1674 (2002)
21. C. Ronda, *Luminescence: From Theory to Applications* (WILEY-VCH Verlag GmbH & Co. KGaA, Weinheim, 2008). ISBN: 978-3-527-31402-7
22. G. Liu, Advances in the theoretical understanding of photon upconversion in rare-earth activated nanophosphors. Chem. Soc. Rev. **44**(6), 1635–1652 (2015)
23. D.R. Gamelin, H.U. Gudel, Excited-state dynamics and sequential two-photon upconversion excitation of Mo^{3+}-doped chloro- and bromo-elpasolites. J. Phys. Chem. B **104**, 10222–10234 (2000)
24. D.R. Gamelin, H.U. Güdel, Two-photon spectroscopy of d^3 transition metals: near-IR-to visible upconversion luminescence by Re^{4+} and Mo^{3+}. J. Am. Chem. Soc. **120**, 12143–12144 (1998)
25. F. Auzel, Upconversion and anti-stokes processes with f and d ions in solids. Chem. Rev. **104**, 139–173 (2004)
26. F. Wang, X.G. Liu, Recent advances in the chemistry of lanthanide-doped upconversion nanocrystals. Chem. Soc. Rev. **38**, 976–989 (2009)
27. Z. Zhang, Q. Han, J.W. Lau, B. Xing, Lanthanide-doped upconversion nanoparticles meet the needs for cutting-edge bioapplications: recent progress and perspectives. ACS Mater. Lett. **2**(11), 1516–1531 (2020)
28. W.T. Carnall, P.R. Fields, B.G. Wybourne, Spectral intensities of the trivalent lanthanides and actinides in solution. I. Pr^{3+}, Nd^{3+}, Er^{3+}, Tm^{3+}, and Yb^{3+}. J. Chem. Phys. **42**(11), 3797–3806 (1965)
29. D. Chen, Y. Wang, Y. Yu, F. Liu, P. Huang, Infrared to ultraviolet upconversion luminescence in Nd^{3+} doped nano-glass-ceramic. J. Rare Earths **26**(3), 428–432 (2008)
30. M.K. Mahata, T. Koppe, H. Hofsäss, K. Kumar, U. Vetter, Host sensitized luminescence and time-resolved spectroscopy of YVO_4: Ho^{3+} nanocrystals. Phys. Proc. **76**, 125–131 (2015)
31. C.M. Trindade, F.G. Rego-Filho, N.G.C. Astrath, C. Jacinto, A.S. Gouveia-Neto, UV–visible-NIR light generation through frequency upconversion in Tm^{3+}-doped low silica calcium aluminosilicate glasses using multiple excitation around 1.2 μm. J. Solid State Chem. **260**, 147–150 (2018)
32. S. Hao, W. Shao, H. Qiu, Y. Shang, R. Fan, X. Guo, L. Zhao, G. Chen, C. Yang, Tuning the size and upconversion emission of $NaYF_4$: Yb^{3+}/Pr^{3+} nanoparticles through Yb^{3+} doping. RSC Adv. **4**(99), 56302–56306 (2014)
33. R. Dey, V.K. Rai, A. Pandey, Green upconversion emission in Nd^{3+}–Yb^{3+}–Zn^{2+}: Y_2O_3 phosphor. Spectrochim. Acta A Mol. Biomol. Spectrosc. **99**, 288–291 (2012)
34. M. Vega, I.R. Martin, J. Llanos, Near-infrared to visible upconversion and second harmonic generation in $BaTiO_3$: Ho^{3+} and $BaTiO_3$: Ho^{3+}/Yb^{3+} phosphors. J. Alloys Compd. **806**, 1146–1152 (2019)
35. P. Singh, R.S. Yadav, P. Singh, S.B. Rai, Upconversion and downshifting emissions of Ho^{3+}-Yb^{3+} co-doped $ATiO_3$ perovskite phosphors with temperature sensing properties in Ho^{3+}-Yb^{3+} co-doped $BaTiO_3$ phosphor. J. Alloys Compd. **855**, 157452 (2021)
36. J.F. Suyver, J. Grimm, M.K. Van Veen, D. Biner, K.W. Krämer, H.U. Güdel, Upconversion spectroscopy and properties of $NaYF_4$ doped with Er^{3+}, Tm^{3+} and/or Yb^{3+}. J. Lumin. **117**(1), 1–12 (2006)
37. R. Shrivastava, D. Fekar, M. Mahilang, Up-conversion phosphors: a short review. Ferroelectrics **614**(1), 112–118 (2023)
38. N. Menyuk, K. Dwight, J.W. Pierce, $NaYF_4$: Yb, Er-an efficient upconversion phosphor. Appl. Phys. Lett. **21**(4), 159–161 (1972)

39. K.W. Kramer, D. Biner, G. Frei, H.U. Gudel, M.P. Hehlen, S.R. Luthi, Hexagonal sodium yttrium fluoride based green and blue emitting upconversion phosphors. Chem. Mater. **16**, 1244–1251 (2004)
40. M. Haase, H. Schafer, Upconverting nanoparticles. Angew. Chem. Int. Ed. **50**(26), 5808–5829 (2011)
41. G.Y. Chen, H.L. Qju, P.N. Prasad, X.Y. Chen, Upconversion nanoparticles: design, nanochemistry, and applications in theranostics. Chem. Rev. **114**, 5161–5214 (2014)
42. A. Maurya, A. Dwivedi, A. Bahadur, S.B. Rai, Enhanced upconversion and downshifting emissions from Tm^{3+}, Yb^{3+} co-doped $CaZrO_3$ phosphor in the presence of alkali ions (Li^+, Na^+ and K^+). J. Alloys Compd. **786**, 457–467 (2019)
43. N. Tripathi, K. Kamada, H. Nakamura, T. Akai, Enhanced near infrared-to-visible upconversion in the $CaTiO_3$: Yb^{3+}/Er^{3+} phosphor via the host lattice modification using co-doping of Mg^{2+} ions. RSC Adv. **13**(50), 35391–35396 (2023)
44. R.S. Yadav, S.J. Dhoble, S.B. Rai, Improved photon upconversion photoluminescence and intrinsic optical bistability from a rare earth co-doped lanthanum oxide phosphor via Bi^{3+} doping. New J. Chem. **42**(9), 7272–7282 (2018)
45. G. Zhou, D. Wang, X. Wang, Z. Shao, M. Jiang, Two-photon absorption and excited state absorption properties of an organic dye PSPI. Opt. Commun. **241**, 215–219 (2004)
46. G. Blasse, B.C. Grabmaier, G. Blasse, B.C. Grabmaier, *A General Introduction to Luminescent Materials* (Springer Berlin Heidelberg, 1994), pp. 1–9

Basic Principle and Mechanism of Upconversion Luminescence in Phosphor Materials

Shreya Choudhary and Jyoti Singh

Abstract Over the past decade, high-quality upconversion phosphors have been synthesized and investigated with the rapid development of nanotechnology. The upconversion materials can efficiently convert near-infrared light into visible or ultra-visible luminescence using a stepwise multiphoton process which make these materials ideal candidates for various applications such as lasers, solar cells, sensors, security inks, and optical imaging. Upconversion can take place in both organic and inorganic materials through a number of different processes. It usually takes place in polycyclic aromatic hydrocarbons (PAHs) in organic matter and in d or f-block elements in inorganic materials. Nowadays lanthanide–doped upconversion materials are proved to be more promising owing to wide range applications in solid state lighting and solar cells. Upconversion materials can efficiently convert sub-band gap photons into above band gap-light in solar cells has led to vast research in upconversion materials and its applications in the energy field. The efficiency of upconversion materials can be elevated through the optimization of the following factors: added impurity, absorption, energy transfer, and emission. The appropriate knowledge of basic principles and mechanism involved in upconversion can lead to novel approaches in enhancing the upconversion luminescence. The proposed chapter thoroughly discusses and focused on the mechanism of upconversion and its differentiation from analogous processes such as simple harmonic generation and two-photon absorption. Various processes possible for upconversion in materials such as energy transfer upconversion (ETU), excited-state absorption (ESA) and photon avalanche (PA) etc. are explained with necessary conditions in detail. This chapter provide the readers an insight to realize how emission energy is high than excitation energy while following the law of conservation of energy. In addition, the chapter elaborates the structure and function of upconversion materials. The efficiency of the upconversion process, fundamental physics involved and atomic level details behind the upconversion process is also discussed. The reported research works in the field over the years are reviewed while simultaneously bringing to the forefront the future scope of the field. The chapter finally summarizes with the limitations,

S. Choudhary · J. Singh (✉)
Department of Applied Physics (SoVSAS), Gautam Buddha University, Greater Noida, Uttar Pradesh, India
e-mail: jyoti.singh.phy@gmail.com

V. Dubey et al. (eds.), *Upconversion Luminescence of Phosphors*, Progress in Optical Science and Photonics 39, https://doi.org/10.1007/978-981-95-1762-6_2

challenges in improving the upconversion luminescence in materials and possible ways to overcome.

Keywords Upconversion · Phosphors · Energy transfer mechanism · Quantum dots · Organic upconversion materials · Cross relaxation mechanism · Non-radiative relaxation · Optical sensing

1 Introduction

The evolution of advanced materials with tunable optical features has significantly impacted a variety of technological domains, ranging from biomedical imaging to energy harvesting. Within this realm, upconversion luminescence (UCL) has surfaced as a significant research focus, especially for uses that necessitate efficient light conversion and manipulation. Phosphor materials, which demonstrate upconversion phenomena, are particularly valuable due to their capacity to convert low-energy infrared light into higher-energy visible or ultraviolet light [1–5].

Upconversion luminescence is a nonlinear optical phenomenon characterized by the sequential absorption of multiple photons, resulting in the emission of light at wavelengths shorter than that of the excitation source. This distinctive phenomenon has garnered considerable attention for its potential applications in fields such as bioimaging, solar energy conversion, optical sensing, and laser technology [6–10]. However, to effectively enhance and optimize the performance of phosphor materials in practical uses, a deep comprehension of the fundamental principles and intricate mechanisms involved is necessary.

The present chapter explores the fundamental principles of upconversion luminescence, focusing on the theoretical aspects of photon absorption and energy transfer in doped phosphor materials. It highlights the significance of rare earth ion-doped phosphors, which demonstrate exceptional upconversion efficiency due to their distinct energy levels and prolonged intermediate states. Additionally, the impact of crystal structure, host matrix influences, and the role of various dopants in shaping luminescence characteristics will also be explained in detail. By exploring the mechanisms that facilitate upconversion luminescence, this chapter aims to provide an in-depth understanding of how these processes can be optimized and adapted for specific applications, while also addressing the ongoing challenges in enhancing upconversion efficiency. The insights derived from this study will form a basis for the creation of novel upconversion materials with superior optical features for state-of-the-art technologies.

2 Introduction to Photon Upconversion

2.1 Foundational Definition and Significance

Photon upconversion is the process of generating a higher-energy photon through the excitation of multiple lower-energy photons. This phenomenon is an example of anti-Stokes emission via a nonlinear optical process [1, 3]. Utilizing appropriate materials, lower-energy photons can be absorbed, elevating an ion to higher-energy levels. Following successive emissions, a photon of higher energy than the initially absorbed one is released. In this context, the ion responsible for emission is termed the activator, while the ion that supplies energy to the activator is referred to as the sensitizer [11–15]. Throughout this process, the particle undergoes multiple metastable states, ultimately resulting in the emission of high-energy photons [1, 3, 5]. The step-wise process associated with photon up conversion mechanism is illustrated in the Fig. 1.

Photon upconversion holds significant promise across various fields, e.g., in biomedicine, where it offers high-energy light ideal for biological tissues, especially in cancer treatments [7, 16–19]. The use of near-infrared excitation wavelengths in upconversion ions provides superior properties and in solar energy conversion by converting low-energy photons from the solar spectrum into higher-energy photons, upconversion materials can potentially increase the overall energy harvested from sunlight [18, 20–24]. Additionally, upconversion contributes in Anti-Counterfeiting and Color Display Technologies because upconversion materials exhibit unique color shifts or luminescent properties that are difficult to replicate. This characteristic makes them ideal for incorporation into banknotes, identification cards, and luxury goods [5, 10, 18].

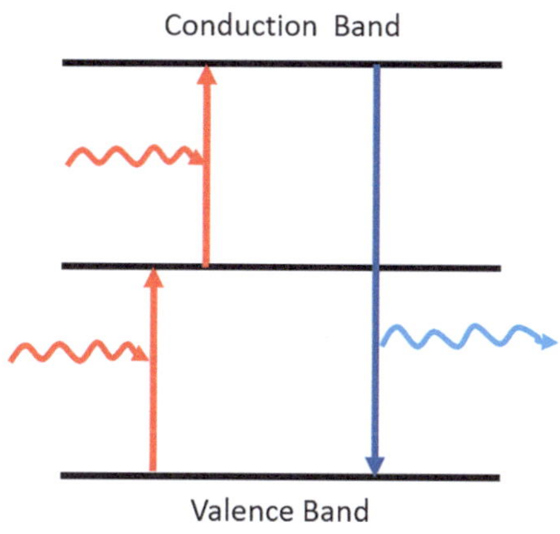

Fig. 1 Schematic diagram of photon upconversion mechanism

2.2 *Historical Foundation and Evolution*

The exploration of stimulated emission began in 1917 when Einstein highlighted its potential during atomic excitation, paving the way for the development of LASER technology. This foundational concept laid the groundwork for understanding different forms of emission, particularly as classified under Stokes' law, which divided emissions in two categories: Stokes emission and anti-Stokes emission. According to Stokes law, the energy of the excitation photon exceeds that of the emitted photon, leading to a phenomenon known as the Stokes shift [2].

While most photoluminescent materials follows Stokes law, anti-Stokes emission presents a compelling contrast, where the emitted photon possesses higher energy than the absorbed photon. Historically, prior to the 1960s, known instances of anti-Stokes emission typically involve energy difference of approximately few kT between emitted photon and absorbed photon as in the case of the Raman effect or from thermal bands [17]. The pivotal moment for photon upconversion arrived in 1966 when F. Auzel proposed that two rare earth (RE) ions in excited states could exchange energy, giving rise to a unique luminescent phenomenon termed "photon upconversion." Unlike other anti-Stokes emissions, the emitted photon in this process exhibits energy elevated by 10–100 kT [8, 17].

Upconversion has predominantly been observed in rare earth ions, where the Laporte selection rule prohibits 4f–4f transitions. However, when integrated into a host matrix, the electronic structure of rare earth ions undergoes modification, allowing for permitted optical transitions due to the mixing of $4f^{n}$ and $4f^{(n-1)}5d$ orbitals [4, 9]. Complementary to upconversion is a similar process known as downconversion, which also boasts numerous applications, particularly in enhancing solar cell efficiency. The unique characteristics of downconversion nanoparticles (DCNPs), including low toxicity, cost-effective material use, and compatibility with diverse ligands or functional groups, further underscore their potential across various domains [1, 3].

3 Mechanisms of Photon Upconversion

Photon upconversion can take place in both organic and inorganic material through a number of methods resulting in various applications in different fields as discussed below:

3.1 *Types of Upconversion Processes in Inorganic Materials*

3.1.1 Excited State Absorption

Energy State Absorption (ESA) is the simplest and most fundamental process for achieving photon upconversion. In this mechanism, upconversion emission occurs when an ion is excited from its ground state to an intermediate energy level (E_1) by absorbing a photon with energy equal to hv\nhv (where "h" is Planck's constant and "v" is the frequency of the incident photon).

Once the ion reaches the intermediate level E_1, it can absorb a second pump photon, which elevates the ion to a higher-energy state (E2). The ion subsequently decays back to the ground state, resulting in the emission of an upconverted photon. Fig. 2 represented the excited state absorption mechanism.

The important factors impacting ESA, include Energy Level Spacing, which refers to the distances between the ground state (G) and intermediate energy level (E_1), as well as the distance between (E_1) and (E_2), respectively. The resemblance in these distances enhances the absorption of photons. Additionally, Reservoir Capacity is noteworthy; the intermediate energy level (E_1) has a considerable reservoir capacity attributed to its ladder-like energy configuration and the prolonged duration of its excited states, which can persist from milliseconds to microseconds. ESA can occur in various singly doped lanthanide ions, including: Er^{3+}, Tm^{3+}, Ho^{3+}, Pr^{3+}, etc.[1, 3, 4].

ESA typically exhibits low upconversion efficiency. This is largely because only a single ion undergoes the complete ESA process, and energy transfer between activators is minimal. As a result, optimizing the conditions and materials used for ESA is essential for improving its efficiency in practical applications [1, 3].

Fig. 2 Schematic diagram of excited state absorption

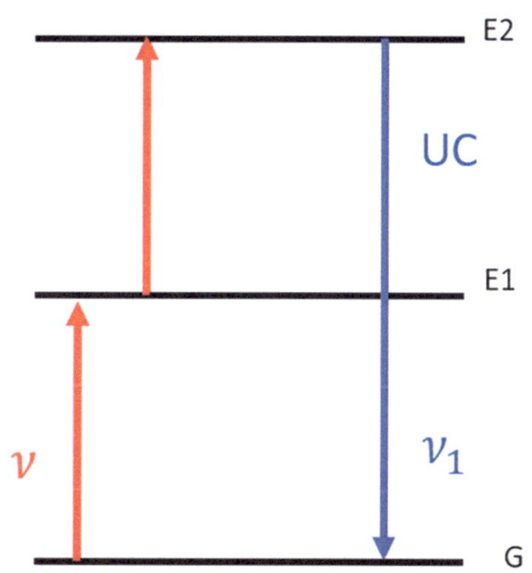

3.1.2 Energy Transfer Upconversion

In contrast to Energy State Absorption (ESA), which operates within a single lanthanide ion, Energy Transfer Upconversion (ETU) involves two nearby ions, referred as the sensitizer and the activator. This mechanism relies on the interaction between these two types of ions to facilitate upconversion [4, 5]. The process of ETU involves three different stages.

1. **Excitation of Ions**: Sensitizers (S) are the first ions to absorb energy from the incident pump photons, getting excited to an intermediate state (E_1). Simultaneously, activators (A) are also excited to the same or a similar intermediate state (E_1) by the energy from the pump photons.
2. **Energy Exchange**: Once both ions are in the E_1 state, they can exchange energy. The sensitizer ion, upon reaching a lower-energy state (G), transfers its excitation energy to the activator ion. This process involves one ion decaying to its ground state while the other is elevated to a higher-energy state (E_2).
3. **Emission of Upconverted Photon**: Following this energy transfer, the activator ion transitions from state E2 back to the ground state (G). During this radiative transition, a photon of energy ($=h\nu 1$) is emitted, which has energy greater than that of the incident photon ($=h\nu$) as depicted in Fig. 3.

In energy transfer upconversion (ETU), energy can be transferred through three primary methods:

(a) **Radiative Energy Transfer**

In this method, energy transfer occurs when a photon is emitted by the sensitizer and subsequently absorbed by the activator. The probability of this transfer as a function of distance R is given by:

$$P_{SA}(R) = \frac{\sigma}{4\pi R^2 \tau_S} \int g_s(v) g_A(v) \mathrm{d}v \tag{1}$$

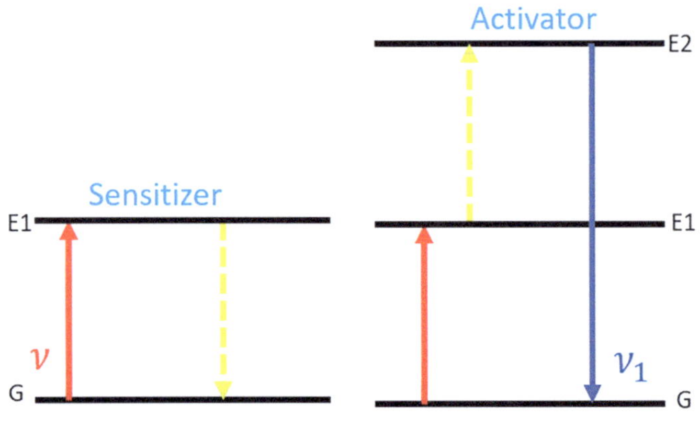

Fig. 3 Schematic diagram of energy transfer upconversion

where τ_s is the sensitizer lifetime, σ_A is the absorption cross section, and the integral is the spectral overlap between S and A. The R^{-2} distance dependence between ions helps determine concentration levels, as well as determining the host based on interatomic distances [8].

(b) **Non-radiative**—A contact between two ions with approximately equal energy between the ground state and an excited state is known as non-radiative energy transfer. The excitation will jump from one ion to another if there is sufficient ion contact [8]. The probability for energy transfer goes as,

$$P_{SA} = \frac{(R_0/R)^n}{\tau_s} \tag{2}$$

where τ_s is the sensitizer lifetime and R_0 is the critical transfer distance for which excitation transfer and spontaneous deactivation of the sensitizer have equal probability. The exponent, n, is an integer that goes as

n = 6 for dipole–dipole interactions.

n = 8 for dipole-quadrupole interactions.

n = 10 for quadrupole–quadrupole interactions.

(c) **Phonon assisted**—Two ions with distinct energy separations can transfer energy with the help of phonons. The energy integral crossover should cause the likelihood of energy transfer to go to zero, however small energies (~ 0.01 eV) can still experience energy transfer if energy is conserved through creation and destruction of a small number of phonons as can be seen in Fig. 4 [8].

The efficiency of the ETU upconversion process is significantly affected by Dopant Concentration because this determines the average distance between neighboring ions. Higher concentrations can enhance the likelihood of energy transfer between sensitizers and activators and Interaction Distance ETU is most effective when these

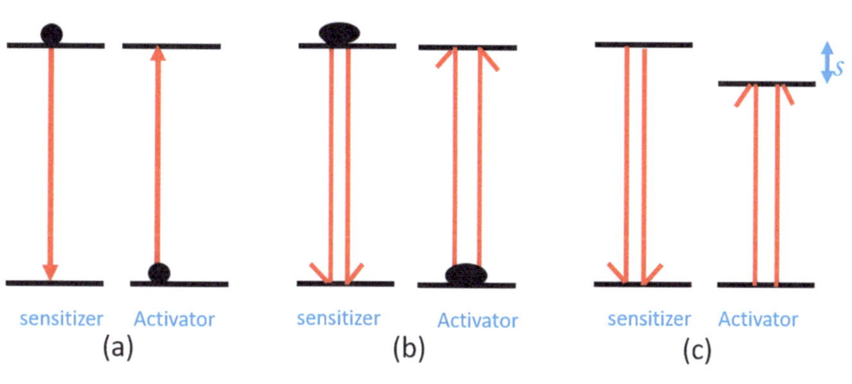

Fig. 4 Energy transfer mechanism in ETU **a** radiative, **b** non-radiative, **c** photon assisted energy transfer

ions are in close proximity, as energy transfer becomes less likely with increased distance. In ETU, sensitizer and activator ions can either be the same or different lanthanide ions. The flexibility in ion choice allows for tailored applications and optimizations in various materials [4].

Advantages of ETU

Due to its larger absorption cross section, ETU is currently considered the most suitable process for upconversion applications. The ability to utilize multiple ions and facilitate efficient energy transfer makes ETU particularly advantageous for enhancing the efficiency of devices relying on upconversion phenomena.

3.1.3 Cooperative Sensitization Upconversion

Cooperative Sensitization Upconversion (CSU) involves three ions, where two adjacent sensitizer ions (e.g., ion 1 and ion 3) absorb pump photons and reach the intermediate excited state (E1). These sensitizer ions then simultaneously interact with a third ion, the activator (ion 2), which is excited to a higher-energy state (E2). Following this excitation, ion 2 can return to its ground state and emit an upconverted photon, as illustrated in Fig. 5.

The excitation of sensitizers is crucial for creating the energy reservoir necessary for the subsequent transfer to the activator. Once both sensitizer ions are in the E1 state, they interact with the activator ion (ion 2). The key feature of CSU is that the energy transfer occurs cooperatively, meaning that both sensitizer ions contribute their excitation energy simultaneously to elevate the activator ion to a higher-energy state (E2) [1, 5].

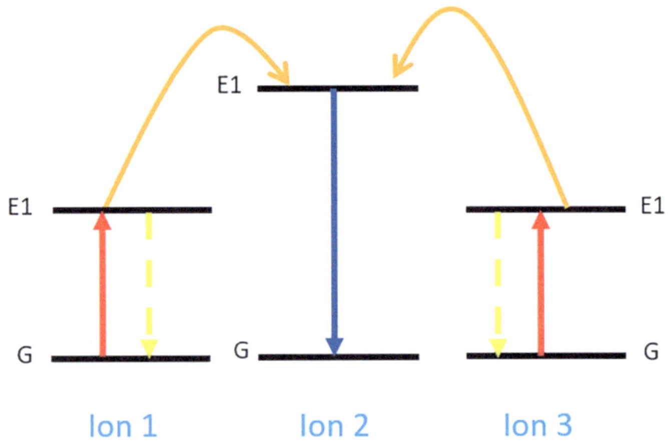

Fig. 5 Schematic diagram of cooperative sensitization upconversion

The activator ion, now receiving energy from both sensitizers, transitions to the higher-energy level E2. This cooperative mechanism enhances the likelihood of exciting the activator compared to individual ion excitation processes. After reaching the higher-energy state, the activator ion (ion 2) eventually relaxes back to its ground state, emitting a photon with energy greater than that of the incident pump photons. This photon represents the upconverted light characteristic of the CSU process. Ion 1 and ion 3 in CSU are typically of the same type, ensuring effective interaction during the energy transfer process [3].

Efficiency Considerations: CSU relies on quasi-virtual pair levels throughout the transitions, which must be maintained. As a result, its efficiency is generally lower than that of the Energy State Absorption (ESA) or Energy Transfer Upconversion (ETU) processes.

Despite its lower efficiency, the limited excitation requirements in CSU can provide advantages in specific applications, particularly in high-resolution bioimaging. This capability allows CSU to achieve imaging results that may not be possible with other upconversion methods, making it a valuable technique in the field of biomedical research [3, 7].

3.1.4 Cross-Relaxation

In the Cross-Relaxation Upconversion (CRU) process, upconversion emission results from ion-ion interactions. This mechanism involves a cross-relaxation effect between two ions, leading to a phenomenon known as self-quenching, where the fluorophore and quencher are the same molecules which dramatically reduces the intensity of emissions. This quenching occurs because the interactions between ions can lead to non-radiative decay processes, where energy is lost as heat rather than emitted as light [1, 4].

Mechanism of Cross-Relaxation

Energy Levels: In CRU, both ions possess intermediate states that have the same energy level, designated as E1. This equality in energy levels facilitates the exchange of excitation energy between the two ions.

Energy Exchange Process: The energy transfer can be represented by the following equation:

$$\text{E2(ion 1)} + \text{G(ion 2)} \rightarrow \text{E1(ion 1)} + \text{E1}'\text{(ion 2)} \tag{3}$$

In this process, the first ion (ion 1) shares some of its excitation energy (initially in the state E2) with the second ion (ion 2), which is in the ground state (G).The excitation energy from ion 1 is effectively redistributed to ion 2, resulting in both ions transitioning to their excited states E1 and E1′ simultaneously.

Dynamics of Cross-Relaxation: The efficiency of the cross-relaxation process can be influenced by several factors, including.

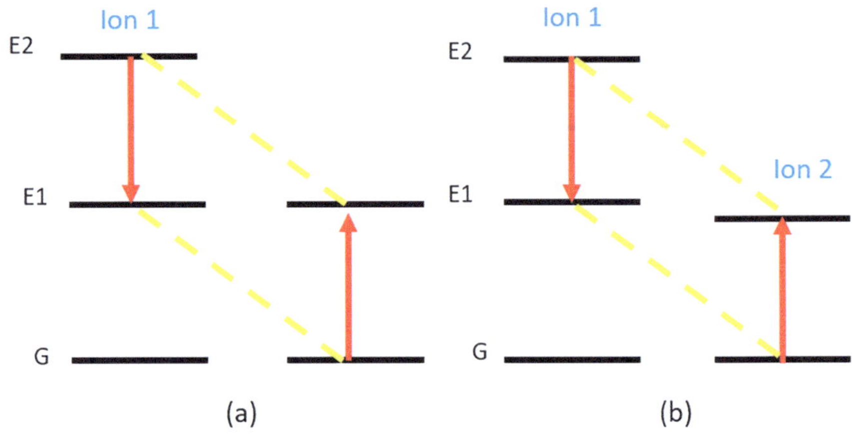

Fig. 6 Schematic diagram showing cross-relaxation in **a** same ions and **b** different ions

Ion Concentration: Higher concentrations can increase the probability of cross-relaxation events but can also exacerbate self-quenching effects.

Distance between Ions: The interaction distance between the ions plays a crucial role; optimal distances enhance energy transfer while maintaining efficient emission.

Ion Combinations: The ions involved in the cross-relaxation process can be either of the same type or different types, allowing for versatility in material design, as illustrated in Fig. 6. Common ion pairs include Yb^{3+} and Er^{3+}, which are frequently used in applications due to their complementary energy levels [1, 3, 5].

Although the CR process can diminish emission intensity, it can also be leveraged to tune the color output in upconversion nanoparticles (UNCPs). By selecting specific combinations of ions, researchers can manipulate the emission wavelength and color, which is beneficial for applications in display technologies and bioimaging, where specific color outputs are desired [1, 14].

3.1.5 Photon Avalanche

The Photon Avalanche (PA) mechanism was first observed in infrared quantum counters based on Pr^{3+} [5]. This mechanism represents a high-intensity upconversion (UC) process resulting from the combination of two distinct processes: Energy State Absorption (ESA) and Cross-Relaxation (CR) as shown in the Fig. 7 [1, 3, 5].

Mechanism of Photon Avalanche

Feedback Loop Creation: The PA mechanism operates through a feedback loop facilitated by CR, linking the excited ion to another nearby ground-state ion. This interaction allows for the rapid accumulation of population in the intermediate energy

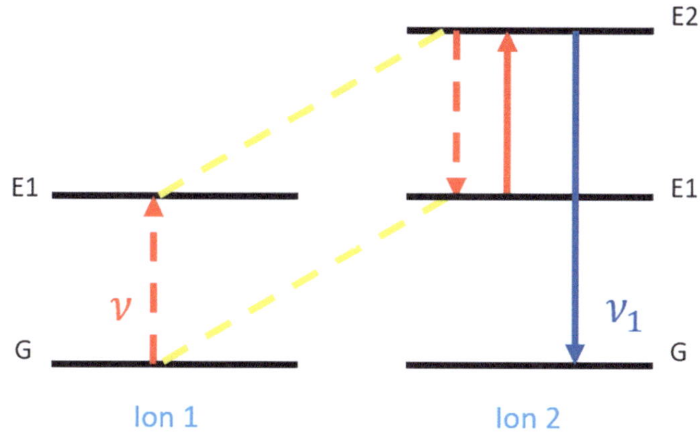

Fig. 7 Schematic diagram of photon avalanche

level, ultimately leading to a powerful upconverted emission through an avalanche effect [4, 5].

The process begins with the population of level E1 of ion 2 via non-resonant weak ground-state absorption. Following this, a resonant ESA process populates the upper visible-emitting level E2. An effective CR process then enables both ions to share the intermediate level E1, as illustrated in Fig. 7. As the feedback loop continues, the population of ion 2 in the E1 state multiplies exponentially. One ion at E1 can excite another, leading to two ions in the E1 state. These two can, in turn, excite four ions, which can then produce eight, and so forth. This rapid multiplication results in a dramatic increase in the intensity of the upconverted emission.

Kinetic Considerations: The PA process is characterized by relatively complex kinetics, where the rates of the involved processes (absorption, ESA, CR, and emission) must be finely balanced to optimize the avalanche effect. The timing of these processes is crucial, as any delay or inefficiency can disrupt the feedback loop.

Role of Dopants: The choice of dopant ions is critical in PA systems. Typically, lanthanide ions are used due to their favorable energy level structures. The interaction between these ions enhances the effectiveness of both ESA and CR, contributing to a more robust avalanche effect [1, 3].

Applications and Limitations

The intense light produced by the PA mechanism has potential applications in various fields, including Solar Energy Conversion (PA can enhance the efficiency of photonic devices by increasing the effective photon flux) and Lighting and Display Technologies (The ability to produce bright, tunable emissions can lead to advancements in LED technology and display systems).

Challenges of the PA Process

- Limited Applicability: The feedback loop is only effective in a relatively small number of upconversion systems, limiting its general applicability across various materials.
- High Pump Intensity Requirement: The PA-induced upconversion requires not only a high intensity of the pump light but also a prolonged response time—often up to several seconds—following the excitation source. This necessity can complicate practical applications, particularly in dynamic environments where rapid response times are crucial.

3.1.6 Energy Migration Upconversion

In the upconversion processes previously discussed, energy transfer has primarily been regarded as a short-range interaction, where energy is exchanged between neighboring ions (sensitizers and activators). However, the Energy Migration Upconversion (EMU) mechanism introduces the possibility of "long-range" energy transfer interactions. In this process, the excitation energy from a sensitizer located at a distance can migrate through other sensitizers before reaching the activators.

Mechanism of Energy Migration Upconversion

EMU suggests that energy transfer is not strictly a local phenomenon rather it is a Long-Range Energy Transfer. It allows for the possibility that activators positioned several nanometers away from sensitizers can still effectively absorb energy. This is facilitated by intermediary sensitizers that mediate the energy transfer process, as illustrated in Fig. 8.

In EMU, the energy migration occurs through a network of sensitizers. When an excitation energy migrates from one sensitizer to another, it can effectively bridge the gap between distant activators, allowing them to participate in the upconversion process even if they are not in immediate proximity.

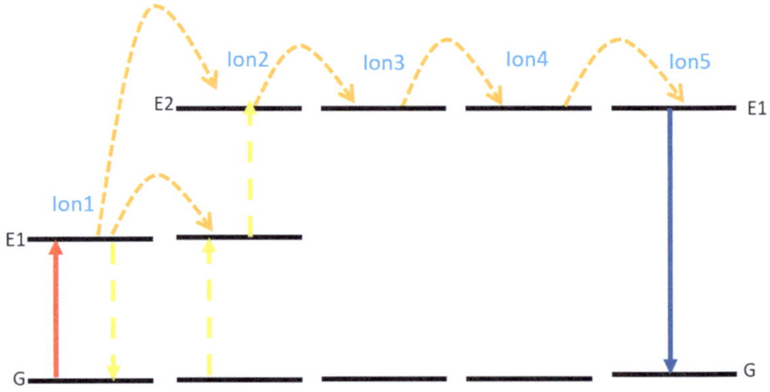

Fig. 8 Schematic diagram of energy migration upconversion

The lack of experimental research on this long-range energy migration process is largely due to the difficulty in distinguishing it from other upconversion mechanisms. The overlapping characteristics of different mechanisms can make it challenging to isolate and study the EMU process independently [1, 3, 4].

Broader Implications: The concept of long-range energy transfer in EMU expands the potential applications of upconversion technologies. By harnessing energy migration over larger distances, it may be possible to develop more efficient photonic devices, improve energy harvesting systems, and enhance the performance of biomedical imaging techniques. All the mechanisms discussed above are type of lanthanide-based upconversion. Upconversion is also possible in organic molecules as discussed below.

3.2 Types of Upconversion Processes in Organic Materials

In organic materials, photon upconversion can take place in organic molecules through processes like as Triple-Triple Annihilation and energy pooling. Organic molecules that can achieve photon upconversion are typically polycyclic aromatic hydrocarbons (PAHs).

3.2.1 Triple-Triple Annihilation

The Triplet–Triplet Annihilation Upconversion (TTA-UC) process relies on a bimolecular system comprising an emitter molecule that undergoes effective triplet–triplet annihilation and a sensitizer molecule that absorbs incident light. This mechanism enables the conversion of low-energy light into higher-energy photons [12, 13].

Mechanism of TTA-UC

Energy Absorption: The sensitizer molecule absorbs low-energy light through its singlet state transition ($S_0 - S_1$). It then quickly undergoes intersystem crossing to the triplet state (T_1), with minimal energy loss (approximately kBT, where B is Boltzmann's constant).The sensitizer, now in its T_1 state, transfers this triplet energy to the emitter molecule [12]. The emitter has a long-lived T_1 state that is energetically higher than half of its own singlet state s_1 but lower than the T_1 state of the sensitizer. When two emitter molecules, both in their T_1 states, come into proximity, they can interact to form complexes with different spin multiplicities (singlet, triplet, and quintet states). This interaction facilitates a transition where the emitter reaches a higher-energy state.

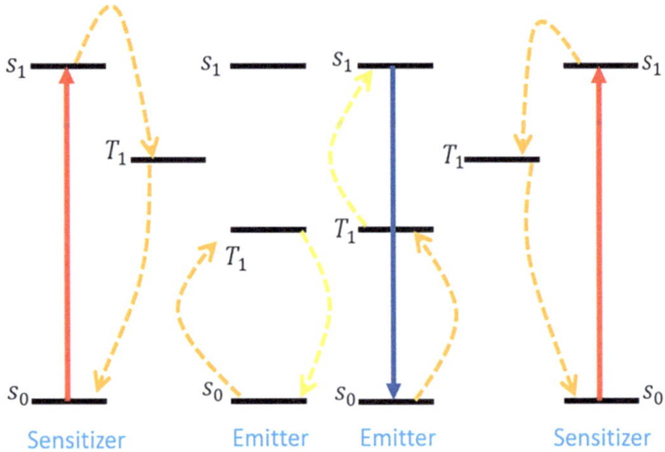

Fig. 9 Schematic diagram of triple-triple annihilation

Once the emitter has accumulated enough energy to exceed the energy of the absorbed photons, it transitions to its singlet state (S_1) and fluoresces. This fluorescence occurs almost immediately upon reaching this higher-energy state, as illustrated in Fig. 9.

The rate equation to express the dynamics of TTA-UC is as follows-

$$\frac{d[^3s^*]}{\mathrm{d}t} = k_\phi[^1s] - k_1^s[^3s^*] - k_{\mathrm{TET}}[^3s^*][^1E] - k_2^{\mathrm{SE}}[^3s^*][^3E^*] - k_2^{SS}[^3s^*]^2 = -\frac{d[^1s]}{\mathrm{d}t} \quad (4)$$

$$\frac{d[^3E^*]}{\mathrm{d}t} = k_{\mathrm{TET}}[^3S^*][^1E] - k_1^E[^3E^*] - k_2^{SE}[^3S^*][^3E^*] - k_2^{EE}[^3E^*]^2 = -\frac{d[^1E]}{\mathrm{d}t} \quad (5)$$

Here $[^3S^*]$ and $[^3E^*]$ are the sensitizer (s) and emitter (e) triplet populations [12].

Upconversion Efficiency: The efficiency of this process can be remarkable. For every N absorbed photons, $N/2$ upconverted photons are generated if all emitter molecules in the triplet states successfully react. This scenario corresponds to what is termed 100% upconversion efficiency. TTA has been proposed as a method for harvesting sub-band gap energy in solar cells. It is particularly effective for solar cells with band gaps above approximately 1.5 eV, where the lifetime of excited states becomes increasingly shorter with higher-energy levels [13].

Challenges and Considerations: While TTA-UC presents exciting possibilities, it also faces challenges, such as the need for optimized sensitizer-emitter pairs and careful management of molecular interactions to minimize energy loss through non-radiative pathways.

3.2.2 Energy Pooling

Energy pooling is a three-electron mechanism where a pair of electrons in excited states transfer energy to a third electron. This process enables the generation of higher-energy photons from lower-energy excitations, as illustrated in Fig. 10 [7].

Mechanisms of Energy Pooling

Energy pooling can be achieved through two different processes.

1. Accretive Energy Transfer (AET): In AET, the first donor molecule becomes virtualized by transferring its energy to a second donor. The second donor then transfers this energy to the acceptor molecule. The dynamics of this process are influenced by various temporal orderings of the interactions, allowing for flexible energy transfer pathways.
2. Cooperative Energy Transfer (CET): In CET, the energy from one donor is directed into the acceptor, transforming it into a virtual state. Here, the acceptor simultaneously receives energy from two donor molecules. This simultaneous energy transfer allows the acceptor to enter one of its eigenstates, leading to the emission of a higher-energy photon.

Advantages of Energy Pooling in Organic Materials

Biocompatibility: Organic materials are generally more biocompatible than inorganic counterparts, making them suitable for biomedical applications.

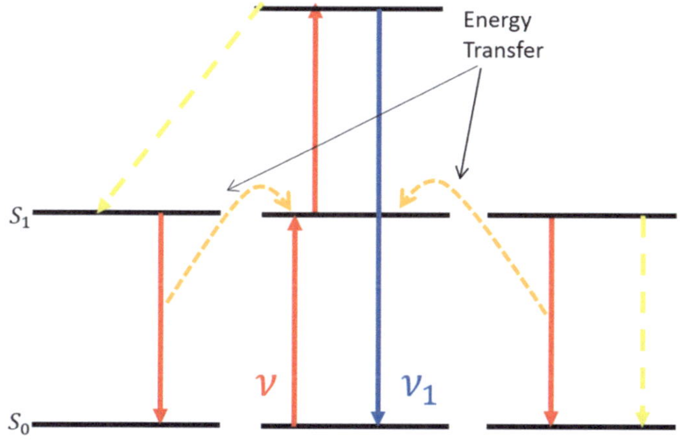

Fig. 10 Schematic diagram of energy pooling

Environmental Safety: Organic compounds often present less toxicity, which is beneficial for both environmental and health considerations.

Cost and Resource Abundance: Organic materials are typically more abundant and less expensive to produce than rare earth materials, enhancing their appeal for various applications.

Applications

Energy pooling mechanisms, including TTA and EP, hold significant promise in fields such as biomedicine, where they can be utilized for: Fluorescence Imaging, Targeted Drug Delivery, Solar Energy Harvesting. Despite their potential, the conditions under which energy pooling will effectively arise in organic condensed matter remain unclear. Further research is needed to fully understand the factors influencing EP in these systems [14].

4 Upconversion Materials

Various materials exhibit upconversion properties due to their unique absorption, emission characteristics, and electronic configurations. Here are some specific materials commonly used for upconversion processes:

4.1 Upconversion Inorganic Materials

Inorganic upconversion materials are widely studied for their ability to convert low-energy photons into higher-energy emissions. Here are some notable inorganic materials that exhibit upconversion properties:

4.1.1 The Lanthanides

Lanthanide-based upconverters are among the most widely studied materials for photon upconversion. This is largely due to the unique electronic properties of lanthanide ions, which possess spectroscopically significant electronic states located in their inner 4f shell. Their f-states exhibit characteristics that are relatively insensitive to the local environment provided by the host lattice. This insensitivity is attributed to the bonding being mediated by the outer 5s and 6p orbitals, allowing the 4f states to remain protected from external perturbations [3, 4]. The energy level diagram of the lanthanide aqua ions are given in Fig. 11.

One of the key advantages of lanthanide ions is that many of their stimulated f-states have long lifetimes, which are essential for upconversion processes. The

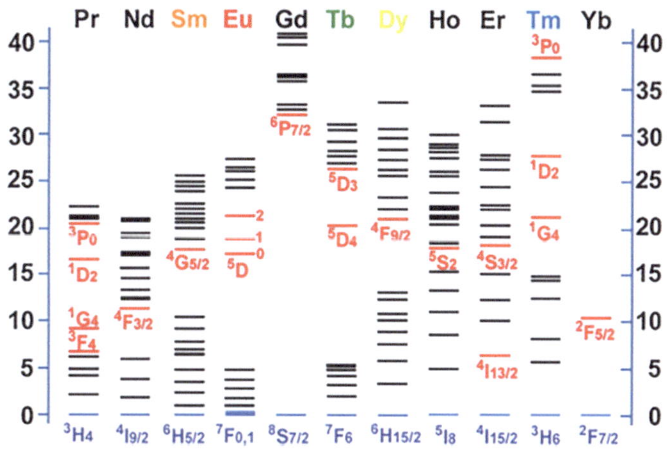

Fig. 11 Partial energy diagrams for the lanthanide aqua ions; the values of the energy levels are given as 1000 cm^{-1}. The red and blue lines indicated the main luminescent and the fundamental levels, respectively. Figures reproduced (adapted) with permission from reference [20], Elsevier

extended lifespan of these states allows for multiple interactions and energy transfers to occur [11].

In lanthanide-based upconversion, the ratio of doping between sensitizer and activator ions plays a pivotal role in achieving high luminescence. An optimal doping ratio not only maximizes the emission intensity but also minimizes self-quenching effect that can significantly diminish the efficiency and brightness of the emitted upconverted light [9].

For effective upconversion using lanthanide ions, two critical components are essential:

1. **Stable Host Material**: The host matrix must be highly stable, providing a robust environment that protects the dopant ions and facilitates their electronic transitions. Common host materials include oxides, fluorides, and phosphates.
2. **Effective Dopant Ion**: The choice of dopant ions is equally crucial. The activator ions must possess suitable energy levels that align well with those of the sensitizers to enable efficient energy transfer and subsequent photon emission [1, 3, 4]. Some common dopant ions are Yb^{3+}, Er^{3+}, Tm^{3+}

Some examples of lanthanide doped materials that undergo upconversion are

Lanthanide-Doped Fluorides: $NaYF_4$.

Lanthanide-Doped Oxide: Y_2O_3.

Lanthanide-Doped Phosphates: $LaPO_4$.

4.1.2 Quantum Dots

Definition: Quantum dots are nanoscale semiconductor particles typically ranging from 1 to 10 nm in diameter. They have discrete energy levels due to quantum confinement effects [21, 22].

Composition: Common materials for quantum dots include cadmium selenide (CdSe), cadmium sulfide (CdS), lead sulfide (PbS), and indium phosphide (InP), among others.

Size-Dependent Properties: The electronic and optical properties of quantum dots can be tuned by altering their size and shape, allowing for precise control over their absorption and emission characteristics.

These core/shell structures of quantum dotes can enhance photoluminescence and stability. Applications of quantum dots include solar energy conversion, Display technologies, etc. [22].

4.2 Upconversion Organic Materials

Organic materials that undergo photon upconversion typically involve specific mechanisms like triplet–triplet annihilation (TTA) and energy pooling. Here are some notable classes of organic materials and examples:

4.2.1 Polycyclic Aromatic Hydrocarbons (PAHs)

PAHs are organic compounds composed of multiple aromatic rings fused together, resulting in a stable π-electron system. This structure contributes to their distinctive optical properties. PAHs can efficiently absorb low-energy photons and transition to excited states [23]. They often have long-lived triplet states, which are essential for participating in upconversion mechanisms. Some examples of PAHs are: Rubrene $C_{18}H_{12}$, Perylene $C_{24}H_{12}$.

Advantages of PAHs

- Biocompatibility: Many PAHs are considered to be less toxic compared to traditional inorganic upconversion materials, making them suitable for biomedical applications.
- Cost-Effectiveness: PAHs are generally more abundant and less expensive to produce than rare earth-based materials

4.2.2 Perovskite Materials

Perovskites typically have the general formula ABX3, Perovskite materials have garnered significant attention for their upconversion capabilities due to their unique electronic properties, tunability, and structural versatility.

Key Features of Perovskite Materials for Upconversion

1. Excitonic Properties:Perovskites exhibit strong excitonic behavior, allowing them to effectively absorb low-energy photons.
2. Energy Level Alignment: The energy levels in perovskites can be finely tuned through composition and structural modifications.
3. Broad Absorption Spectrum: Many perovskite materials possess a broad absorption spectrum, enabling them to capture a wide range of incident photon energies. This characteristic is beneficial for upconversion applications, particularly when targeting near-infrared (NIR) light [24, 25].

Examples of Perovskite Materials in Upconversion: Organometallic Perovskites (e.g., $CH_3NH_3PbI_3$).

4.3 Characteristics of Upconversion Materials

Two components make up a usual upconversion material: doping ions, or emission centers, and an inorganic host matrix [3]. Both the host lattice's variation and the dopant ion's fluctuation inside the lattice can cause a change in the behavior of the material. Modifying the host lattice can have a significant impact on energy as well as radiative and non-radiative (multi-phonon) relaxation characteristics, resulting in completely distinct upconversion luminescence behavior [5, 16].

The following are significant characteristics of the hosts materials:

- High chemical stability
- High transparency in the relevant optical region (UV, visible, and NIR)
- Elevated threshold for optical damage
- Excited state lifetimes
- Photostability
- Doping concentration
- Spectral properties
- Furthermore, the host's vibrational characteristics have a significant impact on the UC emission efficiency, which affects non-radiative relaxations

4.4 Upconversion Performance

Upconversion performance can be determined by its efficiency. Generally, UC efficiency is expressed as a statistic called UC luminescence quantum yield (QY), which is the ratio of absorbed photons to emitted photons in UC [1].

$$QY = \frac{\text{The number of emited photon}}{\text{The numer of absorbed photon}} \tag{6}$$

5 Applications of Upconversion Processes

Not only is the upconversion photoluminescence an intriguing and unique photophysical phenomenon, but it also offers a lot of potential. UCNPs have a wide range of applications. Here are a few of them:

5.1 Chemical Sensors

The UCNPs can be surface-modified to create nanoprobes that can be applied directly to complicated biological samples or living cells, or they can be treated with appropriate polymer materials to create planar sensor layers [6] and thus UCNPs can be used to make many sensors. Some of them are talked about below:

(a) Sensors for enzymatic reactions—In this technique, Self-assembled phospholipid monolayers made of phospholipids modified with PEG and phospholipids tagged with rhodamine B can be applied on the nanoparticles. These nanocomposites have the potential to be employed for tracking PLD activity in living cells. The rhodamine dye is released from the particle surface by PLD-mediated breaking of the phosphodiester bond, which stops the LRET from the nanocrystals to the dye. Ratiometric imaging of PLD function in human malignant cells was conducted using the restored signal of the 540 nm UCL fluorescence and the inert standard signal at 655 nm [6].

(b) Oxygen sensors—By employing a phosphorescent cyclometalated Ir (III) coumarin complex as a probe for oxygen distributed in ethyl cellulose, a UCNP-based oxygen sensor can be created [6].

(c) Temperature sensors—A passive shell of undoped NaYF4 can be borne by core–shell UCNPs to produce nanoprobes with enhanced temperature sensitivity. Peak ratios of the green emissions can be used to calculate temperatures in a roughly linear range from 25 to 60 °C with a resolution of 0.5 °C [6].

5.2 *Bioassays*

A biological test known as a bioassay quantifies the amount of a biomolecule, also called an "analyte," that is present in a solution. Examples of these biomolecules include antibodies and immunoglobulins [5]. In point-of-care theranostics, it is essential for the very sensitive identification of malignant biomarkers in human fluid, for instance samples are necessary for early cancer detection, which improves patient survival rate [15]. There are various benefits to using UCNPs in assays: (i) Due to the special characteristics of UC, backdrop luminescence from the carrier liquid or the assay biochemistry, hence reducing the LOD that is not accessible by traditional assays. (ii) UCNPs' optical properties are immune to the surroundings [5]. The schematic representation of various applications of up conversion materials are summarized in Fig. 12.

(a) **Heterogeneous assay**—The UC heterogeneous assay's removal of autofluorescence has led to a noticeably better LOD when compared to traditional methods [6]. DNA detection, bacterial pathogens (Streptococcus pneumoniae), nucleic acids, human chorionic gonadotropin (a drug of abuse panel), and other analytes can all be detected with UC particles in a variety of heterogeneous bioassays [5].

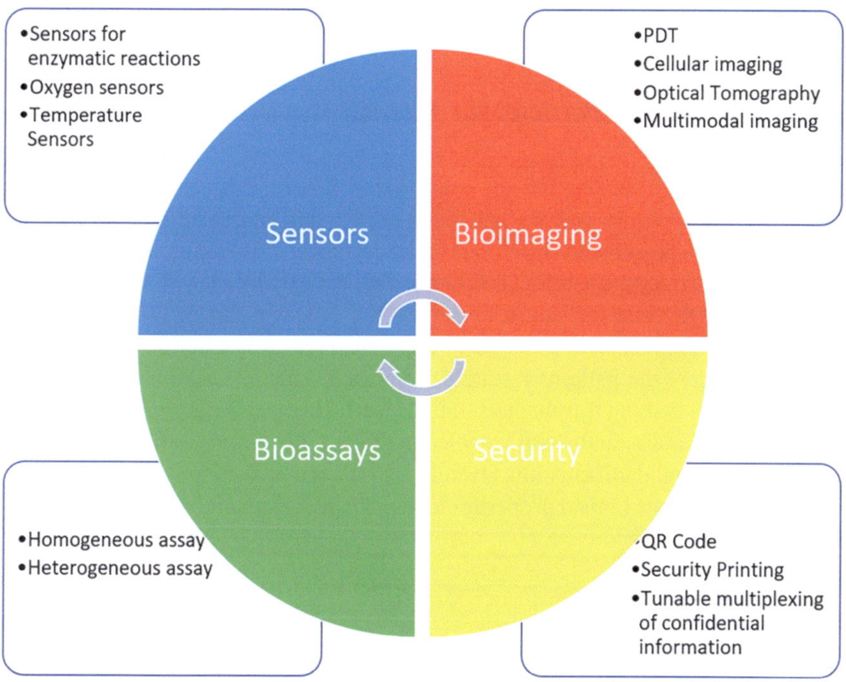

Fig. 12 Schematic diagram showing various applications of upconversions

(b) **Homogeneous assay**—Homogeneous UC assays use long-range dipole–dipole interactions to detect analytes in the solution by utilizing the distance-dependent energy transfer between the donor (UCNPs) and a nearby acceptor molecule [5]. UC-based homogeneous essay can be utilize to detect target DNA [5].

5.3 *Bioimaging*

Photoluminesence imaging is crucial to biomedical research because it can be used to detect, screen, and treat life-threatening diseases with image-guided therapy. But traditional Stokes-shifted fluorophores (such as fluorescent proteins, metal complexes, and organic fluorophores) or quantum dots are not the best choices for bioimaging because their excitation occurs in the UV or visible spectrum, which has poor light penetration and creates a strong imaging background [14].

(a) **Upconversion Photodynamic Therapy:** A clinical procedure known as photodynamic therapy (PDT) uses chemicals that are photo triggered (photosensitizers) to create singlet oxygen, which kills cancer [15]. Because NIR radiation penetrates deeply into biological tissues, there has been a lot of interest recently in PDT due to the introduction of UC nanoparticles that can convert NIR light into the visible spectrum. ROS can be produced by photosensitizers being excited by the nanoparticles' upconverted visual emission [5].

5.4 *Upconversion Particles for Security Application*

Upconversion nanomaterials are not visible to the naked eye but can be seen through NIR excitation. Security printing has found use for this unique attribute of UNCPs. Using UNCPs, a preset read condition can be established by designing them to only render colors correctly at particular excitation power densities [1, 3]. Other than this upconversion QR code tag can be create using e-NIL that can be used for authentication [1]. Moreover, it is possible to implement Tunable Lifetime Multiplexing of UCNPs, where the patterns resulting from each doping. Each layer possesses a unique lifetime, making this method suitable for security applications, ensuring that only authorized individuals with knowledge of the luminescence lifetimes can gain access to confidential information. Overall, UCNPs can be used for security purposes owing to their unique optical properties to create a secure and reliable access control mechanism.

5.5 *Role of UNCPs in Solar Cell*

There are various loss mechanisms in solar cell affecting its efficiency directly such as: Sub-band gap energy photon loss, thermalization of charge carriers, contact voltage loss, junction loss, and reflection loss. Upconverting (UC) materials are inherently opaque to photons that do not match their band gap [3]. This property allows UC material layers to be effectively integrated on the backside of solar cells, eliminating concerns about reflection losses that typically hinder performance. Also, two sub-band gap photons can be converted into one above band gap photon by using UC materials, which offers a solution for the sub-band gap energy photon loss [1, 3].

6 Conclusion

Based on the extensive discussion on photon upconversion, it is evident that this phenomenon holds significant promise across various scientific and technological domains the evolution of photon upconversion has led to the discovery and understanding of diverse mechanisms such as energy transfer upconversion (ETU), cooperative sensitization upconversion (CSU), cross-relaxation, photon avalanche (PA), energy migration upconversion (EMU), and triplet–triplet annihilation (TTA). These mechanisms not only illustrate the complexity and versatility of photon upconversion processes but also underscore their potential applications. In practical terms, photon upconversion offers distinct advantages such as enabling efficient energy transfer from lower-energy photons to higher-energy states, which is particularly useful in fields requiring high-energy light emission like biomedicine, solar cells, and security applications. The ability to convert near-infrared (NIR) light into visible light enhances its utility in bioimaging, photodynamic therapy (PDT), and sensing technologies, where traditional fluorescent dyes or quantum dots fall short due to limited tissue penetration and photodamage concerns. Moreover, the choice of host materials and dopant ions crucially influences the efficiency and applicability of upconversion processes, with lanthanide-based materials emerging as prominent candidates due to their unique electronic transitions and stability. Synthetic strategies such as coprecipitation, solvothermal, and sol–gel methods play pivotal roles in tailoring the properties of upconversion nanomaterials for specific application.

Looking forward, the future of photon upconversion holds exciting prospects. Ongoing research aims to improve UC efficiency, expand material diversity beyond lanthanides to include organic molecules and hybrid nanomaterials, and explore novel UC mechanisms. Advances in nanotechnology and materials science are expected to drive innovations in UC applications, facilitating breakthroughs in fields ranging from quantum computing to environmental monitoring. Moreover, the integration of UC materials into wearable devices, smart sensors, and Internet-of-Things (IoT) technologies could revolutionize data communication and security systems. The ability of

UC particles to emit light in response to specific excitation wavelengths enables their use in encryption, anti-counterfeiting measures, and advanced security applications.

In conclusion, photon upconversion represents a cornerstone of modern photonics and materials science, with profound implications for diverse industries. Continued interdisciplinary research and technological advancements are set to unlock new functionalities and applications, paving the way for a future where UC technologies play a crucial role in addressing global challenges and improving quality of life.

References

1. C. Altavilla (ed), *Upconverting Nanomaterials: Perspectives, Synthesis, And Applications* (CRC Press, 2016)
2. A. Beiser, *Concepts of Modern Physics* (2003)
3. F. Zhang, *Photon Upconversion Nanomaterials*, vol. 416 (Springer, Berlin, 2015)
4. V.K. Rai (ed), *Upconverting Nanoparticles: From Fundamentals to Applications* (Wiley, 2022)
5. G. Chen et al.,Upconversion nanoparticles: design, nanochemistry, and applications in theranostics. Chem. Rev. **114**(10), 5161–5214 (2014)
6. S. Christ, M. Schäferling, Chemical sensing and imaging based on photon upconverting nano- and microcrystals: a review. Methods Appl. Fluoresc. **3**(3), 034004 (2015)
7. M.D. LaCount et al.,Energy pooling upconversion in organic molecular systems. J. Phys. Chem. A **119**(17), 4009–4016 (2015)
8. S.D. Melgaard, Energy transfer upconversion processes. Chem. Rev. **104**(1), 145 (2004)
9. R.-S. Liu (ed.), *Phosphors, Up Conversion Nano Particles, Quantum Dots and Their Applications*, vol. 1 (Springer, Berlin/Heidelberg, Germany, 2017)
10. J. Dong et al.,Plasmon-enhanced upconversion photoluminescence: mechanism and application. Rev. Phys. **4**, 100026 (2019)
11. J. Wang et al., Single-band upconversion emission in lanthanide-doped KMnF3 nanocrystals. AngewandteChemie-International Edition **50**(44), 10369 (2011)
12. T.F. Schulze, T.W. Schmidt, Photochemical upconversion: present status and prospects for its application to solar energy conversion. Energy Environ. Sci. **8**(1), 103–125 (2015)
13. Y.Y. Cheng et al.,On the efficiency limit of triplet–triplet annihilation for photochemical upconversion. Phys. Chem. Chem. Phys. **12**(1), 66–71 (2010)
14. F. Wang et al.,Upconversion nanoparticles in biological labeling, imaging, and therapy. Analyst **135**(8), 1839–1854 (2010)
15. Y.I. Park et al.,Upconverting nanoparticles: a versatile platform for wide-field two-photon microscopy and multi-modal in vivo imaging. Chem. Soc. Rev. **44**(6), 1302–1317 (2015)
16. J. Bergstrand et al., On the decay time of upconversion luminescence. Nanoscale **11**(11), 4959–4969 (2019)
17. F. Auzel, History of upconversion discovery and its evolution. J. Lumin. **223**, 116900 (2020)
18. N.S. Satpute et al., Synthesis and luminescence characterization of downconversion and down-shifting phosphor for efficiency enhancement of solar cells: Perspectives and challenges. ACS Appl. Electron. Mater. **4**(7), 3354–3391 (2022)
19. J.F.-C. Loo et al., Upconversion and downconversion nanoparticles for biophotonics and nanomedicine. Coord. Chem. Rev. **400**, 213042 (2019)
20. S. SeethaLekshmi et al., Lanthanide complex-derived white-light emitting solids: a survey on design strategies. J. Photochem. Photobiol. C Photochem. Rev. **33**, 109–131 (2017)
21. A.J. Nozik, Quantum dot solar cells. Phys. E **14**(1–2), 115–120 (2002)
22. Z. Deutsch, L. Neeman, D. Oron, Luminescence upconversion in colloidal double quantum dots. Nat. Nanotechnol. **8**(9), 649–653 (2013)

23. R. Morales-Cueto et al., Singlet excited-state dynamics of nitropolycyclic aromatic hydrocarbons: direct measurements by femtosecond fluorescence up-conversion. J. Phys. Chem.istry A **111**(4), 552–557 (2007)
24. T. Liang et al., Application of upconversion photoluminescent materials in perovskite solar cells: opportunities and challenges. Mater. Today Energy **21**, 100740 (2021)
25. Z.A. VanOrman et al., Bulk halide perovskites as triplet sensitizers: progress and prospects in photon upconversion. J. Mater. Chem. C **9**(8), 2685–2694 (2021)

A Review Report on Rare Earth Activated Phosphors for Bioimaging Applications

Ananya Chakraborty, Jelena Mitrić, and Vikas Dubey

Abstract This review presents an overview of various synthesis and characterization techniques used for rare earth activated phosphors intended for bioimaging applications. Here, various researchers developed the rare earth activated phosphors using a variety of synthesis techniques, including solid state, sol–gel, hydrothermal, combustion, etc. As reported in their studies, the resulting phosphors have been compared by using various characterization techniques for its structural analysis, morphological analysis, optical analysis, to assess their potential in bioimaging applications. Previously some authors have highlighted the importance of bioimaging phosphor to improve the quality of medical optical imaging and its low-cost effect. We can stimulate the biological window for tissue imaging experiments using rare earth activated phosphors. Herein, some important wavelength ranges relevant for tissue imaging fall within the biological optical window of 650–950 nm, where reduced absorption and scattering enable deeper light penetration into biological tissues. Current phosphors have a significant drawback that they are only ideal for one time use because their persistent brightness fades after a few hours. To address this problem, persistent phosphors with photo-stimulated properties are used. In this article, an in-depth assessment has been provided that surely will be a good resource for researchers in exploring various fields related to biomedical imaging.

Keywords Phosphors · Bioimaging · Persistent behavior

A. Chakraborty (✉) · V. Dubey
Department of Physics, North-Eastern Hill University, Shillong, Meghalaya, India
e-mail: ananyachakraborty4769@gmail.com

J. Mitrić
Department of Physics, Institute of Physics, Belgrade, Serbia

V. Dubey et al. (eds.), *Upconversion Luminescence of Phosphors*, Progress in Optical Science and Photonics 39, https://doi.org/10.1007/978-981-95-1762-6_3

1 Introduction

Over the years, advancements in medical technologies have transformed disease detection and clinical management. Among these innovations, bioimaging has gained recognition as a powerful and rapidly evolving approach in medical diagnostics.

1.1 Bioimaging

Bioimaging is a form of noninvasively imaging that enables real-time viewing of biological functions without the need of disruptive instruments and minimizes disturbance with living processes [1]. Bioimaging becomes progressively significant in the detection of cancer, surgical guidance, and after surgery assessment [2]. The advancements in medical imaging have profoundly impacted how diseases are diagnosed and monitored. Conventional imaging techniques—such as computed tomography (CT), magnetic resonance imaging (MRI), ultrasound, and X-ray radiography—remain central to clinical diagnostics, offering valuable insights into anatomical and structural features. However, each of these methods has inherent drawbacks. CT and X-ray imaging involve exposure to ionizing radiation, which can be harmful when used frequently. Ultrasound, though radiation-free and affordable, suffers from limited resolution and reduced efficacy for deep-tissue visualization. MRI provides high-resolution soft tissue imaging but is costly, time-intensive, and not always practical for real-time or molecular-level studies. Although traditional modalities are part of the broader domain of bioimaging—which encompasses non-invasive visualization of biological structures and processes—recent advances in molecular and optical bioimaging have introduced new dimensions to diagnostics. These emerging techniques offer enhanced sensitivity, improved specificity, and the ability to monitor biological functions at the cellular and molecular level, making them valuable complements to conventional imaging systems [2–4].

Among the diverse range of imaging approaches, luminescence-based techniques have gained considerable attention for their exceptional sensitivity, biocompatibility, and high-resolution capabilities. In particular, rare earth (lanthanide)-doped luminescent phosphors have emerged as promising optical probes, having extensive applications in bioimaging, medication delivery, treatment of cancer, and health monitoring [5].

1.2 Rare Earth-Activated Phosphors

In the late eighteenth century, the scandium, yttrium, and lanthanide (Ln) series (from lanthanum to lutetium) are rare earth (RE) elements that were discovered and are grouped in the periodic table in the IIIB group. In reality, RE elements are not relatively uncommon in terms of their availability on Earth; they are typically found in the trivalent state (RE^{3+}). Over the past 200 years, RE elements have been extensively studied and have been employed in the domains of chemical engineering, metallurgy, optical, electronics, ceramics, glass, textiles, catalysis, and biomedicine since they have their distinct chemical as well as physical characteristics [6]. Mostly rare earth activated phosphors exhibit exceptional luminescent properties including high quantum efficiency. These materials can absorb light in one wavelength and re-emit it at another, which is critical for effective imaging [7].

They furthermore possess two luminescence modes: Up-conversion (UC) and down-conversion (DC) [2]. Figure 1 shows the schematic representation of down-conversion process and up-conversion process.

Up-conversion phosphors (anti-Stokes luminescence) use multi-photon absorption methods to transform near-infrared (NIR) light into high-energy light which comes in the visible or ultraviolet (UV) range [8].

Down-conversion (Stokes shift) phosphors, release photons having less energy in the visible or near-infrared range while absorbing higher-energy photons (UV or visible) [9, 10].

RE-doped nanoparticle possesses a long durability, excellent fluorescence steadiness, tuned emission, and minimal photobleaching. Figure 2 shows the Dieke's diagram of trivalent lanthanoid ions displaying different stages of energy and governing stages of transition. The f-f electronic transitions in lanthanide ions give rise to the distinctive optical features associated with RE activated phosphors, which include narrow emission peaks and lengthened emission durations [12, 13].

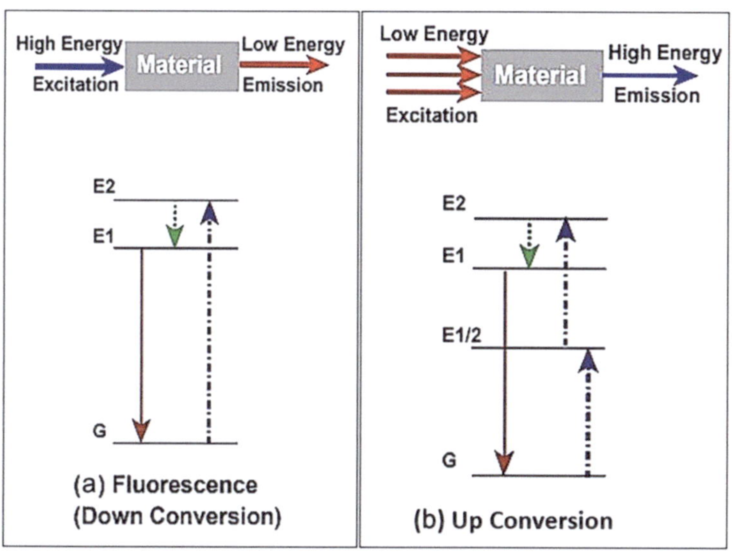

Fig. 1 Schematic diagrams of down-conversion process (left) and up-conversion process (right) (taken from Ref. [11])

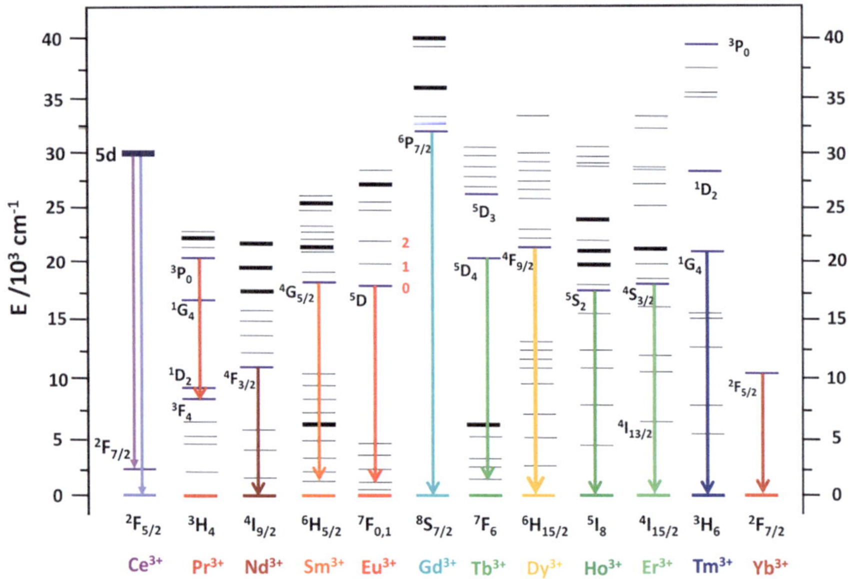

Fig. 2 Dieke's diagram of trivalent lanthanoid ions displaying different stages of energy and governing stages of transition (taken from Ref. [18])

Certain lanthanide-doped materials can store excitation energy in traps (defects) and release it slowly, emitting light for minutes to hours after excitation stops, these particles may keep releasing light for an extended period because of their persistent luminescence property [14–16]. Persistent luminescence is a rare optical phenomenon in which light emission lasts for hours or even days following the excitement is turned off. It is also termed as Long After Glow. Due to its several uses in bioimaging, information storage, energy efficiency, and night vision, this self-sustaining optical characteristic has garnered a lot of interest [17].

With the growing demand in RE phosphors for bioimaging, numerous studies targeting synthesis methodologies, optical features, and therapeutic uses are being benefited. Thus, in order to demonstrate this emerging matter, an in-depth assessment encompassing the latest advancements, existing issues and upcoming perspectives in this domain is necessary. Here, a brief review of elements that contain rare earth elements (which have been shown to boost persistent luminescence in the host materials and also enhance deep-tissue imaging) as doping agents will be mentioned. This study covers their synthesis strategies, characterization processes, luminescence mechanisms, and practical applications in bioimaging. By covering these objectives, this article will be a beneficial resource for researchers exploring various fields related to biomedical imaging.

2 Literature Review

Rare earth activated phosphors have shown immense potential in bioimaging due to their unique optical properties, especially persistent luminescence (PL) and up-conversion luminescence (UCL). These mechanisms offer high contrast, long emission lifetimes, low background interference, and deep tissue penetration, making them highly effective for both in vitro and in vivo biomedical imaging. The following section categorizes the recent research progress based on the luminescence mechanism and host matrix composition.

2.1 Persistent Luminescence-Based Rare-Earth Phosphors

Persistent luminescence (PL), or afterglow, refers to the ability of materials to emit light for extended periods following the removal of excitation. This property eliminates the need for continuous external illumination, enabling background-free imaging and long-duration tracking in vivo.

2.1.1 Silicate and Aluminate-Based Persistent Phosphors

Silicate and aluminate materials have been extensively studied as host lattices for persistent luminescence due to their structural stability, ease of doping with divalent and trivalent rare-earth ions, and their capability to accommodate efficient trap centers for long-duration afterglow emission.

In an important contribution, Homayoni et al. [19] developed water-dispersible $Sr_2MgSi_2O_7:Eu^{2+}$, Dy^{3+} nanoparticles using a sol–gel synthesis approach. To enhance biocompatibility, these nanoparticles were modified with amphiphilic PMMA-b-PEG polymers, which significantly improved their dispersibility in aqueous environments. A remarkable feature of this work was the conjugation of the phosphors with verteporfin, a photosensitizer. This innovative design allowed the persistent afterglow emission of the phosphors to activate verteporfin and generate singlet oxygen without the need for continuous light exposure, thus enabling effective photodynamic therapy (PDT) even in deep-tissue environments where light penetration is typically limited. In a separate study, Sharma et al. [20] explored $Ca_2MgSi_2O_7:Eu^{2+},Dy^{3+}$ phosphors synthesized via combustion. The resulting material exhibited bright green emission at 532 nm under near-UV excitation, attributed to the allowed $4f^65d^1 \rightarrow 4f^7$ transitions of Eu^{2+}. The inclusion of Dy^{3+} served to introduce deep traps that effectively prolonged the afterglow duration, making the material potentially useful for imaging, tissue engineering, and drug delivery applications, though direct bioimaging studies were not performed. In another independent study, Sun et al. [21] introduced $SrAlO_4:Eu^{2+}$, Dy^{3+} nanoparticles coated with polyethylene glycol and pyrophosphoric acid to improve their biological compatibility. When administered in vivo, these particles exhibited strong afterglow for over 30 min post-UV excitation and could be reactivated repeatedly with simple LED light sources, facilitating prolonged imaging sessions over multiple days without the need for repeated nanoparticle injections. Expanding on implantable biomedical devices, Hu et al. [22] developed an implantable "optical battery" by embedding $SrAl_2O_4:Eu^{2+}$, Dy^{3+} persistent luminescence nanoparticles alongside up-conversion nanoparticles (UCNPs) and photosensitizers into a polydimethylsiloxane (PDMS) matrix. This composite system could be charged within seconds using NIR laser excitation, thereby enabling repeated cycles of irradiation-free PDT while minimizing thermal damage, a major advancement for safe and controllable phototherapy. Calatayud et al. [23] contributed to this field by investigating $Sr_4Al_{14}O_{25}:Eu^{2+}$, Dy^{3+} persistent phosphors. Their study confirmed the stable and non-toxic nature of these materials, reinforcing their suitability for biomedical imaging applications that demand long-term biocompatibility. In a distinct study, Li et al. [24] synthesized Sr_2SiO_4 phosphors co-doped with Eu^{2+} and Pr^{3+} ions. This system exhibited tunable luminescence properties, which are particularly advantageous for multi-color bioimaging and multiplexed diagnostic applications that require precise emission wavelength control.

2.1.2 Sulfide-Based Persistent Phosphors

In a significant study, Rodríguez Burbano et al. [25] studied CaS nanoparticles co-doped with Eu^{2+} and Dy^{3+} ions. The resulting phosphors exhibited pronounced red persistent luminescence that remained visible for up to five hours following ultraviolet excitation. An outstanding feature of these nanoparticles was their ability to undergo photo-stimulated luminescence (PSL), whereby trapped charge carriers could be released under near-infrared excitation at 980 nm, producing renewed luminescence. This property enabled multiple extended imaging sessions, with PSL cycles lasting up to 18 min after each reactivation. The combination of long-lasting afterglow and reactivatable luminescence renders these materials particularly well-suited for repeated, non-invasive deep-tissue bioimaging, where sustained signal visibility is critical. In a separate contribution, Gao et al. [26] developed CaS nanocrystals co-doped with Eu^{2+} and Sm^{3+} ions, specifically designed to leverage trap-controlled charge storage mechanisms for photo-stimulated luminescence. These nanocrystals were initially activated using either X-ray or UV radiation, storing energy within deep trap centers created by Sm^{3+} doping. After initial activation by X-ray or UV exposure, the stored energy could be released as NIR emission under low-energy stimulation. Their biocompatibility was verified through successful testing in live animal models, demonstrating suitability for long-term, non-invasive deep-tissue imaging. Importantly, their surface modification provided excellent biocompatibility, and their functional performance was validated through successful imaging experiments in live animal models.

2.1.3 Garnet and Perovskite Structures

Garnet and perovskite-based phosphors are increasingly being explored for their tunable broadband emission and thermal stability, both of which are highly advantageous for biomedical applications.

Jiang et al. [27] employed defect engineering in $Ca_3Sc_2Si_3O_{12}:Eu^{2+}$ garnet phosphors, creating oxygen vacancies that modified the coordination environment around Eu^{2+} ions. This approach resulted in broadband near-infrared persistent luminescence ranging from 650 to 1100 nm, making the material exceptionally suited for bioimaging applications requiring deep penetration and minimal background interference. In another study, Dai et al. [28] studied $Ca_3Ga_2Ge_3O_{12}:Pr^{3+}$, Yb^{3+} phosphors, demonstrating persistent luminescence with improved thermal stability and long-duration emission, attributes highly valuable for in vivo imaging applications. In the domain of perovskite structures, Pellerin et al. [29] synthesized $LaAlO_3$ co-doped with Cr^{3+} and Sm^{3+} ions. The presence of Cr^{3+} provided strong NIR emission at 700 nm, while Sm^{3+} introduced trap states that extended the afterglow duration. The emission wavelength aligns well with the biological transparency window, enhancing suitability for deep-tissue imaging.

2.1.4 Other Persistent Host Systems

Additional innovative host systems have emerged offering unique persistent luminescence behaviors. Xiong et al. [30] investigated $LiGa_5O_8$ doped with Pr^{3+}, which exhibited both visible and NIR afterglow following UV excitation. Importantly, tissue penetration and cytotoxicity assessments confirmed the material's potential for bioimaging applications, highlighting its multifunctionality for both biomedical imaging and optical data storage. Further Rosticher et al. [31] examined $CaMgSi_2O_6$ nanophosphors co-doped with Eu^{2+}, Mn^{2+}, and Dy^{3+}, achieving strong red afterglow emission at approximately 680 nm through synergistic energy transfer from Mn^{2+} and efficient charge trapping by Dy^{3+}. These nanophosphors hold promise for non-invasive in vivo imaging due to their nanoscale size and long-lasting luminescence.

2.2 Up-Conversion Luminescence-Based Rare-Earth Phosphors

Up-conversion nanoparticles (UCNPs) absorb multiple NIR photons, emitting at shorter wavelengths, enabling deep-tissue imaging with negligible autofluorescence and high signal-to-noise ratio. Zijlmans et al. [32] were among the first to explore up-conversion properties for biological applications, pioneering the use of UC nanocrystals as bio-labels. Since this early work, substantial progress has been made in developing UC-based nanoprobes for diverse biomedical and bioanalytical applications, owing to their sharp emission bands, high photostability, low background fluorescence, and superior tissue penetration capabilities.

2.2.1 Fluoride-Based Up-Conversion Phosphors

Fluoride-based host lattices, particularly $NaYF_4$ as host has been widely regarded as the most efficient matrices for up-conversion luminescence due to excellent lattice compatibility with rare-earth dopants.

Nam et al. [33] developed highly efficient $NaYF_4$ nanoparticles co-doped with Yb^{3+} and Er^{3+} ions. These nanoparticles exhibited strong visible emission under 980 nm NIR excitation, making them highly suitable for continuous, real-time imaging with minimal background noise and photobleaching, especially for dynamic biological processes. Further expanding the application of these fluorides, Idris et al. [34] demonstrated the use of $NaYF_4$: Yb^{3+}, Er^{3+} up-conversion nanoparticles for real-time tracking of transplanted cells in live animal models. The deep tissue penetration, stability, and long-term signal retention offered by these nanoparticles position them as powerful tools for monitoring cell-based therapies in regenerative medicine. Cao et al. [35], in their study, developed β-$NaYF_4$:Er^{3+}, Yb^{3+} nanocrystals with uniform size and controlled dopant distribution. By optimizing the Yb^{3+}/Er^{3+} ratio (20% Yb^{3+}, 1% Er^{3+}), they achieved dominant red emission under 980 nm excitation through efficient $Yb^{3+} \rightarrow Er^{3+}$ energy transfer. The enhanced red luminescence, offering superior tissue penetration, demonstrates their potential for deep-tissue bioimaging and bio-labeling. Zhou et al. [36] further contributed to the field by synthesizing water-soluble $NaYF_4$:Yb^{3+},Tm^{3+} up-conversion nanocrystals (UCNPs) via a solvothermal method, exhibiting strong near-infrared (NIR) luminescence at ~ 800 nm upon 980 nm excitation. The NIR-to-NIR emission falls within the biological window, providing deep tissue penetration and minimal autofluorescence. Both in vitro (HeLa cells) and in vivo (C. elegans) models indicated good biocompatibility, excretion behavior and low cytotoxicity, supporting their suitability for NIR bioimaging. In another study, Cho et al. [37] developed multifunctional nanoclusters composed of $NaYF_4$: Yb^{3+}, Er^{3+} up-conversion nanoparticles (UCNPs) coupled with gold nanorods (AuNRs) through PEGylation, designed for simultaneous bioimaging and targeted chemotherapy of bladder cancer. Under near-infrared (NIR) excitation, the UCNP component produced visible luminescence, enabling high contrast, background-free imaging of cancer cells. The AuNRs provided plasmonic enhancement, allowed localized optoporation of cancer cells, and promoted efficient intracellular drug delivery when required. In addition to improving the luminescence intensity, surface modification strategies have been employed to enable targeted bioimaging. Xiong et al. [38] functionalized $NaYF_4$:Yb^{3+}, Er^{3+} up-conversion nanophosphors with amine groups, allowing further conjugation with folic acid (FA) ligands for targeted imaging of folate-receptor-positive cancer cells. Under 980 nm excitation, these nanophosphors exhibited strong green and red up-conversion emissions. Both in vitro and in vivo studies confirmed their excellent targeting ability, low toxicity, good biocompatibility, and high photostability, demonstrating their strong potential for deep-tissue targeted bioimaging.

Expanding beyond conventional $NaYF_4$-based hosts, Janbandhu et al. [39] synthesized Er^{3+}/Yb^{3+} co-doped $KAlF_4$ up-conversion phosphors as an alternative to conventional hosts for bioimaging. Under 980 nm laser excitation, these phosphors exhibited strong red emission at 662 nm $^4F_{9/2} \rightarrow {}^4I_{15/2}$ along with moderate green emissions at 547 and 522 nm weak NIR emissions. A weak near-infrared emission band (~ 834 nm) was also observed. The incorporation of Yb^{3+} as a sensitizer effectively enhanced the visible up-conversion efficiency by promoting energy transfer to Er^{3+} activators. The efficient red emission under NIR excitation, makes $KAlF_4:Er^{3+}$, Yb^{3+} a promising candidate for bioimaging applications.

2.2.2 Oxide-Based Up-Conversion Hosts

Beyond fluoride hosts, oxide-based up-conversion systems offer enhanced structural stability and multifunctionality for biomedical imaging. Sun et al. [40] synthesized La_2O_3 nanoparticles co-doped with Yb^{3+} and Er^{3+} ions, exhibited strong green up-conversion luminescence under 980 nm excitation, attributed to the efficient energy transfer from Yb^{3+} to Er^{3+} ions. In addition to their efficient up-conversion luminescence, this material also exhibited localized optical heating effects which enabled their potential for temperature-responsive biomedical therapies as well as real-time imaging. Furthermore, their low cytotoxicity was confirmed, and confocal microscopy imaging in HeLa cells demonstrated successful cellular uptake and luminescence, highlighting the promising bioimaging capabilities of $La_2O_3:Yb^{3+}/Er^{3+}$ nanocrystals. In another study, Ajithkumar et al. [41] developed Gd_2O_2S phosphors doped with Yb^{3+} and Er^{3+}, which demonstrated strong up-conversion luminescence under 980 nm excitation. Importantly, the inclusion of Gd^{3+} conferred paramagnetic properties, enabling these particles to function as MRI contrast agents in addition to optical imaging probes. Cytotoxicity assessments confirmed their biocompatibility, establishing these materials as promising candidates for multimodal bioimaging platforms. In a significant contribution, Gupta et al. [42] synthesized $La_2Zr_2O_7$ nanoparticles co-doped with various rare-earth ions achieving versatile luminescence behavior. The Er^{3+}/Yb^{3+} co-doped system demonstrated efficient up-conversion luminescence upon 980 nm excitation, emitting strong green (~ 540 nm) and red (~ 660 nm) light through two-photon processes, along with a significant near-infrared downshifting emission at approximately 1540 nm ($^4I_{13/2} \rightarrow {}^4I_{15/2}$ transition of Er^{3+}). The emission at 1540 nm, located within the NIR-II biological window, offers particular promise for deep-tissue bioimaging due to its minimized tissue absorption and scattering. Furthermore, the $Yb^{3+}/Tm^{3+}/Gd^{3+}$ triply doped $La_2Zr_2O_7$ nanoparticles displayed ultraviolet up-conversion emissions (centered around 296–306 nm), facilitated by energy transfer from Tm^{3+} to Gd^{3+} ions. These wide spectral emissions, covering ultraviolet, visible (red and green) and NIR-II regions, $La_2Zr_2O_7$-based nanoparticles emerge as highly adaptable materials with considerable promise for biomedical imaging as well as environmental applications like water disinfection.

2.3 NIR-II Luminescence of Rare-Earth Phosphors

Several studies have demonstrated the advantages of second near-infrared (NIR-II, 1000–1700 nm) luminescence for deeper tissue imaging, including reduced tissue absorption, minimized scattering, low autofluorescence, and improved spatial resolution. Jiang et al. [43] developed $LiYF_4$ nanocrystals doped with Nd^{3+} ions as novel probes for NIR-II bioimaging. The $LiYF_4$ host, optimized with 5% Nd^{3+} doping, produced sharp and efficient emission peaks within the second near-infrared window, offering enhanced tissue penetration and low autofluorescence under biological conditions. Following surface modification, these nanoprobes achieved in vivo penetration depths of approximately 3–5 mm with excellent spatial resolution (~ 0.2 mm), enabling lymphatic and circulatory system imaging with high signal-to-noise ratios. A separate investigation was done by Wang et al. [44] where they synthesized $LuPO_4:Nd^{3+}$ nanophosphors via hydrothermal synthesis method. These nanophosphors exhibited strong Nd^{3+} emission at ~ 1064 nm, achieving remarkable tissue penetration depths up to 15 mm. Additionally, due to the high atomic number of lutetium, these materials demonstrated strong X-ray absorption, making them suitable for combined NIR-II luminescence and X-ray computed tomography (CT) imaging. Cytotoxicity evaluations confirmed favorable biocompatibility, supporting their potential for deep-tissue multimodal imaging applications.

The reviewed literature demonstrates that rare earth activated phosphors, whether based on persistent luminescence or up-conversion mechanisms, hold great promise for biomedical imaging. Both persistent and up-conversion luminescence-based rare-earth phosphors offer distinct and complementary advantages for bioimaging applications. Persistent luminescence phosphors, particularly those based on silicate, aluminate, sulfide, garnet, and perovskite host lattices, have demonstrated exceptional potential for excitation-free, long-term imaging. Meanwhile, up-conversion phosphors excel in real-time, deep-tissue visualization under NIR excitation with negligible background noise. Moreover, NIR-II emitting rare-earth phosphors, such as Nd^{3+}-doped systems, have shown exceptional deep-tissue imaging potential with superior spatial resolution, often enabling multimodal imaging. Table 1 represents a summary of rare earth-based phosphors that are already discussed along with their host types, emission mechanisms, and bioimaging applications. The combination of advanced synthesis techniques, surface functionalization strategies, and intelligent material design continues to expand the versatility and clinical potential of rare earth activated phosphors.

Table 1 Rare-earth phosphors for bioimaging applications

Material system	Authors	Host type	Emission type	Bioimaging applications
$Sr_2MgSi_2O_7:Eu^{2+}, Dy^{3+}$	Homayoni et al. [19]	Silicate	Persistent luminescence	Photodynamic therapy (PDT) activation in deep tissues
$Ca_2MgSi_2O_7:Eu^{2+}, Dy^{3+}$	Sharma et al. [20]	Silicate	Persistent luminescence	Potential for imaging, tissue engineering, drug delivery
$SrAlO_4:Eu^{2+}, Dy^{3+}$	Sun et al. [21]	Aluminate	Persistent luminescence	Long-duration in vivo imaging (repeated LED reactivation)
$SrAl_2O_4:Eu^{2+}, Dy^{3}$ + UCNPs	Hu et al. [22]	Aluminate composite	Persistent luminescence	Implantable "optical battery" for irradiation-free PDT
$Sr_4Al_{14}O_{25}:Eu^{2+}, Dy^{3+}$	Calatayud et al. [23]	Aluminate	Persistent luminescence	Long-term biocompatible imaging probes
$Sr_2SiO_4:Eu^{2+}, Pr^{3+}$	Li et al. [24]	Silicate	Persistent luminescence	Tunable multicolor bioimaging
$CaS:Eu^{2+}, Dy^{3+}$	Rodríguez Burbano et al. [25]	Sulfide	Persistent luminescence	Repeated deep-tissue imaging via PSL
$CaS:Eu^{2+}, Sm^{3+}$	Gao et al. [26]	Sulfide	Persistent luminescence	Deep-tissue bioimaging in live animal models
$Ca_3Sc_2Si_3O_{12}:Eu^{2+}$	Jiang et al. [27]	Garnet	Persistent luminescence	Deep-tissue NIR imaging (650–1100 nm)
$Ca_3Ga_2Ge_3O_{12}:Pr^{3+},Yb^{3+}$	Dai et al. [28]	Garnet	Persistent luminescence	In vivo persistent luminescence imaging
$LaAlO_3:Cr^{3+}, Sm^{3+}$	Pellerin et al. [29]	Perovskite	Persistent luminescence	Deep-tissue imaging (NIR ~ 700 nm)
$LiGa_5O_8:Pr^{3+}$	Xiong et al. [30]	Oxide	Persistent luminescence	Real-time bioimaging with tissue penetration
$CaMgSi_2O_6:Eu^{2+}, Mn^{2+}, Dy^{3+}$	Rosticher et al. [31]	Silicate	Persistent luminescence	Non-invasive red afterglow in vivo imaging (~ 680 nm)
$NaYF_4:Yb^{3+}, Er^{3+}$	Nam et al. [33]	Fluoride	Up-conversion	Real-time dynamic biological imaging with minimal background
$NaYF_4:Yb^{3+}, Er^{3+}$	Idris et al. [34]	Fluoride	Up-conversion	Long-term in vivo cell tracking in animal models

(continued)

Table 1 (continued)

Material system	Authors	Host type	Emission type	Bioimaging applications
β-$NaYF_4$:Er^{3+}, Yb^{3+}	Cao et al. [35]	Fluoride	Up-conversion	Deep-tissue red luminescence bioimaging under NIR
$NaYF_4$:Yb^{3+},Tm^{3+}	Jia-Cai Zhou et al. [36]	Fluoride	Up-conversion	In vitro (HeLa cells) and in vivo (C. elegans); efficient bioimaging
$NaYF_4$:Yb^{3+}, Er^{3+} + AuNRs	Cho et al. [37]	Fluoride Composite	Up-conversion	Bladder cancer imaging + therapy
$NaYF_4$:Yb^{3+}, Er^{3+} + FA	L Xiong et al. [38]	Fluoride	Up-conversion	Targeted bioimaging of folate-receptor-positive cancer cells
$KAlF_4$:Er^{3+}, Yb^{3+}	Janbandhu et al. [39]	Fluoride	Up-conversion	Bioimaging
La_2O_3:Yb^{3+}, Er^{3+}	Sun et al. [40]	Oxide	Up-conversion	Cellular bioimaging with temperature-responsive therapeutic potential
Gd_2O_2S:Yb^{3+}, Er^{3+}	Ajithkumar et al. [41]	Oxide	Up-conversion	Multimodal bioimaging
$La_2Zr_2O_7$:Yb^{3+}, Er^{3+}/ Tm^{3+}, Yb^{3+}, Gd^{3+}	Gupta et al. [42]	Oxide	Up-conversion + NIR-II	Multimodal UV–visible-NIR-II bioimaging and environmental uses
$LiYF_4$:Nd^{3+}	Jiang et al. [43]	Fluoride	NIR-II emission	High-resolution deep-tissue NIR-II bioimaging
$LuPO_4$:Nd^{3+}	Wang et al. [44]	Phosphate	NIR-II emission	Dual-mode deep-tissue imaging

3 Synthesis Techniques

The implications of rare earth activated phosphors in bioimaging have drawn a plenty of interest rates. These phosphors have been tailored for bioimaging adopting a variety of synthesis tactics, with an emphasis on elevating the caliber and affordability of medical imaging with light. A number of parameters, including the expected the size of the particles, configuration, morphological characteristics, the crystallinity, utmost purity, dopant species distribution, synthesis expenditure, and luminescent parameters; govern the synthesis route selection. On the basis of these, RE phosphors can be synthesized by a variety of methods (Fig. 3), among which solid-state technique, sol–gel processes, hydro (solvo) thermal synthesis and combustion synthesis are the most commonly used approaches. The features of phosphors are profoundly affected by the synthetic pathways employed in their synthesis process [18, 45].

Assessment of Synthesis Approaches for Effectiveness

1. *Solid-State Technique*: The ceramic method, additionally referred to as the solid-state synthesis method, is a process with elevated temperatures that is frequently employed for the manufacture of vast quantities of materials. It is reasonably priced, requires barely anything special machinery, and allows for a fast reaction time. This method makes it possible to synthesize a significant amount of the target material in a simple way, usually producing microcrystalline products with a pure phase. Using a ball milling technique or a mortar and pestle, the procedure starts with grinding constituent minerals such oxides and carbonates [45].

The final emission properties are sensitive to numerous parameters including firing temperature, reaction time, and dopant concentration, all of which can significantly affect the efficiency of luminescent emissions [46, 47].

E.g., Ajithkumar et al. in their report on the development of a rare earth-doped $Gd_2O_2S:Yb^{3+}/Er^{3+}$ phosphor designed for multimodal bioimaging applications, used a high temperature solid-state flux fusion method for synthesis process [41].

Here is a Flow Chart for Solid-State method (Fig. 4).

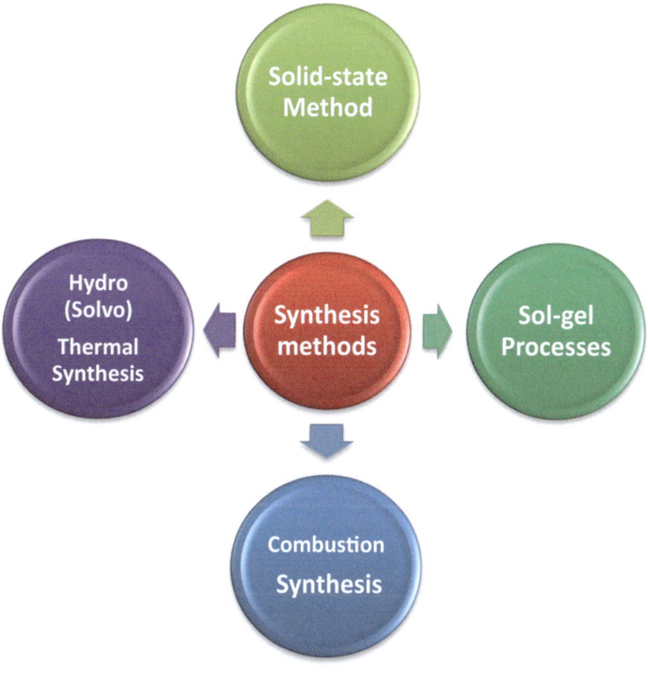

Fig. 3 Various synthesis routes to synthesize the rare earth activated phosphors

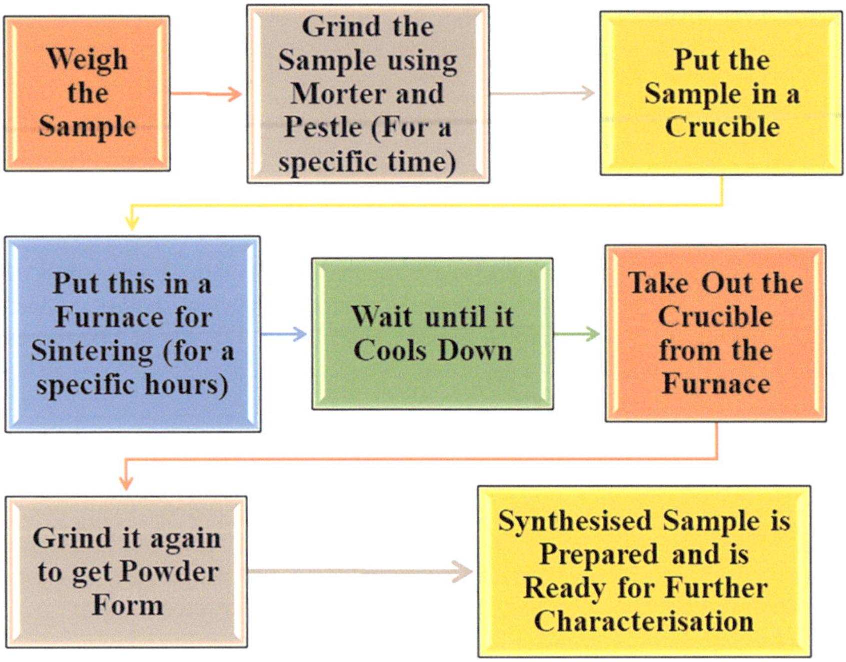

Fig. 4 Flow chart of solid-state method

2. *Sol–Gel Method*: A chemical procedure named sol–gel synthesis is commonly used to yield nanoparticles featuring consistent sizes, superior crystallinity, and ideal purity. In this kind of approach, a solution of chemicals called a sol is used for producing a gel. This process is also often referred by the term citrate-gel process since it typically involves using a chelating chemical, such citric acid, to boost the production of gels. Based upon the vaporization standards, the preexisting elements are submerged in a water-based solution containing the chelating agent followed by being heated to 80–100 °C which leads to the formation of gels. The resultant item is obtained by dehydrating the gel and then heating it under regulated conditions [45].

Highly sensitive rare-earth ions could be added with no disruption to their ionic condition due to the low-temperature conditions necessary for processing[48].

E.g., Homayoni et al. [19] in their report used this sol–gel synthesis route to synthesis $Sr_2MgSi_2O_7$ nanoparticles doped with Eu^{2+} and Dy^{3+} ions. The resulting phosphors were further modified with amphiphilic PMMA-b-PEG polymers to improve water dispersibility and biocompatibility. Pellerin et al. [29] used this synthesis method to create Cr^{3+} (emission center) and Sm^{3+} (trap center) co-doped $LaAlO_3$ nanoparticles.

Here is a Flow Chart for Sol–Gel method (Fig. 5).

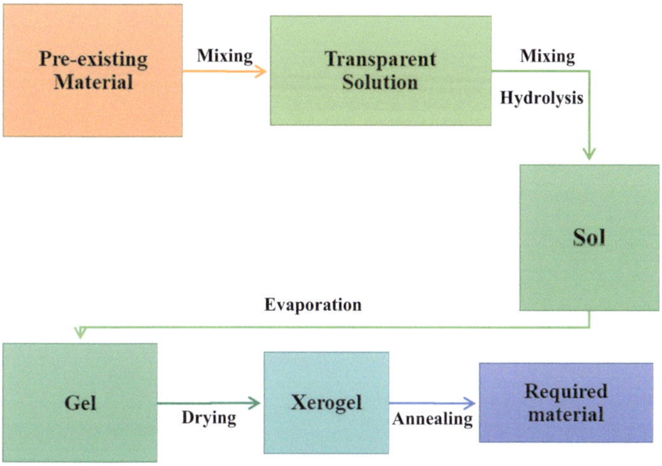

Fig. 5 Flow chart of sol–gel method

3. *Hydrothermal Synthesis*: A low-temperature synthesizing approach for preparing nanoparticles having regulated shape and size of particles is the hydrothermal method of synthesis. Reactive agents get hydrolyzed and condensed in a solvent-filled stainless-steel autoclave. The procedure is known as solvothermal synthesis when another solvent instead of water is utilized. Following that, the closed autoclave undergoes heating for varied lengths of period at several different temperatures in a heating system. The thereby produced material's structure can be altered by varying elements including the pH value, solvent type, surfactants, and templates [45]. It is an excellent technique for creating phosphors with customizable luminous qualities and customizable morphology.

E.g., As demonstrated by Jiang et al. [43] in their study, hydrothermal process is used to synthesis $LiYF_4:Nd^{3+}$ nanophosphors, followed by surface modification to improve biocompatibility and targeting efficiency. Wang et al. [44] in their research work, highlighted this synthesis process. A hydrothermal reaction is used to synthesis $LuPO_4:Nd^{3+}$ nanophosphors for studying their luminescence behavior.

Here is a Flow Chart for Hydrothermal Synthesis method [49] (Fig. 6).

4. *Combustion Synthesis*: A straightforward and reasonably priced low-temperature synthesis technique, the combustion process generates extremely crystalline, homogeneous, and pure nanomaterials. In order to drive an intense reaction that is exothermic, this approach uses both oxidizing and reducing substances, commonly identified as fuels. The mechanism of action works effectively when the valencies of the oxidizing agents and reducing agents are in equilibrium (1:1). A water-based solution containing the fuels and reagents is added to a furnace that has already been heated up. These gases like carbon dioxide and nitrogen are generated throughout the entire procedure, forming an explosion which assists in the synthesis of nanoparticles. Additional calcination is carried

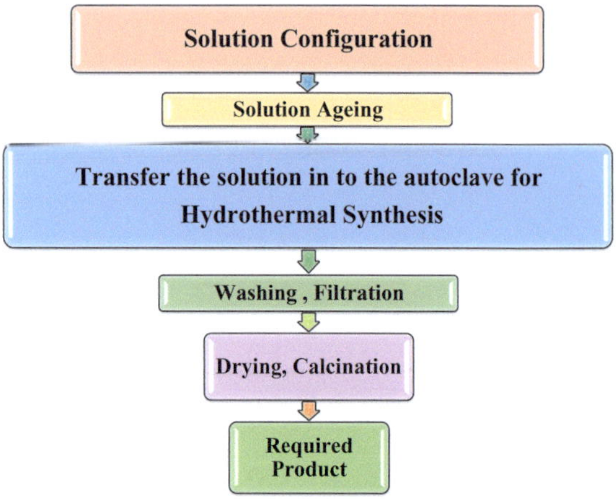

Fig. 6 Flow chart of hydrothermal synthesis method

out to yield a pure-phase material if the starting material has impurities [45]. One notable feature of combustion synthesis is its low energy usage and quick rate of processing.

E.g., Sharma et al. produced $Ca_2MgSi_2O_7$ phosphors co-doped with Eu^{2+} and Dy^{3+} and examined their luminous characteristics by synthesizing the phosphor through combustion method [20].

Here is a Flow Chart for Combustion Synthesis method (Fig. 7).

In a nutshell, the luminous properties of rare earth activated phosphors are substantially affected by their synthesis methodology preference. There are multiple ways

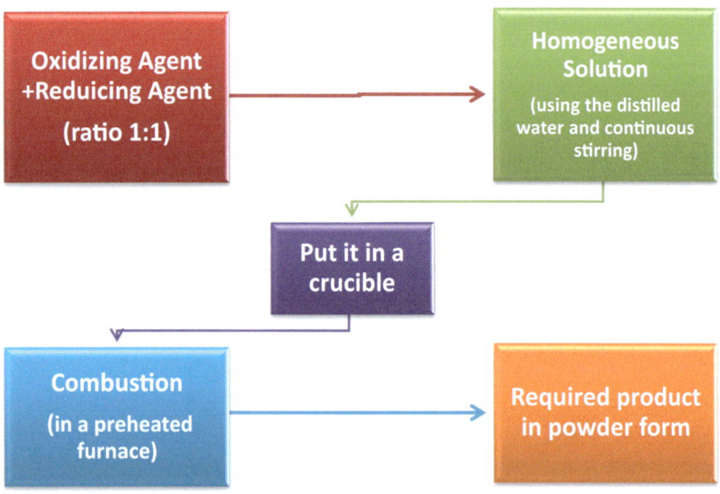

Fig. 7 Flow chart of combustion synthesis method

to improve phosphor performance. The buildup of next-generation phosphors with superior performance criteria compatible with a variety of applications could ensue from further research into these techniques for synthesis and improvements in characterization strategies [18, 48].

4 Characterization Techniques for Phosphors

In order to fully understand the structural, morphological, and optical behaviors of rare earth activated phosphors, an accurate characterization is a must needed process [50]. These methods shed light on the luminescence efficacy, phase pureness, size of particles, crystallinity, along with other functional characteristics of RE phosphor compounds [51]. The following are a few widely utilized phosphor characterization methods for comprehending and enhancing rare-earth phosphors especially for biological imaging applications.

4.1 Structural Analysis

a. *X-Ray Diffraction Analysis (XRD)*—One essential method for figuring out the crystal structure and phase homogeneity of phosphor materials is X-ray diffraction (XRD) [52]. This technique involves letting a collimated monochromatic X-ray beam fall on the specimen, and the detection device receiving the diffracted beam generates a diffraction pattern that follows the Bragg's law [53]. Figure 8 shows the working process of XRD.

Significant details regarding the size, phase, lattice constant, and degree of transparency of phosphor materials can be obtained through patterns generated by XRD

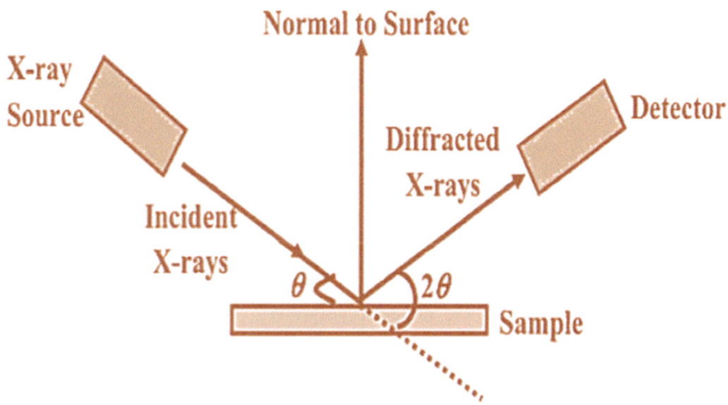

Fig. 8 Representation of the working process of XRD (taken from Ref. [54])

[52]. The crystalline nature of samples that have been synthesized is shown by the strong and intense peaks in XRD patterns, and this is directly related to their luminescence qualities [55].

The crystalline structure of Gd_2O_2S doped with Yb^{3+} and Er^{3+} was verified by X-ray diffraction (XRD) as shown in Fig. 9. The conventional hexagonal phase of Gd_2O_2S (JCPDS Card No. 26-1422) and the recorded XRD pattern matched well, suggesting strong crystallinity with cell parameters a = b = 0.3852 nm, c = 0.6567 nm. The space group is $P_3\overline{m}_1$ and the symmetry is trigonal. In above Fig. 9, The standard diffraction peaks of Gd_2O_2S are represented by vertical lines in the XRD pattern, which are taken from JCPDS File No. 26-1422. As a visual representation of the material's hexagonal lattice arrangement, the inset of the picture shows the crystal structure of Gd_2O_2S, emphasizing the atomic locations of Gd (pink), O (red), and S (yellow) [41].

The crystal structure is similar to that of the A-type rare-earth oxide, but sulfur has been added in place of one oxygen site. ± (1/3, 2/3, u) for Gd (u ≈ 0.28), ± (1/ 3, 2/3, v) for O (v ≈ 0.63), and (0, 0, 0) for S atoms are the atom locations. A seven-coordinated geometry is formed when each Gd atom coordinates with four oxygen and three sulfur atoms. This detailed structural understanding ensures the efficient incorporation of dopant ions into the host lattice, which is essential for maximizing the phosphor's luminescent behavior in bioimaging applications [41].

The following Scherrer equation is being used to figure out the particle size based on the diffraction peak's full width at half maximum (FWHM) [56].

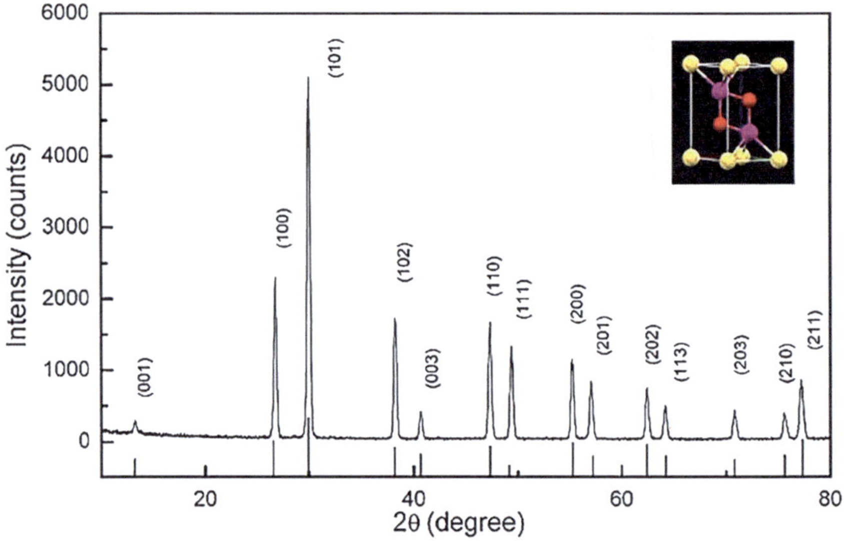

Fig. 9 XRD patterns of Gd2O2S: Yb^{3+}/Er^{3+} (taken from Ref. [41])

$$D = \frac{k\lambda}{\beta \cos\theta}$$

here D is the crystallite size, λ is the wavelength of X-ray, β is the FWHM, θ is the angle known as Bragg's angle, k is known as shape factor which has the value 0.9.

b. *Fourier Transform Infrared (FTIR) Spectroscopy*—Rare-earth phosphors' functional groups are examined by using FTIR spectroscopy, which yields significant insights into their chemical composition as well as bonding [52, 52]. A set of sources, a sample stage, a device for detection, an amplifying device, a converter for analog to digital conversion, and a monitor are components of an FTIR apparatus [58].

The sample both transmits and reflects the radiation that strikes it, and it also absorbs some of the radiation. The signal gets boosted and modified into a digital signal after going via an interferometer, and the FTIR spectra are thereby generated [60]. Figure 10 shows an illustration of a Fourier transform infrared (FTIR) spectrometer's key components. FTIR analysis is highly advantageous as it can identify minor variations in bonding between molecules which other characterization approaches may overlook [57].

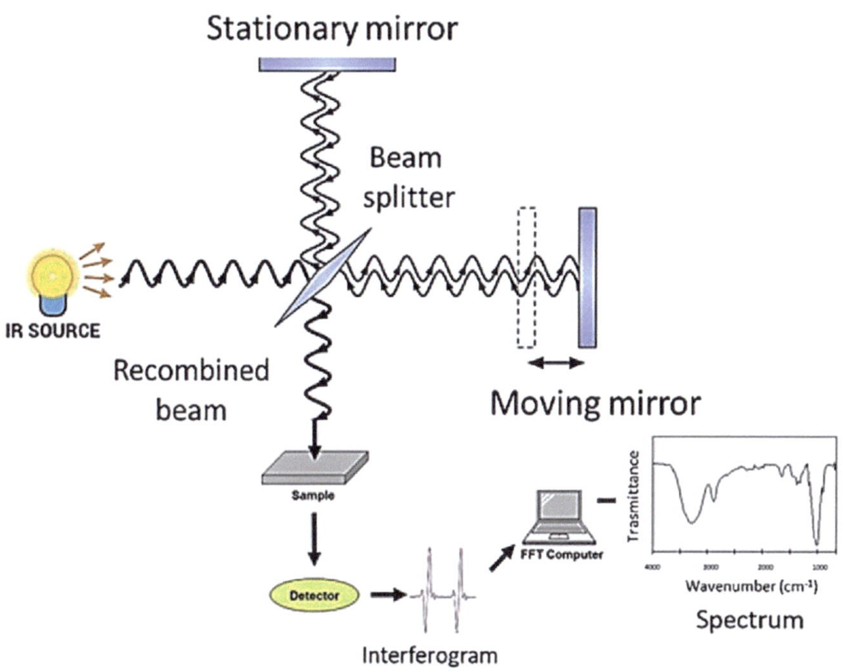

Fig. 10 Illustration of a Fourier transform infrared (FTIR) spectrometer's key components (taken from Ref. [59])

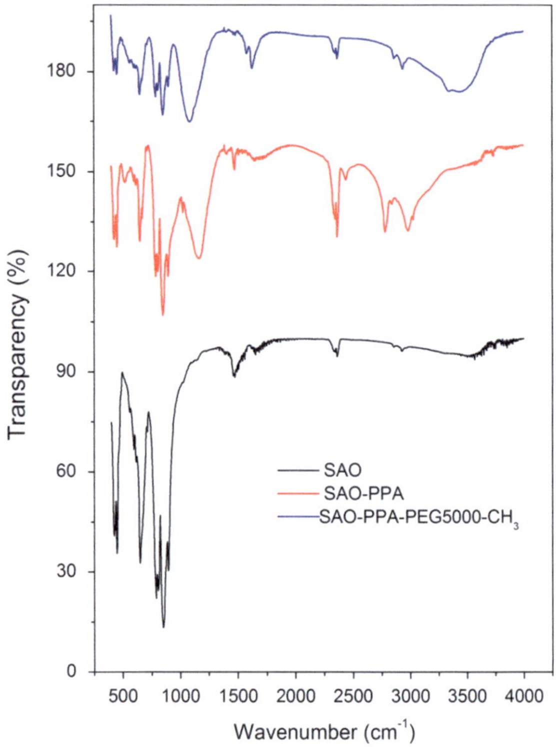

Fig. 11 FTIR spectra of as-synthesized $SrAl_2O_4$ (SAO) nanoparticles, SAO modified with PPA (pyrophosphoric acid), and SAO further grafted with PEG5000-OCH_3. Reproduced by taking permission from Ref. [21]. Licence no. 6020631146975

The surface functionalization of the nanoparticles with PEG5000-OCH_3 was confirmed using FTIR spectroscopy in a work on $SrAl_2O_4$-based (SAO) persistent luminous nanoparticles co-doped with Eu^{2+} and Dy^{3+} is shown in Fig. 11 [21].

The effective grafting of PEG5000-OCH_3 onto SAO-PPA nanoparticles was verified by FTIR spectrum analysis, which showed specific peaks that were associated with P–OH, P=O, C–O–C, and –CO–NH– bonds. Strong absorption bands were seen in the FTIR spectra that was indicated by the appearance of prominent peaks at 2700 cm^{-1} and 1160 cm^{-1}, which corresponded to the stretching vibrations of the P–OH and P=O groups, respectively. Following the attachment of PEG5000-OCH_3, additional peaks appeared at 1083 cm^{-1}, which corresponded to C–O–C bonds from the PEG chains, and 3400 cm^{-1}, which was linked to –CO–NH– bonds, indicating that PEG grafting was successful [21].

The Fourier transform infrared (FTIR) spectrum of pure $KAlF_4$ was recorded in the spectral range of 4000–500 cm^{-1} (Fig. 12) [39].

Multiple absorption bands were observed, indicating the presence of characteristic vibrational modes associated with metal–fluoride and metal–oxygen bonds [39]. Notably, distinct peaks appeared at 869, 813, 709, 597, 548, and 509 cm^{-1}. The absorption band near 597 cm^{-1} is assigned to F–Al stretching vibrations [39, 61, 62]. A broad feature around 709 cm^{-1} is attributed to Al–F stretching modes [39,

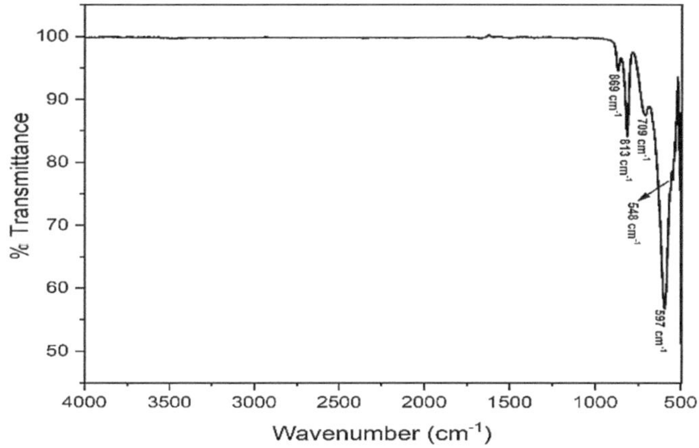

Fig. 12 FTIR spectra of pure $KAlF_4$. Reproduced by taking permission from Ref. [39]. License Number-6054560252333

63]. Furthermore, vibrational modes within the 900–775 cm^{-1} region, along with moderate to strong absorption between 675 and 575 cm^{-1}, are indicative of Al–O stretching vibrations [39, 64]. These observations confirm the presence of both Al–F and Al–O coordination environments in the $KAlF_4$ lattice [39].

These spectral characteristics demonstrate that FTIR is an effective technique to analyze chemical bonding and also surface modifications in phosphor materials, thereby supporting their application in bioimaging.

4.2 Morphological Analysis

a. *Scanning Electron Microscopy (SEM)*—SEM contains vital details regarding the surface morphology of phosphor materials offering insights into their particle dimensions, form, and distribution [65]. These features have an impact on optical behavior of phosphors and interaction with biological tissues thereby making them essential for biomedical imaging purposes. Figure 13 shows the graphical set up for scanning electron microscopy.

Scanning electron microscopy (SEM) is a technique for morphological study of materials utilizing a high-energy electron beam. Electrons generated by an electron gun are focused and directed toward the sample surface. When interacting with the specimen, electron scattering produces a variety of signals that reveal important details about the material's structure and surface morphology [52, 67].

The FE-SEM images of La_2O_3 nanophosphors doped with Yb^{3+} and Er^{3+} ions, synthesized via solvothermal processing are shown in Fig. 14. As shown in Fig. 14a and b at magnifications corresponding to 500 and 300 nm scales, the La_2O_3 nanoparticles exhibit predominantly spherical morphology [40].

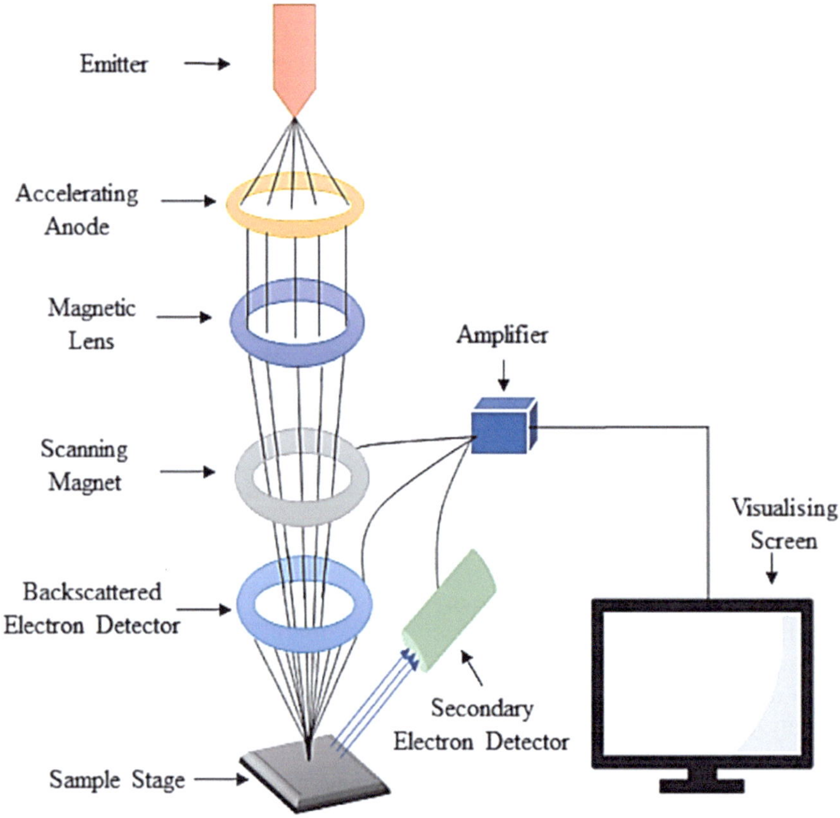

Fig. 13 Scanning electron microscopy setup with schematic diagram. Taken from Ref. [66]

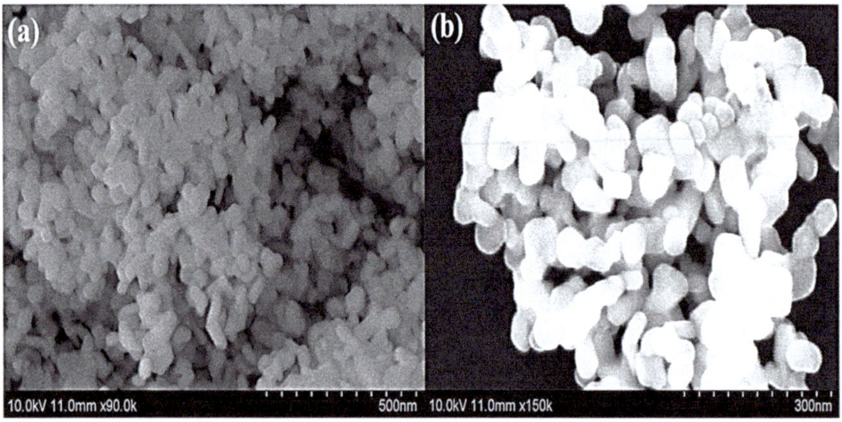

Fig. 14 FE-SEM images of La_2O_3:Yb^{3+}, Er^{3+} nanophosphors at magnifications corresponding to 500 and 300 nm scale bars. Reproduced by taking permission from Ref. [40] License Number 6054570790271

b. *Transmission Electron Microscopy (TEM)*—Transmission electron microscopy (TEM) is an effective technology for studying the microstructure of materials, providing crystallographic information and composition at the nanoscale [68]. Figure 15 shows the schematic diagram of TEM. Compared to SEM, TEM offers even better resolution images, enabling the analysis of the phosphor particles' structural basis and defects [69].

Transmission electron microscopy (TEM) is a high-resolution characterization technique commonly used to investigate the shape, content, and crystallographic properties of phosphor materials. It works by using high voltage (40–100 kV) to accelerate a thermionically produced electron beam (usually from a LaB_6 filament), enabling fine-grained imaging at the nanoscale ($\sim$ 0.2 nm resolution). TEM is particularly beneficial for examining the structural and phase purity of rare-earth-doped phosphors, which is crucial for enhancing their bioimaging performance [52, 71, 72].

The morphological and structural features of $NaYF_4$:Yb^{3+}, Er^{3+} up-conversion nanoparticles were investigated using transmission electron microscopy (TEM),

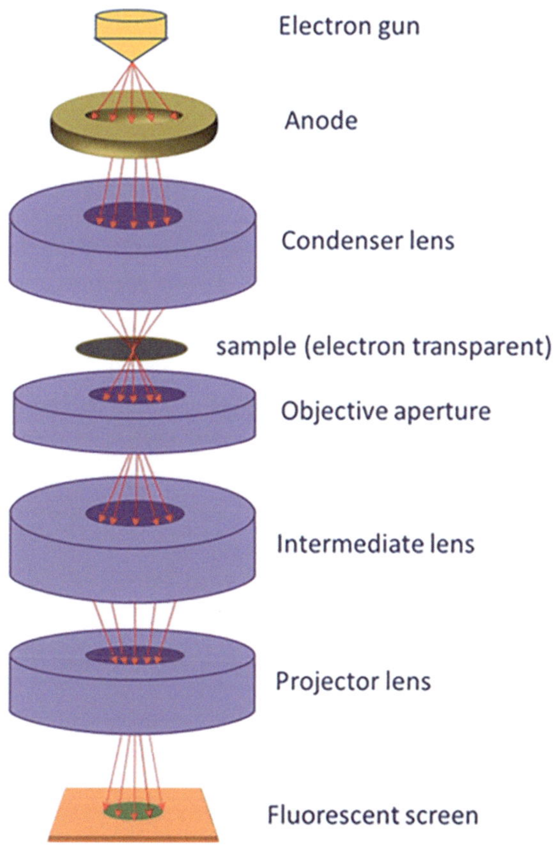

Fig. 15 Schematic diagram of TEM. Taken from Ref. [70]

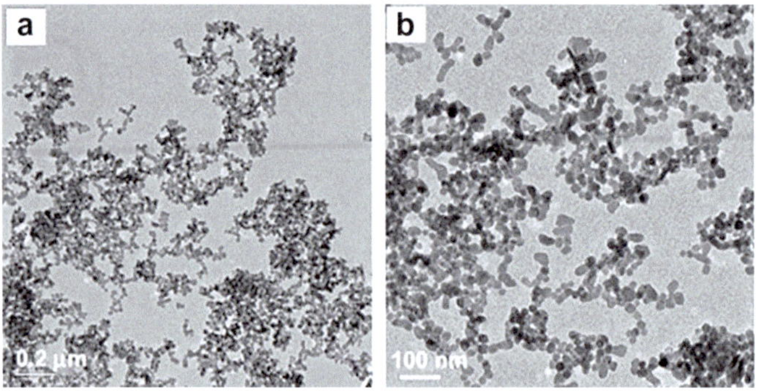

Fig. 16 TEM images of $NaYF_4$ phosphors doped with Yb^{3+} and Er^{3+}. Reproduced by taking permission from Ref. [38]. License Number-6054561244879

obtained after 6 h of hydrothermal treatment. As shown in Fig. 16, the uniform morphology and narrow size distribution indicate controlled growth during the synthesis process [38].

4.3 Optical Analysis

a. *UV–Visible Spectroscopy*: UV–visible spectrophotometry is a widely utilized analytical method that helps in understanding the optical absorption behavior of phosphor materials. It is based on the Beer-Lambert Law [73]. Figure 17 shows the working process of UV–visible spectroscopy through a schematic diagram of single beam UV–visible spectrometer.

The instrument uses a light source that spans the ultraviolet and visible regions of the spectrum. When light interacts with a material, certain wavelengths are absorbed

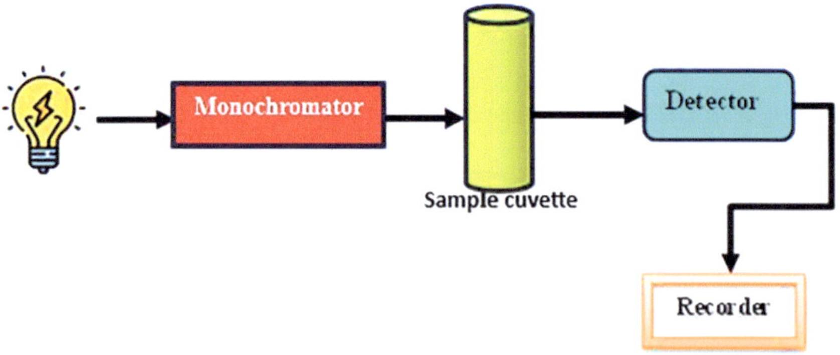

Fig. 17 Schematic diagram of single beam UV–visible spectrometer

based on the arrangement of electrons and the nature of chemical bonds within the atoms or molecules. This absorption causes electrons to transition from their ground state to higher energy levels. The output of this process is generally shown as a spectrum, where absorbance is plotted against wavelength. This spectral data reveals details about electronic transitions and helps in identifying or estimating the presence of specific components in the sample [74, 75].

The optical absorption behavior of $KAlF_4{:}Er^{3+}$, Yb^{3+} up-conversion phosphor was investigated by analyzing UV–Vis spectrum as shown in Fig. 18. Several distinct absorption peaks are observed at approximately 380, 400, 497, 547, and 662 nm, which correspond to characteristic 4f–4f electronic transitions of Er^{3+} ions. Specifically, these transitions are assigned to excitations from the $^4I_{15/2}$ ground state to the $^4G_{11/2}$, $^2H_{9/2}$, ($^4F_{3/2}$ + $^4F_{5/2}$), $^4S_{3/2}$, and $^4F_{9/2}$ excited states, respectively. A noticeable broadening of the absorption band near 497 nm is attributed to the overlap of the closely spaced $^4F_{3/2}$ and $^4F_{5/2}$ levels. A weaker absorption feature is also detected around 662 nm, corresponding to the $^4F_{9/2}$ level [39].

The UV–Vis-NIR absorption spectrum of the sample, recorded at room temperature and shown in Fig. 19, displays multiple distinct absorption bands. These bands exhibit spectral widths and positions that align well with the characteristic absorption profile of Yb^{3+}/Er^{3+} ions. The observed absorption peaks for Er^{3+} are attributed to electronic transitions originating from the ground state to various higher-energy excited states, as indicated in the spectrum [41].

b. *Photoluminescence (PL) Spectroscopy*—Photoluminescence (PL) spectroscopy is widely regarded as a fundamental optical characterization technique because it is non-destructive and provides important insights into both intrinsic and extrinsic electronic transitions within a material [76].

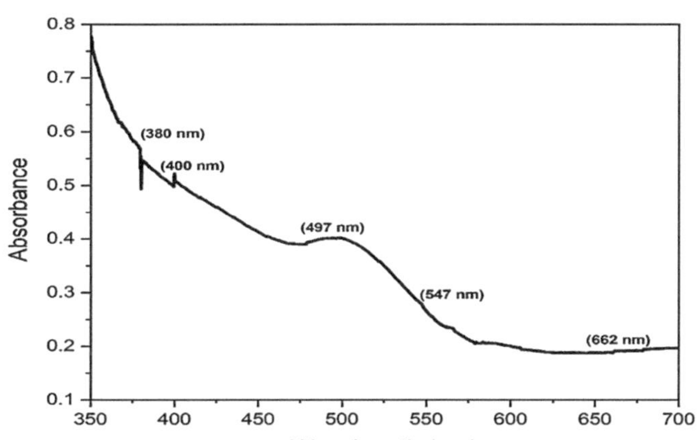

Fig. 18 UV–Vis absorption spectrum of $KAlF_4$ doped with Er^{3+} and Yb^{3+}. Reproduced by taking permission from Ref. [39]. License Number-6054560252333

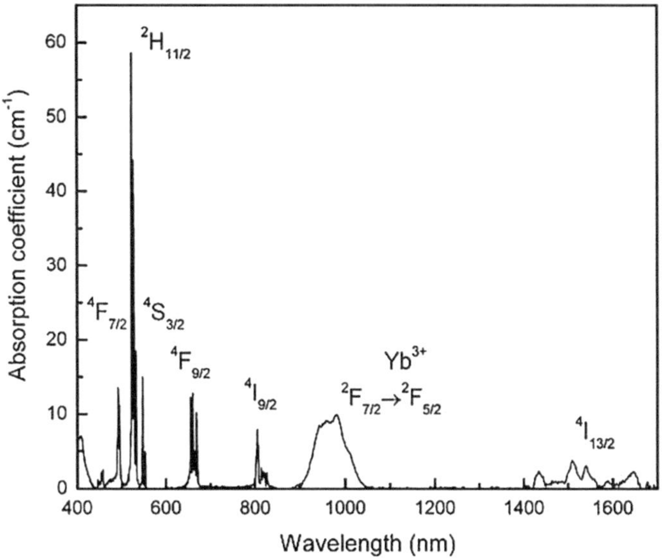

Fig. 19 UV–Vis-NIR absorption spectrum of Gd2O2S:Yb^{3+}/Er^{3+} with all observed spectral transitions. Taken from Ref. [41]

Figure 20 represented the working of photoluminescence (PL) spectroscopy. This spectroscopy involves, light (commonly uses a laser or LED) hits on a material, causing electrons to move to higher energy levels (photoexcitation). After excitation, the electrons relax by losing energy and return to lower energy states, emitting photons (PL emission). The emitted light is detected and amplified by a photomultiplier tube (PMT), with the wavelength and intensity revealing the material's optical properties [52, 52].

The up-conversion luminescence behavior of $KAlF_4$:Er^{3+} and $KAlF_4$:Er^{3+}, Yb^{3+} phosphors was examined under 980 nm laser excitation, as depicted in Fig. 21 [39]. The recorded PL spectra exhibit distinct emission peaks distributed across the visible and near-infrared regions. For $KAlF_4$:Er^{3+} samples with 0.5 mol% Er^{3+} doping, the emission intensities were relatively weak, with the red emission centered at 662 nm barely observable. Increasing the Er^{3+} concentration to 1 mol% resulted in a significant enhancement of the red emission at 662 nm, alongside a noticeable rise in the near-infrared (NIR) band located at approximately 834 nm, as highlighted in the inset of Fig. 22.

Further incorporation of 5 mol% Yb^{3+} ions as sensitizers substantially enhanced the up-conversion emission intensity. In addition to the prominent red emission at 662 nm, strong green emission bands emerged at 522 and 547 nm, with the latter exhibiting higher intensity. The inset illustrates the corresponding enhancement in NIR emission, showing a broad peak at 834 nm accompanied by a shoulder around 805 nm [39].

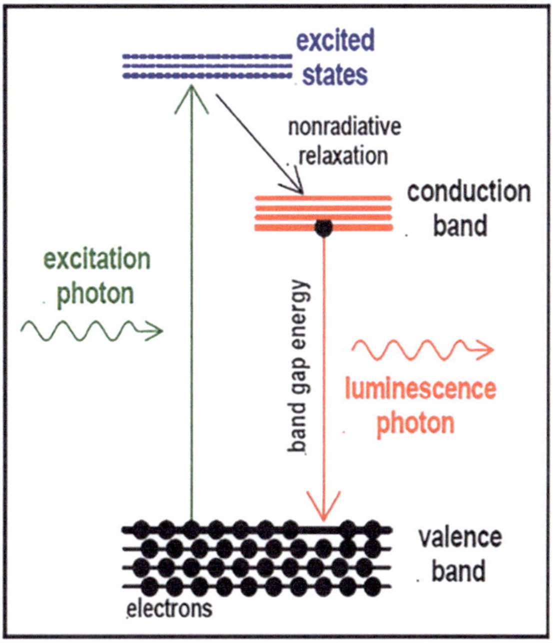

Fig. 20 Working of photoluminescence (PL) spectroscopy. Taken from Ref. [77]

In another study, the up-conversion emission behavior of $NaYF_4$:Yb^{3+}, Tm^{3+} nanocrystals was investigated under 980 nm continuous-wave laser excitation [36]. As shown in Fig. 22, characteristic PL emission peaks of Tm^{3+} ions appear at 405, 475, 650, 700, and 800 nm, corresponding to the $^1D_2 \rightarrow {}^3H_4$, $^1G_4 \rightarrow {}^3H_6$, $^1G_4 \rightarrow {}^3H_4$, $^3F_{2,3} \rightarrow {}^3H_6$, and $^3F_4 \rightarrow {}^3H_6$ transitions, respectively. The incorporation of Yb^{3+} ions, serving as efficient sensitizers, enhances the up-conversion process by absorbing photons and transferring energy to Tm^{3+} ions, thereby significantly increasing the population of excited Tm^{3+} states. As a result, the NIR emission at ~ 800 nm dominates the spectrum, demonstrating efficient NIR-to-NIR up-conversion suitable for bioimaging within the biological transparency window [36].

The up-conversion luminescence mechanism of rare-earth-doped phosphors can be well explained by considering the energy level interactions. Figure 23 shows the schematic energy level diagram illustrating the up-conversion excitation and emission processes for Er^{3+} and Tm^{3+} ions sensitized by Yb^{3+} [79]. As illustrated in Fig. 23, Yb^{3+} ions serve as efficient sensitizers by absorbing 980 nm near-infrared photons and subsequently transferring the absorbed energy non-radiatively to activator ions such as Er^{3+} and Tm^{3+}. The energy transfer from Yb^{3+} populates various excited states of Er^{3+} and Tm^{3+}, which subsequently relax through radiative transitions, resulting in characteristic emissions across the visible and near-infrared regions. The schematic diagram illustrates all possible radiative (solid arrows), non-radiative (dashed/wavy arrows), and energy transfer (thin solid arrows) processes involved in the up-conversion mechanism. Although multiple emission pathways

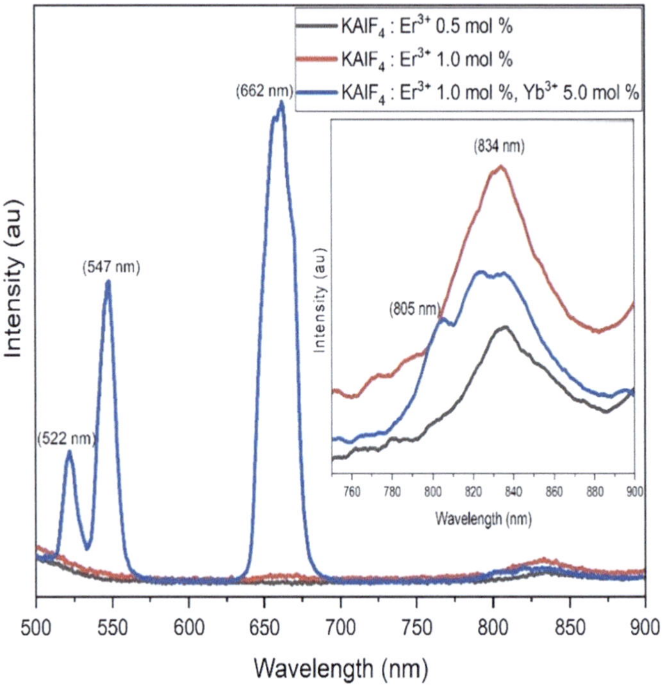

Fig. 21 Photoluminescence spectra of $KAlF_4$:Er^{3+} and $KAlF_4$:Er^{3+}, Yb^{3+} phosphors under 980 nm laser excitation. Inset: highlighting NIR region. Reproduced by taking permission from Ref. [39]. License Number-6054560252333

Fig. 22 Photoluminescence spectra of $NaYF_4$:Yb^{3+}, Tm^{3+} phosphors. Reproduced by taking permission from Ref. [36]. License Number-6054570328495

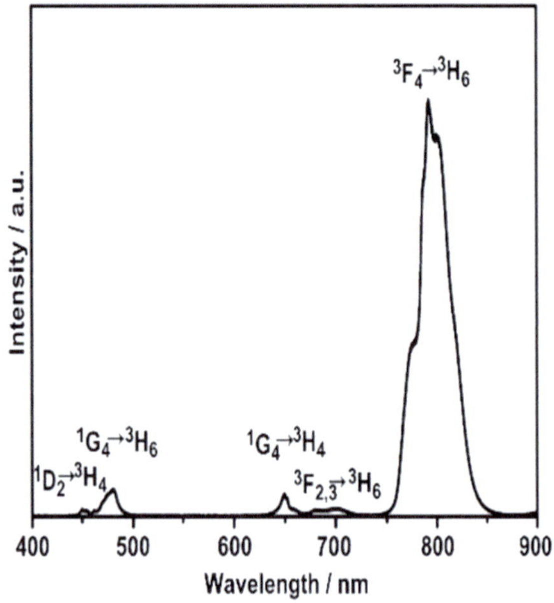

are theoretically possible, only specific transitions are typically observed depending on the host matrix, dopant concentration, and experimental conditions [79].

In this study, the optical characteristics of $SrAl_2O_4$ doped with Eu^{2+} and Dy^{3+} nanoparticle's (modified with PPA and further grafted with PEG5000-OCH_3) were thoroughly investigated to evaluate their potential for bioimaging applications (Fig. 24a–d). These nanoparticles exhibit a broad emission spectrum in the visible region (400–650 nm), with peak intensity around 520 nm (Fig. 24a), which is well within the optical window suitable for biological imaging. The corresponding excitation spectrum extends from the ultraviolet into the visible range (400–475 nm) and shows strong spectral overlap with the emission band of white-light LEDs, making them excitable by clinically relevant light sources (Fig. 24b). One of the notable advantages of these PLNPs is their re-excitability. Although the initial afterglow diminishes within ~ 100 min, luminescence can be fully restored by brief

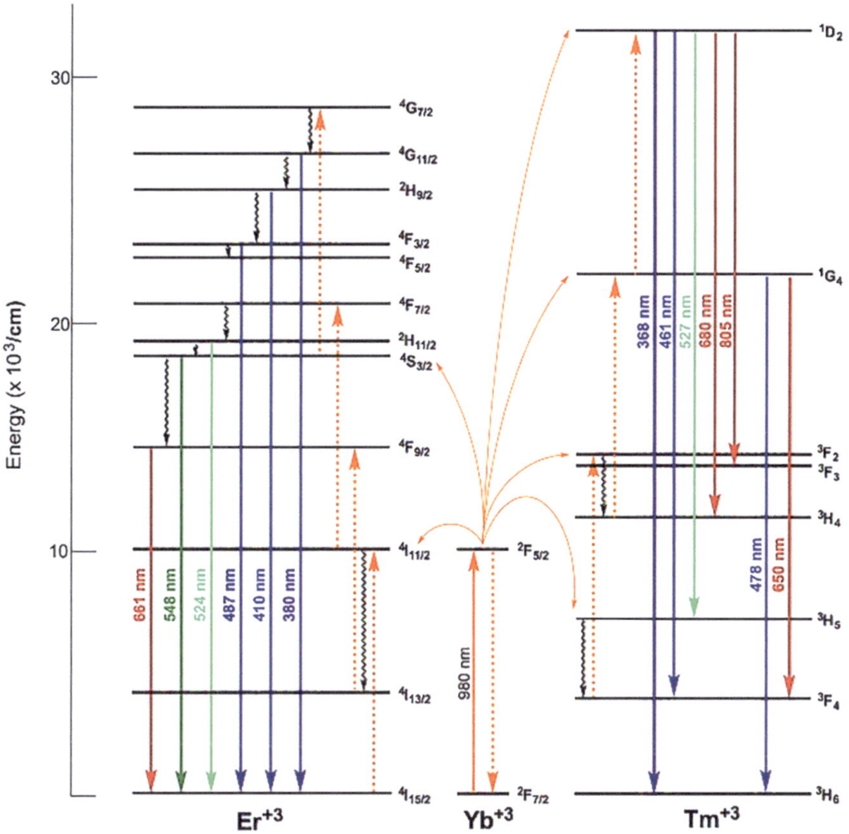

Fig. 23 schematic energy level diagram illustrating the up-conversion excitation and emission processes for Er^{3+} and Tm^{3+} ions sensitized by Yb^{3+}. Reproduced by taking permission from Ref. [79]. License Number-6054580576989

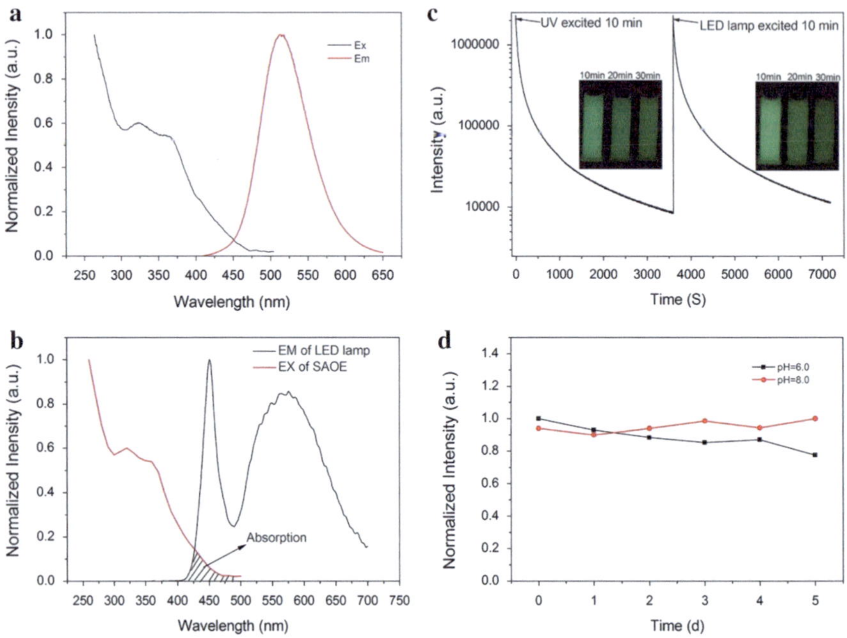

Fig. 24 As-synthesised $SrAl_2O_4$: Eu^{2+}/Dy^{3+} (SAO) nanoparticle's (modified with PPA and further grafted with PEG5000-OCH_3) photoluminescence and afterglow characteristics include the following: **a** the excitation and emission spectra highlighting the optical response of the material, **b** the spectral overlap between the excitation profile and the emission spectrum of a white-light LED source, **c** the afterglow decay curve illustrating the persistence of luminescence over time, and **d** the stability of luminescent intensity under different buffer conditions, demonstrating the material's potential reliability in varied environments. Reproduced by taking permission from Ref. [21]. Licence no. 6020631146975

re-irradiation with white-light LEDs, enabling repeated imaging cycles (Fig. 24c). Green afterglow remains detectable up to 30 min post-irradiation, with strong luminescence efficiency. Additionally, the nanoparticles exhibit excellent chemical and photostability under physiological pH conditions over five days (Fig. 24d). These characteristics highlight their potential for non-invasive, long-term bioimaging, particularly in scenarios where sustained signal monitoring is required [21].

5 Applications in Bioimaging

Rare-earth (RE) activated phosphors are increasingly at the forefront of biomedical innovations due to their exceptional luminescent properties, biocompatibility, and stability under physiological conditions. Biomedical imaging plays a crucial role in both research and clinical settings by providing non-invasive ways to visualize, analyze, and monitor biological processes and structures.

Here's a breakdown of some key applications of biomedical imaging:

a. **Therapeutic Applications**: Rare earth elements (REE)-based nanomaterials are employed in photothermal and photodynamic therapies, particularly for cancer treatment, leveraging their optical properties for targeted therapy. Lanthanide nanoparticles can be integrated with photosensitizers and photothermal agents to create multifunctional platforms for cancer therapy [80].

    I. *Photodynamic Therapy (PDT)*: Up-conversion phosphors doped with rare-earth ions can absorb near-infrared (NIR) light and emit higher-energy visible light, which is used to activate photosensitizing agents. This activation leads to the formation of reactive oxygen species (ROS) that can effectively trigger cell death in cancerous tissues, with minimal harm to healthy surrounding cells [81].

Figure 25 presents a conceptual illustration of a near-infrared (NIR) light-powered "optical battery" implant, designed for tumor therapy via persistent luminescence-driven photodynamic treatment (PDT). Green-emitting persistent luminous materials (GPM), photosensitizers (PSs), up-conversion (UC) components, and electrical structural elements such as the valence band (VB) and trapping band (TB) are all included in this system. During the charging phase, NIR light excitation at 980 nm prompts UC materials to emit UV or blue light, which is absorbed by the persistent phosphors. Energy capture in deep traps is made possible by this process, which excite electrons. The persistent afterglow is caused by radiative recombination involving activator ions, which releases the stored energy gradually after the excitation source is removed. This prolonged emission can activate photosensitizers to initiate PDT without requiring continuous light irradiation [22].

Figure 26 presents tumor images captured 15 days of post-treatment, illustrating the therapeutic efficacy of the PDT.

    II. *Photothermal Therapy (PTT)*: Rare earth-based phosphors integrated into multifunctional nanocomposites can efficiently absorb NIR radiation and convert it into localized heat. This photothermal conversion causes a temperature rise in tumor areas, effectively destroying cancer cells while preserving nearby normal [82].

Figure 27 shows photothermal treatment (PTT) in vivo. In this illustration, mesoporous silica nanoparticles co-loaded with indocyanine green (ICG) and persistent luminescent phosphors (PLPs) are composed of three main components:

i. PLPs are utilized to track the nanoparticles inside the body because they generate long-lasting luminescence when activated by external light sources.
ii. ICG is a near-infrared (NIR) absorber that effectively produces heat from NIR light, making it suitable for use in photothermal treatment.

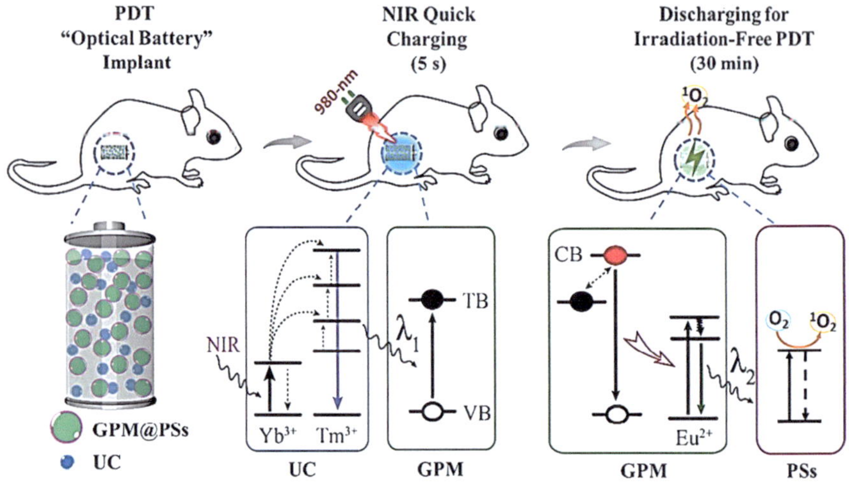

Fig. 25 Schematic representation showing the "optical battery" implant for tumor inhibition that uses persistent luminescence-based photodynamic therapy (PDT). Reproduced by taking permission from Ref. [22]. Licence no. 6020640061888

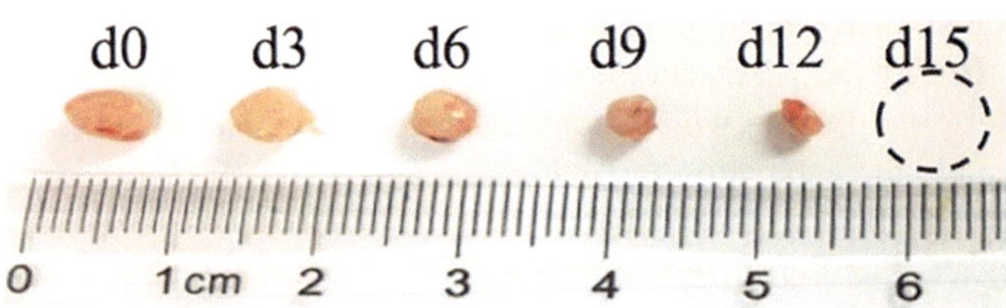

Fig. 26 Images of excised tumor taken at various days following treatment using the PDT. Reproduced by taking permission from Ref. [22]. Licence no. 6020640061888

iii. The mesoporous silica nanospheres serve as a robust three-dimensional scaffold, encapsulating and protecting both the PLPs and ICG, thereby preventing their aggregation and premature release in vivo.

The (ICG + PLPs)@mSiO_2 nanoparticles are injected directly into a tumor in a living mouse following UV pre-irradiation. The tumor spot is then subjected to 808 nm NIR light under the guidance of the persistent luminescence imaging. When exposed to radiation, the ICG in the nanoparticles quickly produces heat, which kills a substantial portion of the tumor cells [82].

b. **Cancer Diagnosis and Monitoring**: Rare earth activated phosphors and rare-earth-doped nanostructures have proven to be valuable tools for cancer diagnosis and monitoring in bioimaging. With their exceptional luminescent and magnetic properties, these materials, enhance imaging capabilities through multimodal techniques, allowing for early-stage detection of tumors [83, 84].

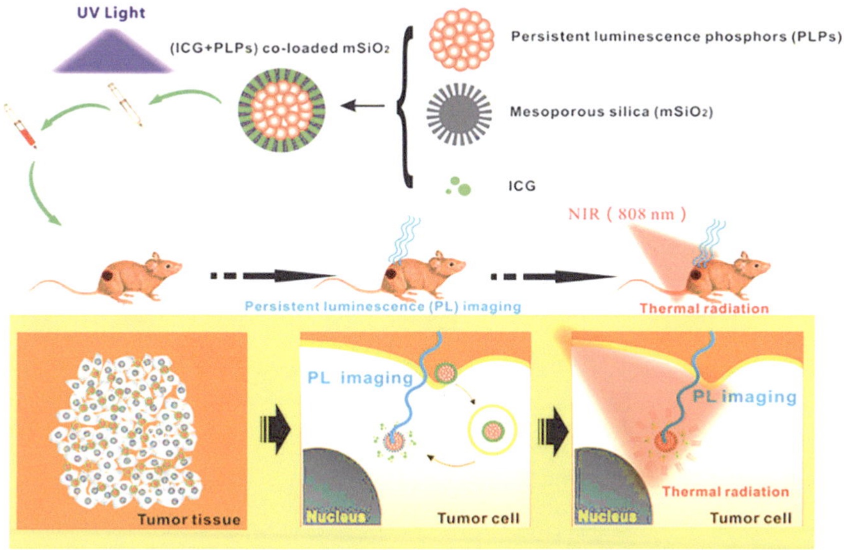

Fig. 27 Use of nanoparticles for persistent luminescence imaging-guided photothermal treatment (PTT) in vivo is shown schematically. Reproduced with permission from [82], Copyright@2016 American Chemical Society

I. For diagnosis, the high brightness, photostability, and low cytotoxicity materials are used to provide excellent contrast in bioimaging, aiding in accurate cancer detection. Additionally, the use of targeting molecules like antibodies or peptides further improves the specificity of tumor identification [83, 84].
II. In monitoring, these nanomaterials enable non-invasive tracking of tumor growth, shrinkage, and treatment response. Their tunable optical and magnetic properties provide continuous, real-time assessments, contributing to better treatment planning and personalized cancer therapies. Rare earth activated phosphors thus play a crucial role in advancing cancer diagnosis and monitoring [83, 84].

Figure 28 shows a graphical representation of multimodal diagnosis and therapy.

c. **Cell and Deep-Tissue Imaging**: Rare earth activated phosphors have emerged as effective probes for imaging at the cellular and tissue scale. Their distinctive luminescence properties—such as characteristic emission wavelengths, strong brightness, and high resistance to photobleaching—allow for detailed visualization of biological structures with minimal background noise. Moreover, their responsiveness to low-energy excitation helps preserve cell viability during live imaging. These materials also show great promise for in vivo applications due to their adjustable emission wavelengths, efficient tissue penetration, and favorable biocompatibility. Functionalization with biological ligands enhances their targeting ability, enabling selective interaction with specific cell types or tissues.

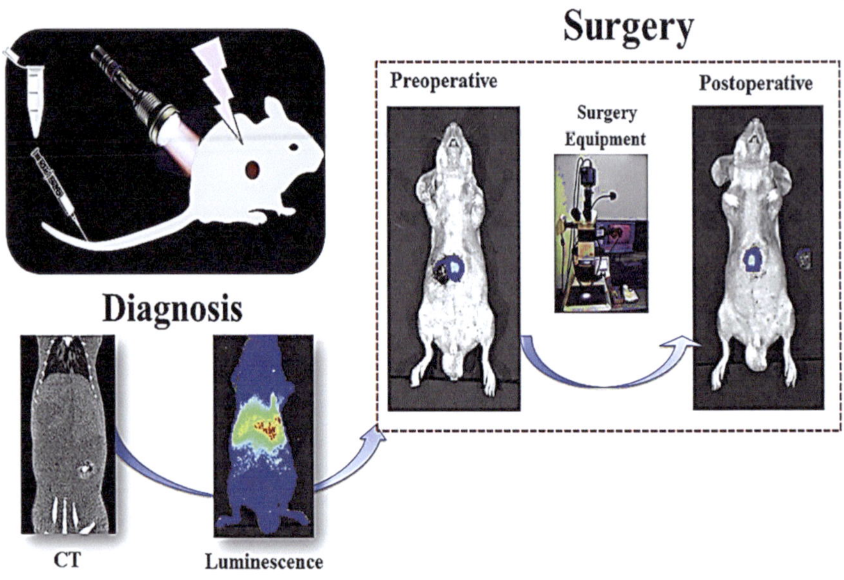

Fig. 28 Representation of rare earth activated luminescent nanoparticles designed for integrated multimodal diagnosis and therapy of hepatocellular carcinoma (HCC). Reproduced by taking permission from Ref. [85]. Licence no. 6020800611171

Such properties make rare-earth-doped nanomaterials suitable for continuous and dynamic observation of cellular processes over extended periods [86, 87].

Figure 29 shows the penetration depth analysis of a Pork Fillet Tissue. The capability of $LuPO_4:Nd^{3+}$ nanoparticles for deep-tissue bioimaging has been demonstrated. To simulate biological tissues, layers of pork fillet with thicknesses ranging from 0.5 to 1.5 cm were placed atop an aqueous dispersion of the nanoparticles. Upon excitation with an 808 nm laser source, the NIR-II luminescence was captured using an EMCCD camera. Although a gradual reduction in emission intensity was observed with increasing tissue thickness, the signal remained clearly detectable even through 1.5 cm of tissue. These findings confirm the excellent tissue penetration ability of $LuPO_4:Nd^{3+}$ nanoparticles, highlighting their strong potential for non-invasive deep-tissue imaging applications within the NIR-II spectral window [44].

d. **In Vivo Targeted Imaging**: Rare-earth-doped phosphors and nanoparticles have become key tools in vivo bioimaging due to their superior characteristics. Rare-earth-doped up-conversion nanoparticles (UCNPs) have emerged as promising candidates for targeted bioimaging. These materials enable high-resolution imaging, offering real-time insights into biological activities within living organisms. Their durable emission properties and resistance to photobleaching make them suitable for prolonged monitoring [88].

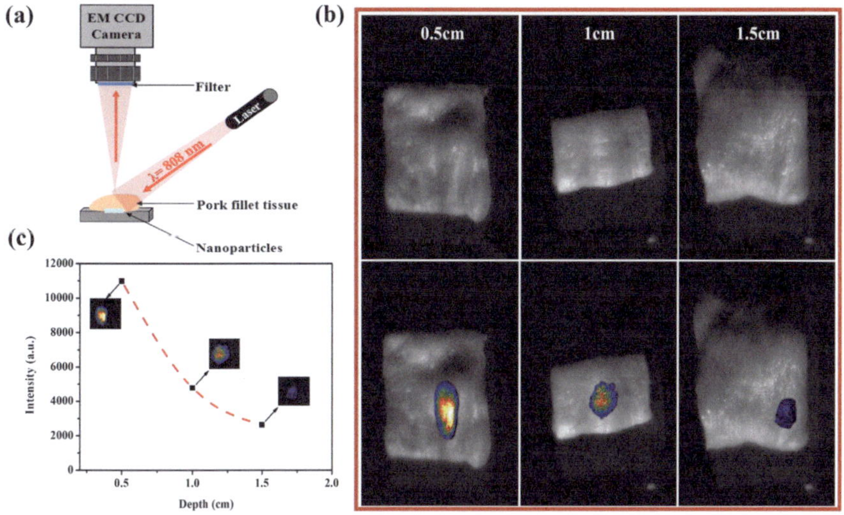

Fig. 29 Penetration depth analysis of $LuPO_4:Nd^{3+}$ NPs in Pork Fillet Tissue **a** A schematic illustration of the experimental setup used for ex vivo imaging studies. **b** NIR-II fluorescence images of $LuPO_4:Nd^{3+}$ nanoparticles captured under 808 nm laser excitation with varying layers of pork fillet tissue (ranging from 0.5 to 1.5 cm) placed above the sample. **c** Plot showing the relationship between NIR-II emission intensity and tissue penetration depth. Reproduced by taking permission from Ref. [44]. Licence no. 6020800317059

Figure 30 shows the uses of rare-earth-doped nanoparticles (RENPs) for in vivo optical imaging of rats [89] and Fig. 31 shows the tumor-targeting capability of folic acid- functionalized up-conversion nanoparticles (UCNPs) in vivo. Mice bearing subcutaneous HeLa tumors in the right rear leg were intravenously injected with UCNPs-FA. From figure, it can be seen that there was a strong luminescence signal at the tumor site post-injection in the UCNPs-FA group [90].

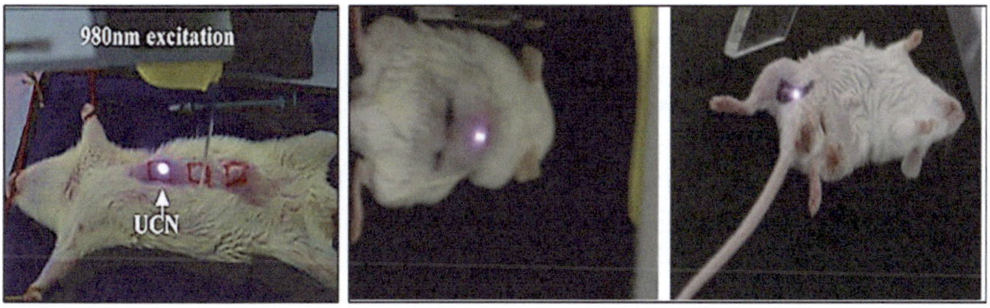

Fig. 30 Near-infrared (NIR) responsive rare-earth-doped nanoparticles (RENPs) for in vivo optical imaging of rats following injection of UCNPs beneath the abdominal skin (left), into the thigh muscles (middle), or under the dorsal skin (right). Reproduced by taking permission from Ref. [89]. Licence no. 6020791490101

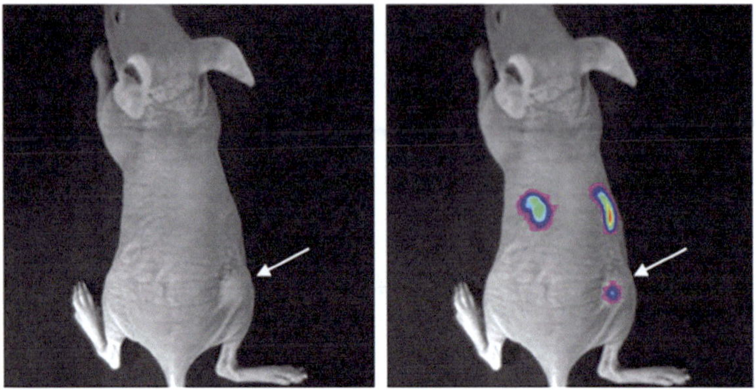

Fig. 31 Upconversion luminescence imaging conducted in vivo on a mice bearing subcutaneous HeLa tumors in the right rear leg (marked by white arrows), following systemic injection of folic acid-conjugated UCNPs. Reproduced by taking permission from Ref. [38]. License Number-6054561244879

e. **Biosensors and Diagnostics**: Rare-earth-doped phosphors and nanoparticles are gaining attention as advanced biosensors for disease detection due to their exceptional sensitivity, stability, and customizable optical properties. These materials, with their precise emission wavelengths and robust photostability, can detect disease-related biomarkers, providing clear and reliable signals for real-time diagnostics [90]. Surface modification further enhances their specificity, transforming these nanomaterials into powerful diagnostic tools suitable for real-time, non-invasive point-of-care applications [91].

6 Challenges in the Bioimaging Field Using Rare Earth Activated Phosphors

The application of rare earth (RE) activated phosphors in bioimaging has gained substantial attention due to their exceptional optical characteristics, such as remarkable stability, tunable emission spectra, and resistance to photobleaching. Despite their great promise, various challenges and limitations hinder their practical use. These challenges impact their performance, safety, and overall feasibility in clinical environments.

Below is a summary of the main issues highlighted in recent research on the use of rare earth activated phosphors in bioimaging:

a. *Biocompatibility and Toxicity*: One of the primary concerns in applying rare earth elements (REEs) in biomedical fields is their possible toxic effects. Although many lanthanide-based materials are considered biocompatible, certain ions or their compounds can be harmful due to their tendency to accumulate in tissues or interfere with cellular functions. To address this, it is crucial to implement effective surface coatings and modifications that enhance their compatibility with biological systems and reduce potential toxicity [92–94].

b. *Aggregation and Dispersion in Biological Environments*: A major issue in bioimaging applications is the tendency of phosphor nanoparticles to aggregate in biological environments. Such aggregation diminishes their dispersion stability, negatively impacting their performance in imaging and targeted delivery. Additionally, it can alter their luminescent properties, resulting in weaker imaging signals [94].
c. *Low Luminescence Efficiency and Photon Conversion*: The limited luminescence efficiency of rare-earth-doped phosphors poses a challenge for their use in bioimaging. Inadequate energy transfer within the phosphor material often leads to weak emission signals, compromising image clarity—especially for deep-tissue imaging. Furthermore, the brightness and emission wavelengths of some rare earth elements may not be ideal for specific biomedical applications [92, 92].
d. *Size Control and Surface Modification*: Controlling the size of rare earth activated phosphors is a key challenge, as size directly influences cellular uptake, tissue penetration, and clearance from the body. The nanoparticle size plays a vital role in their biological behavior—larger particles may be quickly cleared by the immune system, while smaller ones are prone to aggregation and may have reduced energy conversion efficiency. Additionally, effective surface functionalization of rare earth-doped nanoparticles remains a significant challenge. Optimizing surface chemistry is essential for enhancing their stability, dispersion, and interactions with biological systems. Inadequate surface modification can result in poor imaging contrast, increased aggregation, and limited cellular uptake. Tailoring both size and surface characteristics is crucial for improving bioavailability, targeting ability, and overall imaging and therapeutic performance, yet identifying the ideal modification strategy is still a complex task [93, 94].
e. *Vivo Stability and Phototoxicity*: Rare earth nanoparticles often encounter stability challenges within living systems. Once inside the body, they may degrade or dissolve, releasing free ions that can be harmful or shorten their therapeutic effectiveness. For successful application in imaging and therapy, it is crucial that up-conversion nanoparticles maintain their structure and function over time. Although they exhibit some resistance to light-induced damage, extended exposure can still lead to photobleaching, reducing their performance during continuous or long-term imaging. Improving both their chemical durability and resistance to photobleaching is essential for advancing their use in biomedical fields [92, 94].
f. *Scalability of Synthesis and Cost*: Producing rare earth activated phosphors at a scale suitable for biomedical applications remains a significant challenge. The high cost of rare earth elements, combined with complex synthesis and processing steps—including extraction, purification, and surface functionalization—limits their accessibility for clinical use. Additionally, achieving consistent control over particle size, composition, and surface characteristics adds to the difficulty of scaling up production. To enable broader clinical adoption, it is crucial to develop more efficient, reproducible, and cost-effective synthesis methods [93, 94].

g. *Regulatory Challenges and Clinical Translation*: The path to clinical implementation of rare earth activated phosphors is hindered by complex regulatory requirements. As with many emerging nanotechnologies, these materials must undergo comprehensive assessments to ensure their safety, biocompatibility, long-term stability, and absence of cytotoxic effects. The current regulatory framework for nanomaterials remains underdeveloped, resulting in prolonged and costly approval processes. Additionally, ensuring the reproducibility, scalability, and therapeutic efficacy of these phosphors is essential for their successful translation from research laboratories to clinical environments. Overcoming these regulatory and translational challenges is critical to realizing their full potential in biomedical imaging and therapy [92, 94].

7 Future Possibilities

Despite remarkable progress in the development of rare earth (RE) activated phosphors for bioimaging, numerous exciting avenues remain open for exploration.

To enhance the safety of REE-based materials in biological systems, surface modification and biocompatible coatings are key. To improving the biocompatibility and reducing the toxicity of REEs in biological systems, researchers are working on advanced surface modifications, such as biocompatible coatings with biopolymers or PEG, which enhance stability, promote cellular uptake, and minimize cytotoxicity. These coatings will be essential in preventing ion leaching and mitigating immune responses, making REEs safer for in vivo use. Another challenge is nanoparticle aggregation in biological environments, which reduces targeting efficiency. To combat this, stabilizers, surfactants, and functional coatings can be used to improve nanoparticle dispersion and stability. Surface modifications like polymers or silica coatings will help maintain the nanoparticles' long-term stability, ensuring more effective disease targeting and treatment. Low luminescence efficiency of certain REEs also poses a limitation in bioimaging. Advances in energy transfer mechanisms, such as core–shell structures or hybrid materials, will boost up-conversion efficiency and enhance luminescence. This will improve the performance of REEs in deep-tissue imaging and multimodal imaging, where high luminescence is crucial. The size of REE nanoparticles is another key factor in their bioimaging effectiveness. Future research is aiming for precise size control and surface functionalization with targeting ligands like peptides or antibodies, allowing for better cell-specific targeting and improved imaging resolution and sensitivity. This will enhance diagnostic accuracy and therapeutic outcomes. Stability and phototoxicity in vivo remain important challenges. Advanced coatings, such as silica or polymer shells, will help stabilize the nanoparticles, prevent toxic ion release, and maintain their luminescent properties. Scalable and reproducible synthesis methods will be critical to produce nanoparticles at the scale needed for clinical use. Overcoming regulatory challenges is essential for the clinical translation of REE-based nanoparticles. Rigorous preclinical studies will be necessary to assess their long-term stability, and biological interactions. Research

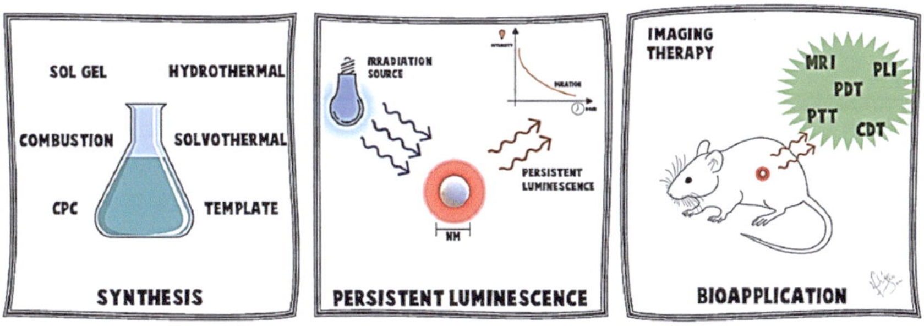

Fig. 32 Graphical representation of rare-earth activated phosphors in biomedical applications. Taken from Ref. [95]

into the long-term effects of these materials will be crucial for regulatory approval and the successful integration of REEs into bioimaging and therapeutic applications. Moreover, on developing cost-effective synthesis methods and finding affordable sources of rare earth materials. Standardizing production processes will make these nanoparticles more accessible for medical applications such as imaging, drug delivery, and cancer therapy [93, 94].

The future of rare earth (RE) activated phosphors in bioimaging is promising, with a range of advancements on the horizon aimed at overcoming current limitations and unlocking their full potential in biomedical applications. Figure 32 shows a schematic illustration of rare earth activated phosphors in biomedical applications. The first panel represents various synthesis strategies which are employed to tailor phosphor morphology and composition. The second panel depicts the principle of persistent luminescence, where phosphors emit light long after excitation, enabling background-free imaging. The third panel highlights the bio-applications of these phosphors in imaging particularly for cancer diagnostics and treatment in vivo.

8 Conclusion

Rare earth activated phosphors represent a transformative class of materials in the field of bioimaging, offering unique optical properties. In persistent luminescence systems, activators such as Eu^{2+}—co-doped with trap-forming ions like Dy^{3+}, Sm^{3+}, Pr^{3+}—store excitation energy and release it gradually, allowing for sustained afterglow emission even after external excitation ceases. These materials typically emit within the visible range (470–700 nm). Also, advancements in host and trap engineering have extended Eu^{2+} -based emissions into the first biological window (NIR-I: 650–950 nm), significantly improving deep-tissue imaging capabilities. A wide range of host matrices, including silicates, aluminates, and sulfides have been employed to stabilize the rare-earth ions, facilitate trap formation, and optimize emission properties for bioimaging. Meanwhile, up-conversion phosphors, utilizing sensitizer-activator pairs such as Yb^{3+}/Er^{3+} or Yb^{3+}/Tm^{3+}, efficiently convert near-infrared excitation (typically around 980 nm) into higher-energy visible or NIR emissions. This mechanism enables real-time imaging with high spatial resolution, excellent tissue penetration, and minimal autofluorescence, as the excitation and emission wavelengths fall within biological optical windows. Host lattices such as fluorides and oxides offer favorable crystal environments that support efficient energy transfer, minimize non-radiative losses, and enhance up-conversion emission intensity. These up-conversion systems have found applications in dynamic cell tracking, multimodal imaging, and therapeutic applications. In addition to that, Nd^{3+} have also been widely explored for their NIR-to-NIR luminescence capabilities. Owing to its strong absorption near 808 nm, Nd^{3+} enables efficient excitation, which falls within the biological transparency window. Moreover, its characteristic emissions in the range of ~ 880, 1060, and 1330 nm further extend bioimaging into both the NIR-I (650–950 nm) and NIR-II (1000–1350 nm) biological windows, offering deeper tissue penetration, reduced scattering, and minimized background autofluorescence.

Advancements in synthesis approaches—ranging from solid-state and sol–gel to hydrothermal and combustion methods—have enabled precise control over dopant incorporation, particle size, crystallinity, emission properties; while surface modifications with biocompatible coatings have enhanced their stability and safety in biological environments.

Despite these considerable advances, several critical challenges remain, including the need to optimize biocompatibility, minimize particle aggregation, enhance luminescence efficiency, ensure long-term in vivo stability, establish scalable synthesis protocols, and achieve regulatory approval for clinical translation. But continuous progress in materials design, surface modification, and biological integration is steadily overcoming these barriers. With ongoing research focused on material innovation, biological safety, and multifunctional capabilities, rare earth activated phosphors are emerging as strong candidates to make significant contributions to next-generation bioimaging and therapeutic technologies.

Acknowledgements The authors gratefully acknowledge the American Chemical Society (ACS), the Royal Society of Chemistry (RSC), Elsevier, Wiley, and other publishers for their kind permission to reproduce selected figures and content included in this work. Appropriate permissions have been obtained for the use of previously published materials, and full credit is given to the original authors and publishers as cited in the respective figure legends.

References

1. M. Mollazadeh-Bajestani, A. Bahmanpour, M. Ghaffari, F. Moztarzadeh, A. Sepahvandi, K. Adebajo, Reviewing the bio-applications of $SrAl_2O_4$: Eu^{2+}, Dy^{3+} phosphor. J. Biol. Med. **7**(1), 044–052 (2023)

2. Q. Fan, X. Cui, H. Guo, X. Yantao, G. Zhang, B. Peng, Application of rare earth-doped nanoparticles in biological imaging and tumor treatment. J. Biomater. Appl. **35**(2), 237–263 (2020)
3. J. Ting, F. Mei, Z. Mengyao, Q. Jianwen, Z. Hu, G. Yong, A novel cholesterol conjugated fluorescence probe for Cu^{2+} detection and bioimaging in living cells. Spectrochim. Acta Part A Mol. Biomol. Spectrosc. **227**, 117530 (2020)
4. L. Wang, X. Yang, X. Chen, Y. Zhou, X. Lu, C. Yan, Y. Xu, R. Liu, J. Qu, A novel fluorescence probe based on triphenylamine Schiff base for bioimaging and responding to pH and Fe^{3+}. Mater. Sci. Eng. C **72**, 551–557 (2017)
5. Y. Ning, M. Zhu, J.L. Zhang, Near-infrared (NIR) lanthanide molecular probes for bioimaging and biosensing. Coord. Chem. Rev. **399**, 213028 (2019)
6. P. Lei, J. Feng, H. Zhang, Emerging biomaterials: taking full advantage of the intrinsic properties of rare earth elements. Nano Today **35**, 100952 (2020)
7. V.B. Pawade, H.C. Swart, S.J. Dhoble, Review of rare earth activated blue emission phosphors prepared by combustion synthesis. Renew. Sustain. Energy Rev. **52**, 596–612 (2015)
8. C. Song, S. Zhang, Q. Zhou, H. Hai, D. Zhao, Y. Hui, Upconversion nanoparticles for bioimaging. Nanotechnol. Rev. **6**(2), 233–242 (2017)
9. D. Chávez-García, K. Juarez-Moreno, I. Calderón-Osuna, P. Navarro, G.A. Hirata, Nanotoxicological study of downconversion Y_2O_3: Eu^{3+} luminescent nanoparticles functionalized with folic acid for cancer cells bioimaging. J. Biomed. Mater. Res. B Appl. Biomater. **108**(6), 2396–2406 (2020)
10. X.Y. Huang, X.H. Ji, Q.Y. Zhang, Broadband downconversion of ultraviolet light to near-infrared emission in Bi^{3+}–Yb^{3+}-codoped Y_2O_3 phosphors. J. Am. Ceram. Soc. **94**(3), 833–837 (2011)
11. Y. Ali, R. Zohre, J. Mostafa, R. Samaneh, Applications of upconversion nanoparticles in molecular imaging: a review of recent advances and future opportunities. Biosci. Biotechnol. Res. Asia **12**, 131–140 (2015)
12. M.K. Chen, Y. Luo, D.P. Tsai, Optical meta-devices for bio-imaging, in *2022 Conference on Lasers and Electro-Optics Pacific Rim (CLEO-PR)* (IEEE, USA, 2022), pp. 1–2
13. X. Liu, R. Lei, F. Huang, D. Deng, H. Wang, S. Zhao, S. Xu, Upconversion luminescence, intrinsic optical bistability, and optical thermometry in Ho^{3+}/Yb^{3+}: $BaMoO_4$ phosphors. Chin. Opt. Lett. **17**(11), 111601 (2019)
14. Q. Le Masne de Chermont, C. Chanéac, J. Seguin, F. Pellé, S. Maîtrejean, J.P. Jolivet, D. Gourier, M. Bessodes, D. Scherman, Nanoprobes with near-infrared persistent luminescence for in vivo imaging. Proc. Nat. Acad. Sci. **104**(22), 9266–9271 (2007)
15. T. Maldiney, D. Scherman, C. Richard, Persistent luminescence nanoparticles for diagnostics and imaging, in *Functional Nanoparticles for Bioanalysis, Nanomedicine, and Bioelectronic Devices*, vol. 2 (American Chemical Society, 2012), pp. 1–25
16. J. Liu, T. Lécuyer, J. Seguin, N. Mignet, D. Scherman, B. Viana, C. Richard, Imaging and therapeutic applications of persistent luminescence nanomaterials. Adv. Drug Deliv. Rev. **138**, 193–210 (2019)
17. Y. Li, M. Gecevicius, J. Qiu, Long persistent phosphors-from fundamentals to applications. Chem. Soc. Rev. **45**, 2090–2136 (2016)
18. S. Singh, D. Singh, P. Siwach, I. Gupta, P. Kumar, Synthesis strategies for rare earth activated inorganic phosphors: a mini review. Appl. Res. **4**(1), e202400190 (2025)
19. H. Homayoni, L. Ma, J. Zhang, S.K. Sahi, L.H. Rashidi, B. Bui, W. Chen, Synthesis and conjugation of $Sr_2MgSi_2O_7$: Eu^{2+}, Dy^{3+} water soluble afterglow nanoparticles for photodynamic activation. Photodiagn. Photodyn. Ther. **16**, 90–99 (2016)
20. S. Sharma, S.K. Dubey, A.K. Diwakar, Luminescence investigation on $Ca_2MgSi_2O_7$: Eu^{2+}, Dy^{3+} phosphor. Int. J. Mater. Sci. **2**(2), 08–15 (2021)
21. M. Sun, Z.J. Li, C.L. Liu, H.X. Fu, J.S. Shen, H.W. Zhang, Persistent luminescent nanoparticles for super-long time in vivo and in situ imaging with repeatable excitation. J. Lumin. **145**, 838–842 (2014)

22. L. Hu, P. Wang, M. Zhao, L. Liu, L. Zhou, B. Li, F.H. Albaqami, A.M. El-Toni, X. Li, Y. Xie, X. Sun, F. Zhang, Near-infrared rechargeable "optical battery" implant for irradiation-free photodynamic therapy. Biomaterials **163**, 154–162 (2018)
23. D.G. Calatayud, T. Jardiel, E. Cordero-Oyonarte, A.C. Caballero, M. Villegas, A. Valle-Noguera, A. Cruz-Adalia, M. Peiteado, Biocompatible probes based on rare-earth doped strontium aluminates with long-lasting phosphorescent properties for in vitro optical imaging. Int. J. Mol. Sci. **23**(6), 3410 (2022)
24. Y. Li, J. Wang, X.M. Wang, F. Pan, T. Zhou, R.J. Xie, Colour tuning via crystalline site-selected energy transfer in a Sr_2SiO_4: Eu^{2+}, Pr^{3+} phosphor. J. Mater. Chem. C **5**(5), 1022–1026 (2017)
25. D.C. Rodriguez Burbano, S.K. Sharma, P. Dorenbos, B. Viana, J.A. Capobianco, Persistent and photostimulated red emission in CaS: Eu^{2+}, Dy^{3+} nanophosphors. Adv. Opt. Mater. **3**(4), 551–557 (2015)
26. Y. Gao, R. Li, W. Zheng, X. Shang, J. Wei, M. Zhang, J. Xu, W. You, Z. Chen, X. Chen, X. Chen, Broadband NIR photostimulated luminescence nanoprobes based on CaS: Eu^{2+}, Sm^{3+} nanocrystals. Chem. Sci. **10**(21), 5452–5460 (2019)
27. C. Jiang, Q. Liu, K. Li, Y. Feng, Y. Fu, Y. Li, Y. Zhang, X. Qian, B. Wei, P. Du, Regulating the broadband near-infrared emission of Eu^{2+}-doped $Ca_3Sc_2Si_3O_{12}$ phosphors through constructing defect for diversified applications. Chem. Eng. J. **496**, 154360 (2024)
28. W.B. Dai, Y.F. Lei, J. Zhou, M. Xu, L.L. Chu, L. Li, P. Zhao, Z.H. Zhang, Near-infrared quantum-cutting and long-persistent phosphor $Ca_3Ga_2Ge_3O_{12}$: Pr^{3+}, Yb^{3+} for application in in vivo bioimaging and dye-sensitized solar cells. J. Alloy. Compd. **726**, 230–239 (2017)
29. M. Pellerin, E. Glais, T. Lecuyer, J. Xu, J. Seguin, S. Tanabe, C. Chanéac, B. Viana, C. Richard, $LaAlO_3$:Cr^{3+}, Sm^{3+}: nano-perovskite with persistent luminescence for in vivo optical imaging. J. Lumin. **202**, 83–88 (2018)
30. P. Xiong, M. Peng, K. Qin, F. Xu, X. Xu, Visible to near-infrared persistent luminescence and mechanoluminescence from Pr^{3+}-Doped $LiGa_5O_8$ for energy storage and bioimaging. Adv. Opt. Mater. **7**(24), 1901107 (2019)
31. C. Rosticher, B. Viana, G. Laurent, P. Le Griel, C. Chanéac, Insight into $CaMgSi_2O_6$: Eu^{2+}, Mn^{2+}, Dy^{3+} nanoprobes: influence of chemical composition and crystallinity on persistent red luminescence. Eur. J. Inorg. Chem. **2015**(22), 3681–3687 (2015)
32. H.J.M.A.A. Zijlmans, J. Bonnet, J. Burton, K. Kardos, T. Vail, R.S. Niedbala, H.J. Tanke, Detection of cell and tissue surface antigens using up-converting phosphors: a new reporter technology. Anal. Biochem. **267**(1), 30–36 (1999)
33. S.H. Nam, Y.M. Bae, Y.I. Park, J.H. Kim, H.M. Kim, J.S. Choi, K.T. Lee, T. Hyeon, Y.D. Suh, Long-term real-time tracking of lanthanide ion doped upconverting nanoparticles in living cells. Angew. Chem. Int. Ed. **50**(27), 6093–6097 (2011)
34. N.M. Idris, Z. Li, L. Ye, E.K.W. Sim, R. Mahendran, P.C.L. Ho, Y. Zhang, Tracking transplanted cells in live animal using upconversion fluorescent nanoparticles. Biomaterials **30**(28), 5104–5113 (2009)
35. T.D. Cao, T.G. Le, T.N. Nguyen, T.N. Dau, V.T. Nguyen, T.T. Van Tran, Investigating the effect of Yb^{3+} and Er^{3+} concentration on red/green luminescent ratio in β-$NaYF_4$: Er, Yb nanocrystals using spectroscopic techniques. J. Mol. Struct. **1210**, 128014 (2020)
36. J.C. Zhou, Z.L. Yang, W. Dong, R.J. Tang, L.D. Sun, C.H. Yan, Bioimaging and toxicity assessments of near-infrared upconversion luminescent $NaYF_4$: Yb, Tm nanocrystals. Biomaterials **32**(34), 9059–9067 (2011)
37. S.K. Cho, L.J. Su, C. Mao, C.D. Wolenski, T.W. Flaig, W. Park, Multifunctional nanoclusters of $NaYF_4$: Yb^{3+}, Er^{3+} upconversion nanoparticle and gold nanorod for simultaneous imaging and targeted chemotherapy of bladder cancer. Mater. Sci. Eng. C **97**, 784–792 (2019)
38. L.Q. Xiong, Z.G. Chen, M.X. Yu, F.Y. Li, C. Liu, C.H. Huang, Synthesis, characterization, and in vivo targeted imaging of amine-functionalized rare-earth up-converting nanophosphors. Biomaterials **30**(29), 5592–5600 (2009)
39. K.S. Janbandhu, V.B. Pawade, S.J. Dhoble, H.C. Swart, Synthesis and upconversion properties of $KAlF_4$: Yb^{3+}/Er^{3+} phosphor for bioimaging application. Infrared Phys. Technol. **126**, 104328 (2022)

40. Z. Sun, G. Liu, Z. Fu, T. Sheng, Y. Wei, Z. Wu, Nanostructured La_2O_3: Yb^{3+}/Er^{3+}: temperature sensing, optical heating and bio-imaging application. Mater. Res. Bull. **92**, 39–45 (2017)
41. G. Ajithkumar, B. Yoo, D.E. Goral, P.J. Hornsby, A.L. Lin, U. Ladiwala, D.P. Dravid, D.K. Sardar, Multimodal bioimaging using a rare earth doped Gd_2O_2S: Yb/Er phosphor with upconversion luminescence and magnetic resonance properties. J. Mater. Chem. B **1**(11), 1561–1572 (2013)
42. S.K. Gupta, M.A.P. Garcia, J.P. Zuniga, M. Abdou, Y. Mao, Visible and ultraviolet upconversion and near infrared downconversion luminescence from lanthanide doped $La_2Zr_2O_7$ nanoparticles. J. Lumin. **214**, 116591 (2019)
43. X. Jiang, C. Cao, W. Feng, F. Li, Nd^{3+}-doped $LiYF_4$ nanocrystals for bio-imaging in the second near-infrared window. J. Mater. Chem. B **4**(1), 87–95 (2016)
44. X. Wang, J. Shi, P. Li, S. Zheng, X. Sun, H. Zhang, $LuPO_4$: Nd^{3+} nanophosphors for dual-mode deep tissue NIR-II luminescence/CT imaging. J. Lumin. **209**, 420–426 (2019)
45. R. Priya, S. Bhatia, S. Kaur, V. Dubey, O.P. Pandey, Synthesis of rare-earth-activated phosphors, in *Rare-Earth-Activated Phosphors* (Elsevier, Amsterdam, 2022), pp. 109–136
46. J. Fu, S. Zhang, T. Ma, Y. Jia, R. Pang, L. Jiang, D. Li, H. Li, W. Sun, C. Li, A convenient and efficient synthesis method to improve the emission intensity of rare earth ion doped phosphors: the synthesis and luminescent properties of novel SrO: Ce^{3+} phosphor. RSC Adv. **5**(114), 93951–93956 (2015)
47. C. Zhang, J. Lin, Defect-related luminescent materials: synthesis, emission properties and applications. Chem. Soc. Rev. **41**(23), 7938–7961 (2012)
48. Y. Karabulut, M. Ayvacıklı, A. Canimoglu, J.G. Guinea, Z. Kotan, E. Ekdal, O. Akyuz, N. Can, Synthesis and luminescence properties of trivalent rare-earth element-doped calcium stannate phosphors. Spectrosc. Lett. **47**(8), 630–641 (2014)
49. G. Yang, S.-J. Park, Conventional and microwave hydrothermal synthesis and application of functional materials: a review. Materials **12**(7), 1177 (2019). https://doi.org/10.3390/ma1207 1177
50. C.H. He, Y.L. Liu, Recent developments of research on rare earth functional materials. Chem. Res. Appl. **13**(3), 232–237 (2001)
51. S. Gai, C. Li, P. Yang, J. Lin, Recent progress in rare earth micro/nanocrystals: soft chemical synthesis, luminescent properties, and biomedical applications. Chem. Rev. **114**(4), 2343–2389 (2014)
52. S. Agrawal, Spectroscopy techniques for rare-earth-activated phosphors, in *Rare-Earth-Activated Phosphors* (Elsevier, Amsterdam, 2022) pp. 147–175
53. R. Elilarassi, G. Chandrasekaran, Structural, optical and magnetic characterization of Cu-doped ZnO nanoparticles synthesized using solid state reaction method. J. Mater. Sci.: Mater. Electron. **21**, 1168–1173 (2010)
54. Y. Xu, Y. Zhou, Y. Li, Y. Zheng, Bridging materials and analytics: a comprehensive review of characterization approaches in metal-based solid-state hydrogen storage. Molecules **29**(21), 5014 (2024)
55. K. He, N. Chen, C. Wang, L. Wei, J. Chen, Method for determining crystal grain size by x-ray diffraction. Cryst. Res. Technol. **53**(2), 1700157 (2018)
56. F. Sanchez-Bajo, A.L. Ortiz, F.L. Cumbrera, Novel analytical model for the determination of grain size distributions in nanocrystalline materials with low lattice microstrains by X-ray diffractometry. Acta Mater. **54**(1), 1–10 (2006)
57. T. Selvalakshmi, P. Venkatesan, S.P. Wu, S. Velmathi, A.C. Bose, Gd_2O_3: RE^{3+} and $GdAlO_3$: RE^{3+} (RE= Eu, Dy) phosphor: synthesis, characterization and bioimaging application. J. Nanosci. Nanotechnol. **17**(2), 1178–1184 (2017)
58. S. Chenna, P.A. Crozier, Operando transmission electron microscopy: a technique for detection of catalysis using electron energy-loss spectroscopy in the transmission electron microscope. ACS Catal. **2**(11), 2395–2402 (2012)
59. B. Faramarzi, M. Moggio, N. Diano, M. Portaccio, M. Lepore, A brief review of FT-IR Spectroscopy studies of sphingolipids in human cells. Biophysica **3**(1), 158–180 (2023)

60. T.E. Weirich, X.D. Zou, R. Ramlau, A. Simon, G.L. Cascarano, C. Giacovazzo, S. Hovmöller, Structures of nanometre-size crystals determined from selected-area electron diffraction data. Found. Crystallogr. **56**(1), 29–35 (2000)
61. N.V. Chukanov, M.F. Vigasina, N.V. Zubkova, I.V. Pekov, C. Schäfer, A.V. Kasatkin, V.O. Yapaskurt, D.Y. Pushcharovsky, Extra-framework content in sodalite-group minerals: complexity and new aspects of its study using infrared and Raman spectroscopy. Minerals **10**(4), 363 (2020)
62. C.V. Duke, J.M. Miller, J.H. Clark, A.P. Kybett, 19F mas NMR and FTIR analysis of the adsorption of alkali metal fluorides onto alumina. J. Mol. Catal. **62**(2), 233–242 (1990)
63. U. Gross, S. Rüdiger, E. Kemnitz, K.W. Brzezinka, S. Mukhopadhyay, C. Bailey, A. Wander, N. Harrison, Vibrational analysis study of aluminum trifluoride phases. J. Phys. Chem. A **111**(26), 5813–5819 (2007)
64. P. Tarte, Infra-red spectra of inorganic aluminates and characteristic vibrational frequencies of AlO_4 tetrahedra and AlO_6 octahedra. Spectrochim. Acta Part A **23**(7), 2127–2143 (1967)
65. A. Dwivedi, A. Roy, S.B. Rai, Photoluminescence behavior of rare earth doped self-activated phosphors (i.e. niobate and vanadate) and their applications. RSC Adv. **13**(24), 16260–16271 (2023)
66. S. Sharma, S. Jaiswal, B. Duffy, A.K. Jaiswal, Nanostructured materials for food applications: spectroscopy, microscopy and physical properties. Bioengineering **6**(1), 26 (2019)
67. A.C. Mohan, B.J.P.T. Renjanadevi, Preparation of zinc oxide nanoparticles and its characterization using scanning electron microscopy (SEM) and X-ray diffraction (XRD). Proc. Technol. **24**, 761–766 (2016)
68. D.S. Rao, K. Muraleedharan, C.J. Humphreys, TEM specimen preparation techniques. Microsc. Sci. Technol. Appl. Educ. **2**, 1232 (2010)
69. S. Zaefferer, A critical review of orientation microscopy in SEM and TEM. Cryst. Res. Technol. **46**(6), 607–628 (2011)
70. https://embiitk.vlabs.ac.in/exp/transmission-electron-microscope/theory.html
71. R.P. Williams, R.D. Hart, A. Van Riessen, Quantification of the extent of reaction of metakaolin-based geopolymers using X-ray diffraction, scanning electron microscopy, and energy-dispersive spectroscopy. J. Am. Ceram. Soc. **94**(8), 2663–2670 (2011)
72. M. Manso, M.L. Carvalho, I. Queralt, S. Vicini, E. Princi, Investigation of the composition of historical and modern Italian papers by energy dispersive X-ray fluorescence (EDXRF), X-ray diffraction (XRD), and scanning electron microscopy energy dispersive spectrometry (SEM-EDS). Appl. Spectrosc. **65**(1), 52–59 (2011)
73. G. Verma, M. Mishra, Development and optimization of UV-Vis spectroscopy-a review. World J. Pharm. Res **7**(11), 1170–1180 (2018)
74. N. Sudharshan, V. Swetha, UV-visible spectroscopy: a comprehensive review on instrumentation. World J. Pharm. Res **12**(19), 1342–1363 (2023)
75. M. Rizwan, C. Gwenin, Nanomaterials in renewable energy: UV-Visible spectroscopy characterization and applications, in *Nano Tools and Devices for Enhanced Renewable Energy* (Elsevier, Amsterdam, 2021), pp. 103–120
76. J.T. Torvik, Dopants in GaN, in *III-Nitride Semiconductors: Electrical, Structural and Defects Properties* (Elsevier, Amsterdam, 2000), pp. 17–49
77. P. Vuong, Optical spectroscopy of boron nitride heterostructures. Doctoral dissertation, Université Montpellier, 2018
78. N.L. Ignjatović, L. Mančić, M. Vuković, Z. Stojanović, M.G. Nikolić, S. Škapin, D.P. Uskoković, Rare-earth (Gd^{3+}, Yb^{3+}/Tm^{3+}, Eu^{3+}) co-doped hydroxyapatite as magnetic, up-conversion and down-conversion materials for multimodal imaging. Sci. Rep. **9**(1), 16305 (2019)
79. M.V. DaCosta, S. Doughan, Y. Han, U.J. Krull, Lanthanide upconversion nanoparticles and applications in bioassays and bioimaging: a review. Anal. Chim. Acta **832**, 1–33 (2014)
80. Q. Zhang, S. O'Brien, J. Grimm, Biomedical applications of lanthanide nanomaterials, for imaging, sensing and therapy. Nanotheranostics **6**(2), 184 (2022)

81. C. Wang, L. Cheng, Z. Liu, Upconversion nanoparticles for photodynamic therapy and other cancer therapeutics. Theranostics **3**(5), 317 (2013)
82. B. Zheng, H.B. Chen, P.Q. Zhao, H.Z. Pan, X.L. Wu, X.Q. Gong, H.J. Wang, J. Chang, Persistent luminescent nanocarrier as an accurate tracker in vivo for near infrared-remote selectively triggered photothermal therapy. ACS Appl. Mater. Interfaces. **8**(33), 21603–21611 (2016)
83. Z. Wei, Y. Liu, B. Li, J. Li, S. Lu, X. Xing, K. Liu, F. Wang, H. Zhang, Rare-earth based materials: an effective toolbox for brain imaging, therapy, monitoring and neuromodulation. Light Sci. Appl. **11**(1), 175 (2022)
84. T. Zako, H. Hyodo, K. Tsuji, K. Tokuzen, H. Kishimoto, M. Ito, K. Kaneko, M. Maeda, K. Soga, Development of near infrared-fluorescent nanophosphors and applications for cancer diagnosis and therapy. J. Nanomater. **2010**(1), 491471 (2010)
85. T. Ai, W. Shang, H. Yan, C. Zeng, K. Wang, Y. Gao, T. Guan, C. Fang, J. Tian, Near infrared-emitting persistent luminescent nanoparticles for hepatocellular carcinoma imaging and luminescence-guided surgery. Biomaterials **167**, 216–225 (2018)
86. E. Hemmer, N. Venkatachalam, H. Hyodo, A. Hattori, Y. Ebina, H. Kishimoto, K. Soga, Upconverting and NIR emitting rare earth based nanostructures for NIR-bioimaging. Nanoscale **5**(23), 11339–11361 (2013)
87. Y. Wei, J. Wang, X-ray/γ-ray/ultrasound-activated persistent luminescence phosphors for deep tissue bioimaging and therapy. ACS Appl. Mater. Interfaces. **16**(42), 56519–56544 (2024)
88. T. Maldiney, A. Bessière, J. Seguin, E. Teston, S.K. Sharma, B. Viana, A.J. Bos, P. Dorenbos, M. Bessodes, D. Gourier, D. Scherman, C. Richard, The in vivo activation of persistent nanophosphors for optical imaging of vascularization, tumours and grafted cells. Nat. Mater. **13**(4), 418–426 (2014)
89. D.K. Chatterjee, A.J. Rufaihah, Y. Zhang, Upconversion fluorescence imaging of cells and small animals using lanthanide doped nanocrystals. Biomaterials **29**(7), 937–943 (2008)
90. D.K. Sardar, K.L. Nash, J.B. Gruber, A. Sayka, Rare-earth doped nanocrystals for biosensing and imaging. Biophotonics Int. **15**(1), 45 (2008)
91. B. Zheng, J. Fan, B. Chen, X. Qin, J. Wang, F. Wang, R. Deng, X. Liu, Rare-earth doping in nanostructured inorganic materials. Chem. Rev. **122**(6), 5519–5603 (2022)
92. Z. Yi, Z. Luo, X. Qin, Q. Chen, X. Liu, Lanthanide-activated nanoparticles: a toolbox for bioimaging, therapeutics, and neuromodulation. Acc. Chem. Res. **53**(11), 2692–2704 (2020)
93. M. Ganguly, S. Agarwal, 14 Recent advances in the biomedical applications of rare earth elements, in *Rare Earth Elements: Processing, Catalytic Applications and Environmental Impact*, vol. 259 (2023)
94. L.D. Sun, Y.F. Wang, C.H. Yan, Paradigms and challenges for bioapplication of rare earth upconversion luminescent nanoparticles: small size and tunable emission/excitation spectra. Acc. Chem. Res. **47**(4), 1001–1009 (2014)
95. D.L. Fritzen, L. Giordano, L.C. Rodrigues, J.H. Monteiro, Opportunities for persistent luminescent nanoparticles in luminescence imaging of biological systems and photodynamic therapy. Nanomaterials **10**(10), 2015 (2020)

Synthesis of Upconversion Phosphors

Ram Gopal

Abstract Our objective in this chapter is to present a thorough discussion of the synthesis methods used for the development of upconversion phosphors. Upconversion phosphors are extensively utilized in a variety of optoelectronic devices, displays, bio-sensors, scintillators, and cathode-ray tubes. Upconversion phosphors can be developed into micro-sized particles, nanoparticles, colloidal nanoparticles, and thin films. Different types of physical and chemical methods are followed to synthesize the nanomaterials and thin films. These phosphors can be synthesized through numerous methods, including solid-state, solvothermal, combustion, co-precipitation, sol–gel, electrospinning, and laser ablation. The chosen synthesis route significantly impacts the luminescent properties of the materials. Therefore, understanding the synthesis methods and their effects is crucial for developing phosphors for practical use. This chapter is structured into three main sections. The first section discusses the impact of different factors on the synthesis processes in the top-down approach. The second section introduces various synthesis methods of the bottom-up approach and emphasizes the precursors, solvents, temperatures, and other synthesis conditions. The final section focuses on the various techniques used to develop thin films of upconversion materials and explores future advancements in synthesis techniques.

Keywords Upconversion phosphor · Synthesis · Nanoparticle · Thin film · Top-down · Bottom-up

1 Introduction

The upconversion phosphors have attracted significant attention over the past few decades due to their unique ability to convert low-energy infrared or visible light into higher-energy visible or ultraviolet light [1]. The upconversion phenomenon is fundamentally different from the more common downconversion process and offers

R. Gopal (✉)
Department of Physics, Panjab University, Chandigarh 160014, India
e-mail: r.gopal1996@gmail.com

V. Dubey et al. (eds.), *Upconversion Luminescence of Phosphors*, Progress in Optical Science and Photonics 39, https://doi.org/10.1007/978-981-95-1762-6_4

many applications such as bio-imaging, photovoltaics, photocatalysis, temperature sensing, and display technologies. The upconversion phosphors consist of a host lattice and one or more dopant ions and the interaction between the host and dopant ions govern the overall performance of the phosphor [2, 3]. Moreover, multiple electronic transitions are involved in the dopant ions, which can be influenced by chemical and electronic structures [4]. The host materials such as oxides, fluorides, molybdates, and tungstates, doped with lanthanide ions, such as erbium (Er^{3+}), holmium (Ho^{3+}), and thulium (Tm^{3+}) [5–8]. These ions are known for their rich energy-level structures and long-lived excited states, which facilitate the sequential absorption of two or more photons, leading to the emission of higher-energy light [9]. This process involves sophisticated energy transfer mechanisms, often facilitated by co-doping with sensitizer ions like Yb^{3+}, which absorb the excitation energy and transfer it to the activator ions (e.g., Er^{3+}, Ho^{3+}, Tm^{3+}) responsible for the visible emission [10].

The ability to manipulate light in such a manner opens up vast possibilities in various scientific and technological domains. In bio-imaging, for instance, upconversion nanoparticles can serve as luminescent probes that provide high contrast and deep tissue penetration with minimal photodamage and auto-fluorescence [11, 12]. In the realm of renewable energy, integrating upconversion materials into solar cells holds the potential to enhance their efficiency by converting sub-bandgap photons into usable light [13, 14]. Additionally, upconversion phosphors find applications in anti-counterfeiting measures, where their unique luminescent properties can be exploited to create security inks and markers [15, 16].

The experimental method used to synthesize a phosphor material plays an important role in deciding its applicability in various fields. The structural parameters (like crystallite size, particle size, and surface morphology) and optical parameters (like optical band gap and luminescence efficiency) of a material strongly depend on the experimental method followed to synthesize it [17]. Various experimental methods used to synthesize upconversion phosphor materials are mechanical ball milling, chemical etching, pulsed laser ablation, sputtering, solid-state reaction, sol–gel, solvothermal, co-precipitation, combustion, colloidal method, thermal decomposition, and so on. The materials prepared by different experimental methods may have different structural and optical properties. For example, the solid-state reaction route is generally used to prepare the bulk materials having particles in the micron range. The bulk phosphors generally have high luminescence efficiency under UV and NIR excitations. Moreover, the solid-state reaction method is facile, user friendly and suitable for the large-scale production of different phosphor materials such as oxide, tungstate, phosphate, silicate, and borate. However, wet chemical methods like sol–gel, solvothermal, and co-precipitation are more stable for developing uniform-size nanoparticles [18]. Several physical vapor deposition and chemical vapor deposition techniques are used to develop thin films.

2 Methods of Synthesis

The synthesis of upconversion phosphors encompasses a wide range of techniques, each with its advantages and challenges. These methods can broadly be categorized into top-down and bottom-up approaches. Another way of classification can be physical and chemical methods. Physical methods for synthesizing phosphors involve processes primarily using physical means such as temperature, pressure, and mechanical forces to transform raw materials into phosphors. These methods do not involve significant chemical reactions but rely on physical processes to achieve the desired material properties. Chemical methods for synthesizing phosphors involve chemical reactions that transform precursor compounds into phosphor materials. These methods often involve solution chemistry and can offer precise control over composition, morphology, and particle size [19, 20] (Figs. 1 and 2).

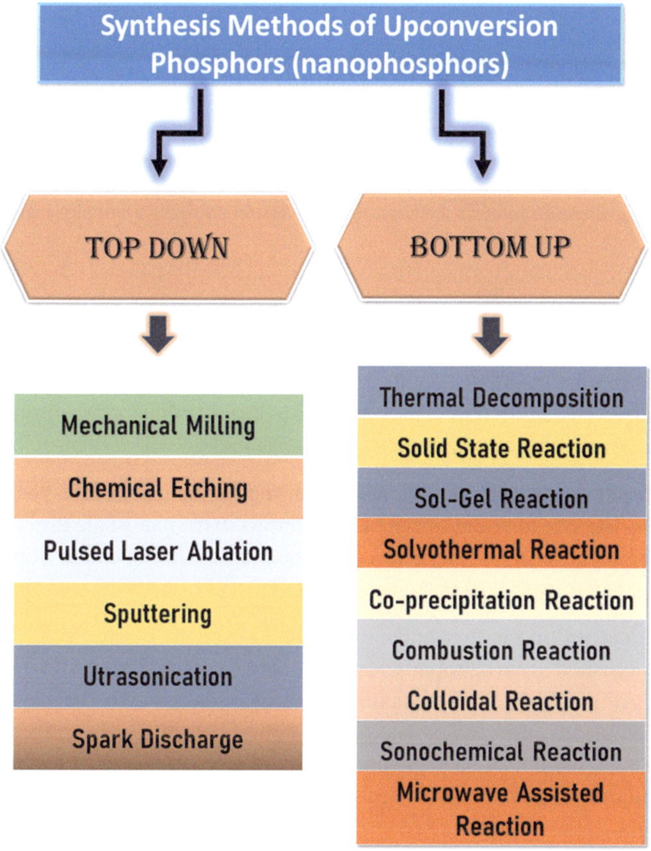

Fig. 1 Classification of synthesis methods of upconversion phosphors into top-down and bottom-up approaches

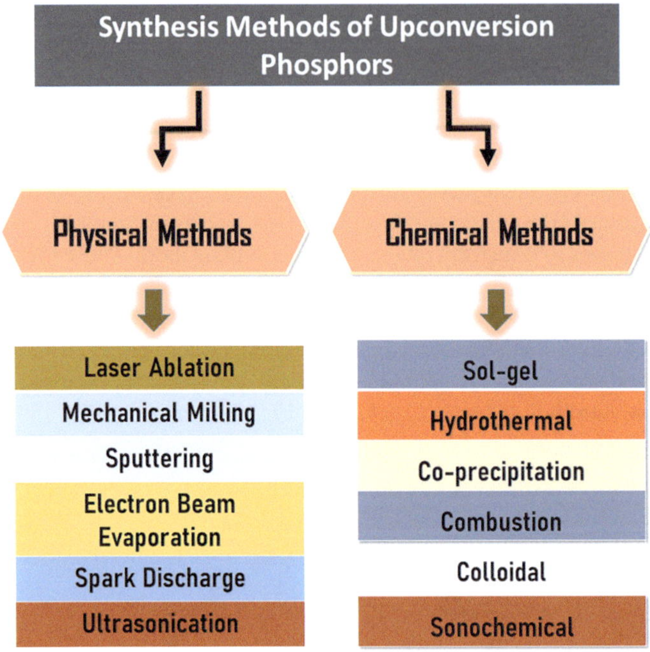

Fig. 2 Classification of synthesis methods of upconversion phosphors into physical and chemical methods

In this chapter, the synthesis methods of upconversion nanoparticles will be discussed following the two major approaches: top-down and bottom-up. Each method has unique advantages and challenges, and selecting a method between the two depends on several parameters such as the desired nanoparticle properties, size, shape, biocompatibility, and application requirements.

1. **Top-Down Approach**: In this approach, the nanoparticles are synthesized by reducing the size of bigger-sized bulk materials to the nanoscale through mechanical, chemical, or physical means. This method is similar to carving or shaping a bigger size material to make smaller particles.

Mechanical milling is a popular top-down approach, where bulk materials are ground or crushed into nanoparticles using high-energy ball milling or attrition milling. During milling, the mechanical energy breaks down the larger particles into smaller pieces, eventually yielding nanoparticles with desired size distributions. However, there may be some probability of contamination of nanoparticles by milling media, and careful optimization is required to achieve the desired nanoparticle properties.

Chemical etching is another top-down approach of synthesizing nanoparticles where bulk materials are selectively removed using chemical reactions. For example, in electrochemical etching, an electric field is applied to dissolve material from the surface, leading to the formation of nanoparticles. This method provides precise control over nanoparticle size and shape but requires complex equipment and careful control of reaction conditions.

Physical methods such as laser ablation and sputtering also fall under the top-down category, where nanoparticles are produced by ablating or sputtering material from a bulk target. The ablated material then condenses into nanoparticles through nucleation and growth processes. The laser ablation and sputtering methods yield high-purity nanoparticles with controlled composition, but they have low output capacity and cannot be used for large-scale production [19].

2. **Bottom-Up Approach**: In this approach, the nanoparticles are synthesized from smaller building blocks, such as atoms, molecules, or ions, through chemical reactions or self-assembly processes. This method is similar to constructing a structure from individual bricks or molecules [20].

Wet chemical synthesis is the most common bottom-up method, where nanoparticles are synthesized by the reduction or nucleation of precursor molecules in a solution. For example, the sol–gel method involves hydrolysis and condensation reactions of metal alkoxides to form nanoparticles in a colloidal suspension. This method offers precise control over nanoparticle size, shape, and composition and can be easily scaled up for large-scale production.

Another bottom-up technique is template-directed synthesis, where nanoparticles are synthesized within a template structure, such as micelles or mesoporous materials. The template provides a confined environment for nanoparticle formation, leading to the controlled growth of nanoparticles with tailored properties. However, template removal and synthesis optimization may pose challenges in this method.

Thus, both top-down and bottom-up approaches have their own benefits and limitations. The top-down methods are well-suited for producing nanoparticles from bulk materials with controlled size and shape, while bottom-up methods provide the precise control of nanoparticle composition at the molecular level. The choice of synthesis approach depends on factors such as the desired nanoparticle characteristics, scalability, and application requirements.

Furthermore, the upconversion phosphors can also be prepared in the form of thin films for various applications and the preparation methods of thin films can be categorized into two well known categories called physical vapor depositions and chemical vapor depositions. Both the methods are popularly used depending on the circumstances and requirements.

2.1 Top-Down Approach

2.1.1 Mechanical Milling

Mechanical milling, also known as mechanical alloying or mechanical attrition, is a solid-state powder processing technique used for synthesizing nanoparticles through repeated mechanical deformation, fracturing, and cold welding of powder particles. This method enables the production of nanoparticles with controlled size, shape, and composition by subjecting precursor materials to high-energy mechanical forces in a milling device such as a ball mill, attritor mill, or planetary mill.

The process of synthesizing nanoparticles using mechanical milling typically involves the following steps:

1. **Selection of Precursor Materials:** The precursor materials, which can be in the form of elemental powders, alloy powders, or composite powders, are selected based on the desired composition of the nanoparticles to be synthesized. The precursor powders are often finely divided and chemically pure to ensure homogeneity and reactivity during milling.
2. **Loading of Precursor Powders:** The precursor powders are loaded into a milling container, typically made of hardened steel or ceramic, along with milling balls or beads. The ratio of powder to milling balls, as well as the size and material of the milling balls, influence the milling efficiency and nanoparticle properties.
3. **Mechanical Milling:** The milling container is sealed and mounted on a milling machine, where it undergoes repeated cycles of mechanical deformation and impact caused by the rotation and movement of the milling balls. The kinetic energy transferred from the milling balls to the precursor powders leads to particle fracturing, cold welding, and alloying, promoting the formation of nanoparticles.
4. **Controlled Milling Parameters:** The milling process parameters, including milling time, speed, ball-to-powder ratio, and milling atmosphere (e.g., inert gas or reactive gas), are carefully controlled to optimize nanoparticle synthesis. Longer milling times and higher milling speeds typically result in smaller nanoparticles but may also lead to increased contamination and grain growth.

Mechanical milling offers several advantages for nanoparticle synthesis, including scalability, versatility, and the ability to produce nanoparticles from a wide range of materials, including metals, ceramics, and polymers. It is particularly suitable for synthesizing upconversion nanoparticles with high melting points, immiscible phases, or metastable phases that are difficult to obtain by conventional chemical synthesis methods. Additionally, mechanical milling is a cost-effective and environmentally friendly approach, as it typically does not require harsh reaction conditions or toxic chemicals. However, challenges such as contamination, agglomeration, and control over nanoparticle size distribution must be carefully addressed during milling to achieve the desired nanoparticle properties [21].

2.1.2 Chemical Etching

The chemical etching method is a widely used technique for synthesizing upconversion nanoparticles with precise control over size, shape, and composition. This method involves the selective dissolution of material from the surface of a bulk substrate through chemical reactions, leading to the formation of nanoparticles. The method starts with the selection of a suitable substrate material, typically a metal or semiconductor, which serves as the precursor for nanoparticle synthesis. The substrate is immersed in a chemical etchant solution that contains reactive species capable of dissolving the substrate material. The choice of etchant solution and reaction conditions depends on the desired nanoparticle characteristics and the chemical properties of the substrate material.

As the substrate is exposed to the etchant solution, chemical reactions occur at the surface, resulting in the preferential removal of material from certain crystallographic planes or defects. This selective dissolution process leads to the formation of nanoscale pits, pores, or features on the substrate surface, which serve as nucleation sites for nanoparticle growth. Nanoparticles are formed through the redeposition of dissolved material onto the substrate surface or as detached particles dispersed in the etchant solution. The size, shape, and morphology of the nanoparticles can be controlled by adjusting parameters such as the composition of the etchant solution, temperature, reaction time, and agitation.

The chemical etching method offers several advantages for nanoparticle synthesis, including high scalability, uniformity, and the ability to produce nanoparticles with tailored properties. By controlling the etching conditions, it is possible to achieve precise control over nanoparticle size, shape, and surface chemistry. Additionally, the chemical etching method can be easily integrated with lithographic techniques to pattern substrates and create ordered arrays of nanoparticles.

Various modifications of the chemical etching method have been developed to enhance its versatility and applicability. For example, template-assisted etching involves using a sacrificial template to create nanoporous structures, which can then be filled with a different material to form nanoparticles. Additionally, photochemical etching utilizes light to induce chemical reactions at the substrate surface, enabling spatially controlled nanoparticle synthesis.

The chemical etching method finds numerous applications in areas such as catalysis, sensing, electronics, and biomedical engineering. Metal nanoparticles synthesized via chemical etching exhibit unique catalytic properties and are used as catalysts in chemical reactions. Semiconductor nanoparticles produced by this method are employed in optoelectronic devices, solar cells, and sensors [22].

2.1.3 Pulsed Laser Ablation

Laser ablation is a versatile technique for synthesizing nanoparticles with precise control over size, shape, composition, and crystallinity. In laser ablation, a high-energy laser pulse is focused onto a target material immersed in a liquid or gas environment, causing rapid vaporization and the formation of a plasma plume. The plasma plume consists of atoms, ions, clusters, and nanoparticles ejected from the target surface due to the extreme heating and vaporization caused by the laser pulse. One of the key advantages of laser ablation is its ability to generate nanoparticles directly from solid bulk materials without the need for additional chemical precursors or reaction steps. This makes it a clean and efficient method for nanoparticle synthesis, particularly for materials that are difficult to synthesize using conventional chemical methods.

The properties of the nanoparticles produced by laser ablation can be controlled by adjusting various parameters, including laser fluence, pulse duration, wavelength, repetition rate, target material composition, and ambient environment. By tuning these parameters, researchers can tailor the size, morphology, crystallinity, and chemical composition of the nanoparticles to meet specific application requirements.

The laser ablation process can be performed in various environments, including liquids (such as water, organic solvents, or surfactant solutions) and gases (such as inert gases or reactive atmospheres). The choice of environment influences the formation and properties of the nanoparticles, as well as the stability and dispersibility of the resulting nanoparticle colloids or suspensions.

In liquid-phase laser ablation, the laser pulse generates a cavitation bubble near the target surface, which rapidly collapses and generates shock waves that contribute to the fragmentation and dispersion of the target material into nanoparticles. The liquid environment also helps to quench and stabilize the nanoparticles, preventing agglomeration and promoting uniform particle size distribution.

In gas-phase laser ablation, the laser pulse directly vaporizes the target material, producing a plasma plume that expands rapidly into the surrounding gas. The nanoparticles formed in the plasma plume can be collected on a substrate or quenched in a cold trap to prevent further growth or agglomeration. Gas-phase laser ablation is particularly suitable for synthesizing nanoparticles in high-purity environments and for studying the fundamental mechanisms of nanoparticle formation and growth.

Laser ablation has been successfully used to synthesize upconversion nanoparticles of various materials, including metal oxides, semiconductors, ceramics, and carbon-based materials. These nanoparticles exhibit unique properties and find applications in diverse fields such as catalysis, sensing, electronics, photonics, biomedicine, and environmental remediation [23].

2.1.4 Sputtering

Sputtering is a versatile and widely used technique for depositing thin films of material onto substrates. Still, it can also be employed to synthesize nanoparticles through a process called sputter deposition or sputter erosion. In sputtering, energetic ions bombard a target material, causing atoms or molecules to be ejected from the target surface. These ejected particles then deposit onto a substrate, forming a thin film or, in the case of nanoparticle synthesis, nanoparticles.

The process of sputter deposition for nanoparticle synthesis typically involves the following steps:

1. **Target Preparation:** A solid target material, often a metal or compound, is prepared and mounted in a vacuum chamber. The choice of target material determines the composition of the nanoparticles to be synthesized.

2. **Substrate Preparation:** The substrate, which can be made of various materials such as silicon, glass, or polymers, is cleaned and placed in the vacuum chamber opposite the target.
3. **Vacuum Chamber Evacuation:** The vacuum chamber is evacuated to create a low-pressure environment, typically in the range of 10^{-6} to 10^{-3} m bar, to minimize gas collisions and ensure efficient ion transport.
4. **Argon Ion Bombardment:** An inert gas, such as argon, is introduced into the vacuum chamber to serve as the sputtering medium. High-energy ions, typically generated by a radio frequency (RF) or direct current (DC) power source, are directed toward the target material.
5. **Sputtering Process:** When the energetic ions collide with the target surface, they transfer momentum to the target atoms, causing them to be ejected or sputtered from the target surface. These sputtered atoms then travel through the vacuum chamber and deposit onto the substrate surface.
6. **Nanoparticle Formation:** Upon deposition onto the substrate, the sputtered atoms undergo nucleation and aggregation processes, eventually forming nanoparticles. The size, shape, and composition of the nanoparticles can be controlled by adjusting parameters such as sputtering power, target material, substrate temperature, and deposition time.

Sputtering offers several advantages for nanoparticle synthesis, including excellent control over nanoparticle size, composition, and uniformity, as well as compatibility with a wide range of materials. Additionally, the process can be easily scaled up for large-scale production and is suitable for the synthesis of nanoparticles with diverse properties for applications in electronics, catalysis, sensing, and biomedical fields [24].

2.1.5 Ultrasound (Ultrasonication)

The ultrasound method, also known as ultrasonic irradiation or sonication, is a powerful and versatile technique for synthesizing nanoparticles through the application of high-frequency acoustic waves (ultrasound) to a reaction mixture. This method facilitates the formation of upconversion nanoparticles by inducing cavitation, acoustic streaming, and micro-turbulence within the reaction medium, leading to enhanced mass transport, mixing, and chemical reactions.

The process of synthesizing nanoparticles using ultrasound typically involves the following steps:

1. **Preparation of Reaction Mixture:** The precursor materials, which can be in the form of metal salts, metal complexes, or organic compounds, are dissolved or dispersed in a suitable solvent or solvent mixture. Additional reagents, stabilizers, or surfactants may be added to control nanoparticle size, shape, and stability.
2. **Introduction of Ultrasound:** The reaction mixture is placed in a sonication chamber or vessel equipped with an ultrasonic transducer. High-frequency

acoustic waves, typically in the range of 20 kHz to several MHz, are applied to the reaction mixture using the ultrasonic transducer.

3. **Cavitation and Micro-turbulence:** The acoustic waves induce cavitation, which is the formation, growth, and collapse of microbubbles within the reaction medium. During bubble collapse, extreme temperatures and pressures are generated locally, leading to the formation of highly reactive species such as free radicals, ions, and hot spots. These reactive species promote chemical reactions and nucleation events, facilitating the formation of nanoparticles.
4. **Nucleation and Growth:** The precursor species in the reaction mixture undergo nucleation and growth processes induced by the reactive species generated during cavitation. Nanoparticles form through the aggregation of atoms, ions, or molecules, followed by further growth and stabilization through surface interactions.
5. **Controlled Synthesis:** The size, morphology, and composition of the nanoparticles can be controlled by adjusting parameters such as ultrasound intensity, frequency, duration, precursor concentration, and reaction temperature. By optimizing these parameters, it is possible to achieve nanoparticles with tailored properties for specific applications.

The ultrasound method offers several advantages for upconversion nanoparticle synthesis, including rapid reaction kinetics, high efficiency, scalability, and versatility. It is particularly suitable for the synthesis of a wide range of nanoparticles, including metal nanoparticles, metal oxide nanoparticles, semiconductor nanoparticles, and organic nanoparticles. Additionally, the ultrasound method is environmentally friendly, as it typically does not require harsh reaction conditions or toxic chemicals, making it an attractive option for green synthesis approaches in nanotechnology [25].

2.1.6 Spark Discharge

The spark discharge method is a unique and versatile technique for synthesizing upconversion nanoparticles through the application of high-voltage electrical discharges in a gas or liquid medium. This method is particularly effective for producing nanoparticles of metals, metal oxides, and carbon-based materials, offering advantages such as high purity, scalability, and control over particle size and morphology.

The process initiates with the introduction of precursor materials, typically in the form of metal electrodes or carbonaceous materials, into a reaction chamber filled with an inert gas or liquid medium. The choice of precursor material depends on the desired composition of the nanoparticles to be synthesized. For example, metal electrodes such as platinum, gold, or silver are commonly used for metal nanoparticle synthesis, while graphite electrodes are employed for carbon-based nanoparticle production.

Once the reaction chamber is prepared, a high-voltage electrical discharge is applied across the electrodes, initiating a series of complex physical and chemical processes. The electrical discharge generates a plasma, characterized by high temperatures, intense electromagnetic radiation, and reactive species such as ions, electrons, and free radicals.

During the discharge process, the precursor materials undergo vaporization, dissociation, and nucleation, leading to the formation of nanoparticles within the plasma. The rapid cooling of the nanoparticles upon exiting the plasma environment helps preserve their size and morphology. Additionally, the presence of the inert gas or liquid medium serves to control the growth and agglomeration of nanoparticles, ensuring uniform dispersion and preventing particle aggregation.

The spark discharge method offers several advantages for nanoparticle synthesis, including high production rates, narrow size distributions, and the ability to produce nanoparticles with controlled crystal structure and surface properties. Furthermore, this method enables the synthesis of nanoparticles in a continuous and scalable manner, making it suitable for industrial-scale production.

One of the key advantages of the spark discharge method is its versatility, allowing for the synthesis of nanoparticles with diverse compositions and functionalities. For example, metal nanoparticles synthesized via this method exhibit catalytic, magnetic, and optical properties, making them useful for various applications such as catalysis, sensing, and biomedical imaging. Similarly, carbon-based nanoparticles produced by spark discharge exhibit unique electronic, mechanical, and thermal properties, enabling their use in energy storage, nanoelectronics, and composite materials [26].

2.2 *Bottom-Up Approach*

2.2.1 Thermal Decomposition Method

The thermal decomposition method is a widely used technique for synthesizing upconversion nanophosphors with precise control over size, shape, composition, and crystallinity. This method comprises the decomposition of precursor compounds at elevated temperatures to form nanoparticles in the gas or solution phase. The process typically begins with the dissolution of the precursor materials in a high-boiling solvent, such as organic solvents or surfactants, to form a clear solution. The choice of precursor compound depends on the desired composition of the nanoparticles, with metal alkoxides, metal carbonyls, and metal complexes being commonly used precursors.

Once the precursor solution is prepared, it is heated to temperatures above the decomposition temperature of the precursor compound, typically ranging from 100 to 400 °C. The thermal decomposition of the precursor compound results in the formation of volatile decomposition products, such as gases or small molecules, and the nucleation of nanoparticles within the solution. The nucleation and growth of nanoparticles occur through a series of complex chemical reactions, including nucleation, growth, coalescence, and surface modification. The kinetics of these processes are influenced by factors such as temperature, precursor concentration, reaction time, and the presence of capping agents or surfactants.

The choice of solvent and reaction conditions plays a crucial role in determining the size, morphology, and crystallinity of the resulting nanoparticles. Organic solvents with high boiling points are often used to facilitate the dissolution of the precursor compound and control the reaction kinetics. Additionally, the presence of surfactants or capping agents can help stabilize the nanoparticles and prevent agglomeration by forming a protective layer around them.

The thermal decomposition method offers several advantages for nanoparticle synthesis, including high purity, narrow size distribution, and the ability to produce nanoparticles with well-defined shapes and crystalline structures. Furthermore, this method can be easily scaled up for large-scale production of nanoparticles with reproducible properties.

The thermal decomposition method has been widely applied for the synthesis of various types of nanoparticles, including metal nanoparticles, metal oxide nanoparticles, semiconductor nanoparticles, upconversion nanoparticles, and magnetic nanoparticles. These nanoparticles find applications in diverse fields, such as catalysis, sensing, biomedical imaging, drug delivery, and electronic devices [17].

2.2.2 Solid-State Reaction Method

The solid-state reaction is the conventional and popular method of synthesis of phosphor materials. The method is suitable for the large-scale production of different phosphor materials such as oxide, tungstate, phosphate, silicate, and borate due to its simple and straightforward methodology. Reactants in the form of oxides and carbonates are taken in the stoichiometric ratio, ground and mixed homogeneously and heated at high temperatures to get the desired product. Rigorous grinding is required to mix all the reagents homogeneously and bring them in close proximity with each other. The volatile organic mediums such as acetone or ethanol are used for grinding. During the heat treatment, precursors break in the form of ions and ionic inter-diffusion takes place to complete the reaction. The rate of reaction depends on the temperature of reaction, grain boundary contacts, surface area and reactivity of reactants. The initial reaction rate is fast because of short diffusion path of reacting ions and the further reaction proceeds slower. To enhance the reaction rate, the diffusion path must be reduced by crushing the mixture again and again throughout the heat treatment. To decrease the reaction temperature and increase the reaction rate, fluxes such as NH_4Cl and H_3BO_3 are used. Fluxes enhance the crystallinity, morphology, and luminescence of the material. The solid-state reaction is usually carried out at high temperatures (800–1600 °C). Sometimes, impurity phases like unreacted intermediates appear, which can be removed by repeated grinding and heating. This method is environment-friendly, simple, relatively inexpensive, and non-toxic. However, the completion time of the method may be relatively long, and a high temperature is required. The obtained powder has a non-uniform and relatively larger particle size. All the oxide-based upconversion phosphors, such as tungstate, phosphate, molybdate, and phosphate, can be easily synthesized by solid-state reaction method [27].

2.2.3 Sol–Gel Reaction Method

The sol–gel method is suitable for the synthesis of nano-size phosphors with uniform size and controlled morphology. In this method, metal nitrates are used as precursors; citric acid serves as a chelating agent and polyethylene glycol is used as a surfactant. All the precursors taken in the appropriate ratio are mixed in deionized water along with citric acid and polyethylene glycol. The resulting solution is heated at 80 to 100 °C with continuous stirring to form a solution with colloidal particles known as sol and later on condensation takes place to form an elastic solid (with three-dimensional cross-linked network) within the liquid known as gel. The obtained gel is dried under different conditions like superficial drying, thermal drying, and freeze-drying, and finally, it is heated at a suitable temperature to get the final product. Factors that affect the nature, shape, and size of the synthesized particles include the nature of the solvent used, the pH of the solution, the nature of additives added, and their amount and temperature of heat treatment of the material. Different forms of phosphors such as thin film, nanomaterial, glass and ceramic, can by synthesized using this method at a low cost [28].

2.2.4 Solvothermal Reaction Method

The solvothermal synthesis process operates within a sealed system, necessitating elevated temperatures and pressures to initiate a chemical reaction or break down precursor materials, directly forming a desired compound from a solution. This method is applicable to both aqueous and non-aqueous solvents and commonly employs solvents such as water, ethanol, methanol, ammonia, carbon dioxide, hydrochloric acid, and hydrofluoric acid. When water is the solvent used, the term shifts to "hydrothermal."

The procedure involves blending the precursors within the solvent and then sealing them in an autoclave. Subsequently, the autoclave is heated to temperatures surpassing the solvent's boiling point. The autoclave functions as a closed system, elevating the solution's temperature and pressure, thereby causing the dissolved material to crystallize. By carefully selecting precursor composition and adjusting reaction conditions, it is feasible to attain nanoparticles with high purity and very narrow size distribution that are uniformly dispersed. Unlike high-temperature methods, this technique does not necessitate high-temperature treatments. Despite being a low-temperature synthesis, it results in phosphors with high crystallinity. The size, shape, and phase of the phosphors are greatly influenced by factors such as reaction temperature, pH, solvent type, and precursor composition. One notable drawback of this approach is its incapacity to generate bulk quantities of phosphors in a single operation, primarily because the size of the autoclaves used limits mass production [23].

2.2.5 Co-Precipitation Method

Co-precipitation methods are most appropriate for the synthesis of silicate and fluoride-based phosphors. Precursors in the form of nitrates, chlorides, and acetates are used in this method. Precursors taken in appropriate ratios are dissolved in a solvent such as deionized water or ethanol using continuous stirring. A precipitating agent such as ammonium hydroxide is added dropwise in the solution to make it transparent and result in the formation of the precipitate. The derived precipitates are filtered and washed several times with ethanol and deionized water to remove the unwanted impurities. The washed precipitate is further dried at about 60–80 °C to decompose the hydroxides and carbonates and then annealed at high temperature for better crystallization of the material. This method is suitable for synthesizing uniform-sized nanoparticles with the appropriate use of precursors, precipitating agents and solvents and by regulating the pH of the solution [29].

2.2.6 Combustion Method

The combustion method is a solution method that does not require high-temperature calcination. It is a simple, low cost, and energy-efficient method. The obtained final product through this method is highly crystalline, pure, and uniform. It is an exothermic reaction that occurs with the use of different oxidizing and reducing agents. Metal nitrates serve as an oxidizing agent and urea is a reducing agent. The reducing agents serve as fuels and provide the necessary conditions for an exothermic reaction to take place. The oxidizer and fuel must be taken in an appropriate ratio to obtain maximum heat during the reaction. All the precursors (metal nitrates and fuel) taken in appropriate proportion are added in an aqueous medium and heated at 80 to 100 °C with continuous stirring to form a homogeneous solution. The obtained solution is further ignited with the help of a flame. The flame temperature is so high, around 1200 °C, that the desired product is formed in very little time, i.e., < 10 min. In some cases, organic impurities and mixed phases are observed, which can be eliminated by further calcination at higher temperatures [30].

2.2.7 Colloidal Method

The colloidal method of synthesizing nanoparticles is a versatile and widely utilized approach that enables the production of nanoparticles with precise control over size, shape, composition, and surface properties. This method involves the formation of nanoparticles through the reduction or precipitation of precursor species in a liquid medium, typically called the colloidal solution or dispersion.

The process begins with the dispersion of the precursor materials, such as metal salts or metal–organic compounds, in a solvent or solvent mixture. These precursor species may be in the form of ions, molecules, or clusters, depending on the specific synthesis route and desired nanoparticle characteristics. The choice of solvent and reaction conditions plays a crucial role in determining the nucleation and growth kinetics of the nanoparticles.

Once the precursor species are dispersed in the solvent, the reduction or nucleation step initiates the formation of primary nanoparticles. This step involves the addition of a reducing agent or a nucleating agent, which triggers the conversion of precursor species into nanoscale particles. The selection of the reducing agent and reaction conditions can influence the size, morphology, and crystallinity of the resulting nanoparticles.

As the reaction progresses, the primary nanoparticles undergo further growth through the addition of precursor species onto their surfaces. The growth process can be controlled by adjusting parameters such as temperature, reaction time, precursor concentration, and the presence of stabilizing agents or surfactants. These additives help prevent agglomeration and stabilize the nanoparticles by forming a protective layer around them, thereby preventing undesirable interactions and promoting uniform growth.

The colloidal method offers several advantages for nanoparticle synthesis, including scalability, reproducibility, and the ability to produce nanoparticles with well-defined properties. Additionally, the colloidal approach allows for the synthesis of nanoparticles with complex architectures, such as core–shell structures, Janus particles, and hierarchical assemblies, by introducing additional synthetic steps or modifying reaction conditions.

Furthermore, the colloidal method can be tailored to produce nanoparticles with specific functionalities for various applications, including catalysis, sensing, drug delivery, and nanoelectronics. Surface modification techniques, such as ligand exchange, surface functionalization, and post-synthesis treatments, can be employed to tailor the surface chemistry of nanoparticles and impart desired properties such as stability, biocompatibility, and reactivity [23].

2.2.8 Sonochemical Method

The sonochemical method is a novel method of synthesis that uses high-intensity ultrasound to develop nano to bulk-sized phosphors. The prime advantage of the method is that it does not require high temperature and high pressure. The sonochemical method is also termed the co-precipitation method, which uses ultrasound waves. The solution of precursors (metal nitrates) is prepared at a particular pH value and then subjected to high-intensity ultrasonic radiation. At the same time, a precipitating agent is added slowly to the solution. The physical and chemical effects of ultrasonic radiation reduce the time of reaction and facilitate the development of nanophosphors. The properties of the final product depend on the frequency and duration of the ultrasound used. Phosphors with different morphologies and dimensions can be developed by the sonochemical method [31].

2.2.9 Microwave-Assisted Method

The method is not as economical as it uses delicate microwave reactors. However, microwave-assisted synthesis techniques have emerged as a valuable alternative to many conventional methods used for creating inorganic phosphors, organic phosphors, or polymers. This method proves highly efficient in crafting precise nanomaterials. The principle involves using microwaves to heat the precursors, prompting a chemical reaction that leads to the formation of the final product. What sets this approach apart from traditional heating methods is its ability to attain very high temperatures under normal conditions. Microwaves specifically target molecules with dipolar characteristics, primarily water molecules, to generate heat. Dipolar molecules possess the ability to efficiently absorb microwaves and convert them into thermal energy. When these dipoles become excited by microwaves, they align themselves with the external field, resulting in vigorous molecular movement and, consequently, substantial internal heating. Therefore, it is crucial to dissolve the precursors in a polar solvent to effectively produce phosphors using this technique. It is worth noting that different solvents will absorb microwave radiation to varying degrees due to variations in their molecular polarity. The greater the polarity of the solvent, the more heat it generates [32].

3 Thin Film Deposition Techniques for Upconversion Phosphors

Thin film deposition techniques are methods adopted to apply thin layers of material onto a substrate. These techniques are extensively used in industries and research fields such as semiconductors, optics, electronics, and materials science. Thin films can range from a few nanometers to several micrometers in thickness, and the deposition method often determines the film's properties, such as thickness, uniformity, composition, and crystallinity. Thin film deposition techniques can be classified into physical and chemical vapor deposition techniques on the basis of the involved mechanism. The classification is shown (Fig. 3) [33]. A detailed discussion of thin film deposition techniques is presented below:

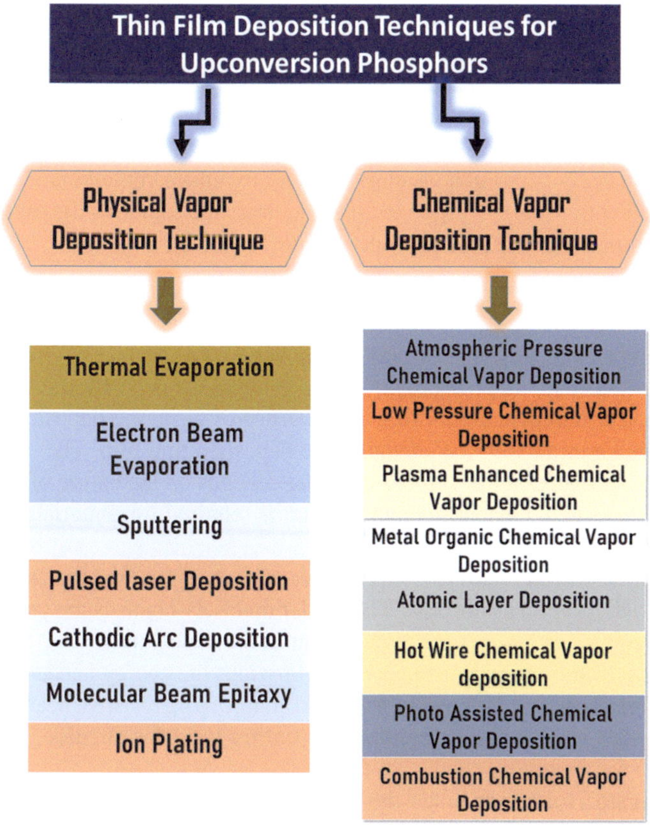

Fig. 3 Classification of thin film deposition techniques

3.1 Physical Vapor Deposition Technique (PVD)

Physical vapor deposition (PVD) is a versatile and widely used method for the synthesis of thin films. It involves the physical transfer of material from a solid or liquid source to a substrate in a vacuum environment. PVD encompasses several techniques, including evaporation, sputtering, and molecular beam epitaxy. Here, we provide an overview of the PVD process, its working principle, instrumentation, and applications. The basic principle of PVD involves the following steps: The material to be deposited (target) is prepared in a solid or liquid state with the required properties. Then, the target material is vaporized using thermal energy (evaporation) or by bombarding it with high-energy ions (sputtering) or photons (pulsed laser deposition). The vaporized atoms or molecules travel through a vacuum or low-pressure gas environment in a vacuum chamber to reach the substrate surface. The vaporized material finally condenses on the substrate, forming a thin film [34].

In this process, the material to be deposited is physically removed from a source (usually a solid or liquid) and deposited onto a substrate. The material is vaporized by physical means such as heat (thermal evaporation), light (pulsed laser deposition), or ion bombardment (sputtering) in a vacuum environment. The vaporized atoms or molecules then travel to the substrate, where they condense to form a thin film. These processes normally occur in a vacuum chamber to reduce contamination and allow for the efficient transport of vaporized material from the source to the substrate [14]. The key components of a PVD system include:

I. **Vacuum Deposition Chamber**: A sealed environment where the deposition process occurs, typically maintained at a high vacuum to minimize contamination and ensure a mean-free path for the vaporized species. The vacuum in the deposition chamber provides better control over the deposition rate, thickness, and uniformity of the thin film. The absence of air in the deposition chamber will also minimize the oxidation of the deposited thin film. The vacuum level to be maintained in the vacuum chamber depends on the nature of the material to be deposited. Generally, high vacuum is required for materials sensitive to oxygen.
II. **Source Material Holder**: Holds the target (source) material that will be vaporized to deposit on a substrate. The holder precisely controls the position of the target and rotates as per the requirement of the deposition.
III. **Heating Element or Plasma Source**: Provides the energy needed to vaporize the target material. This can be a thermal source (e.g., resistive heating, electron beam) for evaporation, a high-energy pulsed laser for ablation, or a plasma source for sputtering.
IV. **Substrate Holder**: It holds the substrate layer, on which the thin film will be deposited. The substrate may be heated or cooled to maintain the desired temperature of the substrate. It can also be rotated to achieve uniform film deposition on the substrate.
V. **Gas Inlet System**: Introduces inert or reactive gases into the chamber if needed (commonly in reactive sputtering).
VI. **Control Systems**: Monitors and regulates vacuum levels, power supply, temperature, and gas flow.

Physical vapor deposition (PVD) encompasses a variety of techniques used to synthesize thin films. Each method offers unique advantages and is suited to different materials and applications. Here are the main PVD techniques used for thin film synthesis:

3.1.1 Thermal Evaporation

Thermal evaporation is the simplest and frequently preferred technique for depositing thin films. In this technique, the target material is heated in a high vacuum environment using a resistive heater, electron beam, or other heat sources to evaporate it. The vaporized atoms travel through the vacuum in the deposition chamber and condense on the substrate, forming a thin film. This technique is known for depositing high-purity thin films with minimal contamination. The instrumental setup is cost-effective and suitable for large-scale production. Thermal evaporation is ideal for materials with low to moderate melting points, such as organic compounds and some oxides. They are commonly used in the production of organic electronic devices and optical coatings. R. Balakarthikeyan et al. fabricated SnS thin films using the thermal evaporation technique to analyze the structural and optical properties for photo-sensing applications [35]. NiO thin films were prepared by Davoud Dastan et al. for applications in H_2S gas sensing [36]

3.1.2 Electron Beam Evaporation

Electron beam evaporation is a PVD technique where a beam of electrons is used to vaporize the source material. A focused electron beam is directed at the target material, providing high energy to the target to vaporize it. High-melting-point materials can also be vaporized and deposited on different substrates by this technique. The vaporized material condenses in a high vacuum chamber on the substrate. This method is suitable for refractory metals, ceramics, and complex oxides. Particularly, used in semiconductor manufacturing, optical coatings, and metal films. The conduction and switching performance of electron beam deposited polycrystalline Nb2O5 based nano-ionic memristor were studied by Satyajeet Sahoo for non-volatile memory applications [37].

3.1.3 Sputtering

The basic idea behind sputtering is the ejection of atoms from a source material by the bombardment of energetic ions in a plasma. The ions from the plasma are accelerated toward the target material in a vacuum chamber having an inert gas like Argon. Subsequently, the ions hit the target causing atoms to be ejected (sputtered) from the target. The sputtered atoms then condense on the substrate, forming a thin film. The sputtering can be DC sputtering, RF sputtering, and magnetron sputtering. In DC sputtering, a direct current power source is used to form plasma, and is usually used for thin film preparation of conductive materials. RF sputtering uses radio frequency power to generate the plasma, and is primarily suitable for both conductive and insulating materials. Magnetron sputtering improves the sputtering process using magnetic fields to confine the plasma near the target, increasing deposition rates and efficiency. The sputtering techniques are versatile for developing thin films of a broad range of materials such as metals, alloys, semiconductors, and insulators. The sputtering techniques are capable of depositing thin films over a large area and on substrates of complex geometry. Sputtering techniques are superior to the evaporation technique in terms of material choice, deposition rate, substrate geometry, and scalability. So, this technique is preferred by several researchers for depositing thin films. K. Naveen Kumar et al. analyzed the electrochemical properties of WO_3 thin films by DC magnetron sputtering [38]. A series of Ni thin films was successfully prepared by Thomas R. Koenig et al. by the sputtering method [39].

3.1.4 Pulsed Laser Deposition (PLD)

A high-power pulsed laser usually emitting in UV and blue regions with a pulse of 20–30 ns and pulse energy of 0.5–0.8 J is used. The laser pulse ablates the target material from its surface, creating a plasma plume of ejected atoms/molecules. The plume expands and deposits the material onto the substrate in a vacuum chamber. A major advantage of PLD is that the source material can be a metallic solid, ceramic, pellet, or even liquid. This method is effective for complex oxides, perovskites, and materials that are difficult to evaporate or sputter. PLD is used in the fabrication of high-quality thin films for electronic, magnetic, and optical applications. A drawback of this technique is that the beam area of the laser is quite small, restriction the deposition on a large area [40]. The PLD is frequently used by several researchers for depositing high-quality thin films of different materials. Mariangela Curcio et al. successfully deposited TiO_2 thin films for utilization in Li-ion batteries [41]. Ayman M. Mostafa et al. studied the linear and nonlinear optical properties of Ag/Zn/ZnO nanocomposite thin films deposited by pulsed laser deposition technique [42].

3.1.5 Cathodic Arc Deposition

Cathodic arc deposition is a PVD technique in which an electric arc is used to evaporate material from a cathode and deposit it on a substrate. A high-current, low-voltage arc discharge is created on the surface of a cathode (made of target material), producing a highly ionized vapor of the target material. The ionized material emitted from the target (source) material travels to the substrate, where it condenses to form a thin film. The method is known for producing dense, adherent coatings. The technique is used in hard coatings for tools, wear-resistant coatings, and decorative finishes. The nanocrystalline titanium nitride thin films are deposited by Jayaraman Senthilselvan et al. on fluorine-doped tin oxide-coated glass substrate by cathodic arc deposition to study the structural, mechanical, optical, and electrical properties [43].

3.1.6 Molecular Beam Epitaxy (MBE)

Molecular beam epitaxy is a sophisticated and highly controlled deposition technique used when atomic-level control is required during thin film deposition. The primary idea behind MBE is the slow and precise deposition of the atoms and molecules on a hot substrate to form a crystalline structure layer by layer. High-purity source materials are evaporated or sublimated in ultra-high vacuum to form molecular beams. The molecular beams are directed at the heated substrate, where they condense and form an epitaxial layer with controlled composition and thickness. The technique provides precise control over film thickness and composition due to the slow deposition rate, making it suitable for research and development of advanced semiconductor and quantum devices. Nowadays, the utility of MBE is being explored to deposit 2-dimensional materials. The single-crystal GaAs was deposited by Brelon J. May et al. on NaCl thin films using molecular beam epitaxy [44].

3.1.7 Ion Plating

Ion plating is a PVD technique in which the substrate is bombarded with energetic ions during the deposition process. The target material to be deposited is evaporated using any of the PVD techniques, like evaporation, arc deposition, or sputtering, and allowed to move toward the substrate. The ionized material is also accelerated towards the substrate for its continuous and periodic bombardment with the help of a bias voltage, ensuring good adhesion and uniform coverage. The process produces coatings with excellent adhesion and uniformity. The method is commonly used for decorative coatings, corrosion-resistant coatings, and thin-film optical filters. TiSiN coatings with various silicon concentrations were deposited on 316 L steel by Q. Wan et al. using cathodic arc ion plating [45].

3.1.8 Advantages and Limitations of PVD

The vacuum environment reduces contamination, producing high-purity films. The technique is versatile and suitable for a wide range of materials and substrates. The technique offers controlled deposition, allowing precise control over film thickness and composition. Techniques like ion plating enhance the adhesion of the thin films. PVD processes do not produce hazardous chemical byproducts and are environmentally friendly. In addition to several advantages, the PVD techniques have some limitations and challenges. The PVD instruments are costly, requiring a high initial investment for vacuum equipment and maintenance. The technique is not suitable for materials that are difficult to vaporize or sputter. They cannot be deposited by this technique. Some techniques, like thermal evaporation and other high-energy processes, can induce stress in the films. PVD techniques can be slower compared to some chemical deposition methods. So, large-scale production can be challenging and cost-intensive.

3.2 Chemical Vapor Deposition Technique (CVD)

Chemical vapor deposition (CVD) is a widely used technique for thin film synthesis, involving the chemical reaction of vapor-phase precursors on a substrate to form a solid material. This process occurs at elevated temperatures in a controlled environment, often leading to high-purity and high-quality films. Various CVD techniques have been developed to optimize the deposition process for different materials and applications. The CVD process generally consists of the following steps:

1. **Introduction of Precursors**: Gaseous reactants (precursors) are introduced into the reactor chamber. These precursors can be elemental gases, metal–organic compounds, or other volatile compounds.
2. **Chemical Reactions**: The precursors undergo thermal decomposition, reduction, or other chemical reactions at or near the heated substrate surface, forming a solid film.
3. **Deposition**: The reaction products deposit on the substrate, forming a thin film. By-products of the reaction are typically volatile and are removed from the chamber by the gas flow.

CVD involves a chemical reaction between gaseous precursors in the vicinity of the substrate. The reaction products form a solid material that deposits onto the substrate as a thin film. The process usually takes place at elevated temperatures to facilitate the chemical reactions. It can take place at atmospheric pressure or under low pressure, depending on the specific process. The reaction chamber may contain various gases, which contribute to the chemical reactions that form the film [46]. Here are the main types of CVD techniques:

3.2.1 Atmospheric Pressure Chemical Vapor Deposition (APCVD)

It is a CVD technique in which the deposition process occurs at atmospheric pressure. Precursors are introduced into the reactor where they decompose and react at the substrate surface to form a thin film. Elevated temperatures facilitate the reaction and deposition. This method is used for depositing doped and undoped semiconductor materials such as silicon dioxide (SiO_2) and silicon nitride (Si_3N_4). The method is commonly employed in semiconductor device fabrication and solar cell manufacturing. We demonstrated the large-scale deposition of high-quality graphene using atmospheric pressure chemical vapor deposition on Cu and illustrated how this technique eliminates major difficulties associated with the low-pressure CVD approach while allowing direct extension of this technology to large-scale graphene production [47].

3.2.2 Low-Pressure Chemical Vapor Deposition (LPCVD)

In a low-pressure CVD, the low pressure (0.1 to 10 Torr) is required for depositing thin films. The reduced pressure improves the uniformity and conformality of the film. Here, high temperatures (500–900 °C) are used to break down the precursors to drive the reaction. This technique is ideal for depositing polysilicon, silicon nitride, and silicon dioxide. The technique is also widely used in microelectronics for gate oxides, diffusion barriers, and insulating layers. Yuxuan Zhang and his team illustrated the growth of unintentionally doped semiconductor indium oxide with ultrafast growth rates on c-plane sapphire by low-pressure chemical vapor deposition in his publication [48].

3.2.3 Plasma-Enhanced Chemical Vapor Deposition (PECVD)

PECVD can work at lower temperatures compared to conventional CVD techniques by employing a plasma to activate the chemical reaction. The high-quality depositions can be performed by this technique on temperature-sensitive substrates at relatively lower temperatures. The plasma plays a vital role in enhancing the rate of the chemical reaction by supplying additional energy to the reactants. The gaseous precursors are introduced into the vacuum chamber and a plasma is generated by a DC or RF power source. The electrons and ions inside the plasma dissociate the gaseous precursors into reactive species. These reactive species reach the substrate and form the film. The lower operating temperature of this technique makes it suitable for depositing materials that can degrade at high temperatures, such as polymers. This method is intensely used for depositing silicon-based thin films such as silicon dioxide, silicon nitride, and amorphous silicon. So, this technique is popular in semiconductor device fabrication industries such as solar cells, thin film transistors, and protective coatings. Farnaz Tabarkhoon et al. deposited silicon carbide thin films by plasma-enhanced chemical vapor deposition and tested its quality by FTIR and EDS analysis [49].

3.2.4 Metal–Organic Chemical Vapor Deposition (MOCVD)

The deposition technique utilizes metal–organic compounds as precursors. Precursors decompose and react on the heated substrate to form the desired thin film. Typically occurs at temperatures between 300 and 800 °C. Widely used for the growth of compound semiconductors such as gallium arsenide (GaAs) and indium phosphide (InP). Critical for manufacturing optoelectronic devices, LEDs, and high-frequency transistors, SiN_x passivated MOCVD-grown AlScN/GaN heterostructures with improved structural quality have been developed by Christian Manz et al. [50].

3.2.5 Atomic Layer Deposition (ALD)

ALD is a sophisticated technique used for single or multi-atomic layer depositions. Ultrathin films with high uniformity can be deposited using this technique. Precursors are introduced sequentially in a cycle, with each cycle forming a monolayer of the material. Reactions are self-limiting, ensuring precise control over film thickness and composition. The deposition can be performed at relatively low temperatures (typically 100–300 °C). ALD technique is ideal for depositing ultra-thin films and nanostructures. The method is extensively used for the fabrication of high-k dielectrics, gate oxides, and barrier layers in semiconductor devices. Hyun Gu Kim and his group studied the effects of Al precursors on deposition selectivity of atomic layer deposited Al_2O_3 using ethanethiol inhibitor [51].

3.2.6 Hot-Wire Chemical Vapor Deposition (HWCVD)

In this technique, a hot wire (usually tungsten) is used to decompose the precursor gases. High deposition rates at relatively low substrate temperatures can be achieved by this technique. The hot wire directly decomposes the precursor gases without requiring a plasma. This technique is used for depositing hydrogenated amorphous silicon (a–Si) and microcrystalline silicon (μc–Si). HWCVD is commonly employed in thin-film solar cells and display technology. A soft ionization laser-based technique using 10.5 eV photon energy has been used to probe radical growth precursors in the hot-wire chemical vapor deposition of a-Si:H by H. L. Duan and his team [52].

3.2.7 Photo-Assisted Chemical Vapor Deposition (PACVD)

In this CVD technique, ultraviolet (UV) light is used to activate the chemical reactions. It can facilitate deposition at lower temperatures compared to thermal CVD. The precise control can be achieved in this technique. UV light allows for selective area deposition and enhanced control over film properties. The technique is used for depositing oxides, nitrides, and other compounds on temperature-sensitive substrates. PACVD is frequently used for optical coatings, microelectronics, and surface modification. Feng Teng and his team reported the preparation of black TiO_2 by hydrogen plasma-assisted chemical vapor deposition and its photocatalytic activity in their publication [53].

3.2.8 Combustion Chemical Vapor Deposition (CCVD)

A combustion flame is used to generate the precursor species and drive the chemical reactions in this technique. Rapid deposition of thin films can be performed by this technique. This technique does not require sophisticated vacuum systems. This technique is suitable for large-area coatings and high-throughput production. It is also used in glass coatings, protective films, and catalytic surfaces. A highly sensitive and fast-responding SnO_2 sensor fabricated by combustion chemical vapor deposition by Ying Liu and his team [54].

3.2.9 Advantages and Limitations of CVD Techniques

The CVD techniques can yield high-purity films with negligible contamination. The CVD techniques are capable of producing uniform and conformal coatings over large areas and complex geometries. These techniques are also suitable for different types of materials, such as metals, semiconductors, insulators, and complex compounds. Moreover, the CVD techniques can be used for both small-scale research applications and large-scale industrial production. These methods offer precise control over film thickness, composition, and microstructure. However, CVD techniques have some limitations. Many CVD techniques require high temperatures, which can limit the choice of substrate materials. Some CVD techniques, such as MOCVD and ALD, require complex equipment and precise control over process parameters. The availability and cost of suitable precursors can be a limiting factor for certain materials. The use of hazardous chemicals and gases requires stringent safety measures and environmental controls.

4 Final Remarks

Upconversion phosphors are important materials that can be applied in different technologies. The advancement in synthesis methods and materials engineering has enabled the development of upconversion phosphors with refined properties and enhanced performance. The synthesis of upconversion phosphors is an important aspect in various optoelectronic applications, where various methods produce materials with tailored properties and functionalities. Starting with the top-down approach of synthesizing nanoparticles from bulk materials to the bottom-up approach of nanoscale structures, each synthesis approach offers unique advantages and challenges. The upconversion phosphors prepared in the form of thin films are continuously used in several devices. Physical vapor deposition techniques are used for the fabrication of thin films with precise control over composition and morphology. PVD techniques offer a range of methods to deposit thin films with varying properties and applications. High-quality thin films for several industrial and scientific applications can be prepared by PVD techniques. Each technique has its own strengths and is selected on the basis of the specific requirements of the material and the desired properties of the thin film.

Chemical vapor deposition techniques are primarily used for the controlled growth of upconversion phosphors from gaseous precursors. CVD encompasses a diverse array of techniques for thin film synthesis, each with unique mechanisms and applications. These techniques enable the fabrication of high-quality coatings for a wide range of industries. The continued development and refinement of CVD technologies promise to expand their applications and capabilities, driving innovation in materials science and engineering. As the demand for upconversion phosphors continues to escalate across diverse applications, the synthesis strategies elucidated in this chapter serve as indispensable tools for researchers and engineers striving to unlock the full potential of these luminescent materials. By harnessing the insights gleaned from this chapter, future endeavors in upconversion phosphor synthesis are poised to transcend existing boundaries, ushering in a new era of innovation and advancement in materials science and technology.

References

1. M. Matakgane, T.P. Mokoena, M.R. Mhlongo, Recent trends of oxides heterostructures based upconversion phosphors for improving power efficiencies of solar cells: a review. Inorg. Chem. Commun. **156**, 111202 (2023). https://doi.org/10.1016/j.inoche.2023.111202
2. G. Liu, Advances in the theoretical understanding of photon upconversion in rare-earth activated nanophosphors. Chem. Soc. Rev. **44**, 1635–1652 (2015). https://doi.org/10.1039/C4CS00187G
3. I. Gupta, S. Singh, S. Bhagwan, D. Singh, Rare earth (RE) doped phosphors and their emerging applications: a review. Ceram. Int. **47**, 19282–19303 (2021). https://doi.org/10.1016/j.ceramint.2021.03.308
4. Y. Zhou, Y. Cheng, Q. Huang, J. Xu, H. Lin, Y. Wang, Abnormal thermally enhanced upconversion luminescence of lanthanide-doped phosphors: proposed mechanisms and potential applications. J. Mater. Chem. C. **9**, 2220–2230 (2021). https://doi.org/10.1039/D0TC05759B
5. R. Priya, S. Bhatia, S. Kaur, V. Dubey, O.P. Pandey, Synthesis of rare-earth-activated phosphors, in: *Rare-Earth-Activated Phosphors* (Elsevier, 2022, pp. 109–136). https://doi.org/10.1016/B978-0-323-89856-0.00004-3
6. K. Narita, Methods of phosphor synthesis and related technology, in: *Phosphor Handbook* (CRC Press, 2006). https://doi.org/10.1201/9781420005233.pt3
7. G. Paramasivam, V.V. Palem, T. Sundaram, V. Sundaram, S.C. Kishore, S. Bellucci, Nanomaterials: synthesis and applications in theranostics. Nanomaterials **11**, 3228 (2021). https://doi.org/10.3390/nano11123228
8. V.B. Pawade, N.R. Pawar, S.J. Dhoble, Upconversion in some fluoride crystal system—a review. Infrared Phys. Technol. **123**, 104148 (2022). https://doi.org/10.1016/j.infrared.2022.104148
9. S. Kaur, H. Kaur, A.S. Rao, G.V. Prakash, A review on photoluminescence phosphors for biomedical, temperature sensing, photovoltaic cell, anti-counterfeiting and white LED applications. Phys. B Condens. Matter. **690**, 416224 (2024). https://doi.org/10.1016/j.physb.2024.416224
10. P. Zhang, L. Liang, X. Liu, Lanthanide-doped nanoparticles in photovoltaics—more than just upconversion. J. Mater. Chem. C. **9**, 16110–16131 (2021). https://doi.org/10.1039/D1TC02441H
11. M.V. DaCosta, S. Doughan, Y. Han, U.J. Krull, Lanthanide upconversion nanoparticles and applications in bioassays and bioimaging: a review. Anal. Chim. Acta **832**, 1–33 (2014). https://doi.org/10.1016/j.aca.2014.04.030
12. G. Liang, H. Wang, H. Shi, H. Wang, M. Zhu, A. Jing, J. Li, G. Li, Recent progress in the development of upconversion nanomaterials in bioimaging and disease treatment. J. Nanobiotechnology. **18**, 154 (2020). https://doi.org/10.1186/s12951-020-00713-3
13. A. Ghazy, M. Safdar, M. Lastusaari, H. Savin, M. Karppinen, Advances in upconversion enhanced solar cell performance. Sol. Energy Mater. Sol. Cells **230**, 111234 (2021). https://doi.org/10.1016/j.solmat.2021.111234
14. A. Khare, A critical review on the efficiency improvement of upconversion assisted solar cells. J. Alloys Compd. **821**, 153214 (2020). https://doi.org/10.1016/j.jallcom.2019.153214
15. H. Suo, Q. Zhu, X. Zhang, B. Chen, J. Chen, F. Wang, High-security anti-counterfeiting through upconversion luminescence. Mater. Today Phys. **21**, 100520 (2021). https://doi.org/10.1016/j.mtphys.2021.100520
16. M. Gao, J. Li, D. Xia, L. Jiang, N. Peng, S. Zhao, G. Li, Lanthanides-based security inks with reversible luminescent switching and self-healing properties for advanced anti-counterfeiting. J. Mol. Liq. **350**, 118559 (2022). https://doi.org/10.1016/j.molliq.2022.118559
17. X. Zheng, R.K. Kankala, C.-G. Liu, S.-B. Wang, A.-Z. Chen, Y. Zhang, Lanthanides-doped near-infrared active upconversion nanocrystals: upconversion mechanisms and synthesis. Coord. Chem. Rev. **438**, 213870 (2021). https://doi.org/10.1016/j.ccr.2021.213870
18. H.S. Naher, B.A.H. Al-Turaihi, S.H. Mohammed, S.M. Naser, M.A. Albark, H.A. Madlool, H.A.M. Al-Marzoog, A. Turki Jalil, Upconversion nanoparticles (UCNPs): synthesis methods, imaging and cancer therapy. J. Drug Deliv. Sci. Technol. **80**, 104175 (2023). https://doi.org/10.1016/j.jddst.2023.104175

19. I. Ijaz, E. Gilani, A. Nazir, A. Bukhari, Detail review on chemical, physical and green synthesis, classification, characterizations and applications of nanoparticles. Green Chem. Lett. Rev. **13**, 223–245 (2020). https://doi.org/10.1080/17518253.2020.1802517
20. N. Abid, A.M. Khan, S. Shujait, K. Chaudhary, M. Ikram, M. Imran, J. Haider, M. Khan, Q. Khan, M. Maqbool, Synthesis of nanomaterials using various top-down and bottom-up approaches, influencing factors, advantages, and disadvantages: a review. Adv. Colloid Interface Sci. **300**, 102597 (2022). https://doi.org/10.1016/j.cis.2021.102597
21. Y. Jiao, C. Ling, J. Wang, H. Amanico, J. Saczek, H. Wang, S. Sridhar, B. Bin Xu, S. Wang, D. Wang, Controllable synthesis of upconversion nanophosphors toward scale-up productions. Part. Part. Syst. Charact. **37**(2020). https://doi.org/10.1002/ppsc.202000129
22. A.A. Leonardi, M.J. Lo Faro, A. Irrera, Silicon nanowires synthesis by metal-assisted chemical etching: a review. Nanomaterials **11**, 383 (2021). https://doi.org/10.3390/nano11020383
23. D.N. Karimov, P.A. Demina, A. V. Koshelev, V. V. Rocheva, A. V. Sokovikov, A.N. Generalova, V.P. Zubov, E. V. Khaydukov, M. V. Koval'chuk, V.Y. Panchenko, Upconversion nanoparticles: synthesis, photoluminescence properties, and applications. Nanotechnologies Russ. **15**, 655–678 (2020). https://doi.org/10.1134/S1995078020060117
24. A. Das, K. Bae, W. Park, Enhancement of upconversion luminescence using photonic nanostructures. Nanophotonics. **9**, 1359–1371 (2020). https://doi.org/10.1515/nanoph-2020-0159
25. A.A. Ansari, V.K. Thakur, G. Chen, Functionalized upconversion nanoparticles: new strategy towards FRET-based luminescence bio-sensing. Coord. Chem. Rev. **436**, 213821 (2021). https://doi.org/10.1016/j.ccr.2021.213821
26. G. Galbács, A. Kéri, A. Kohut, M. Veres, Z. Geretovszky, Nanoparticles in analytical laser and plasma spectroscopy—a review of recent developments in methodology and applications. J. Anal. At. Spectrom. **36**, 1826–1872 (2021). https://doi.org/10.1039/D1JA00149C
27. R. Gopal, J. Manam, $SrWO_4$: Er^{3+}; an efficient green phosphor for LED and optical thermometry applications. J. Mater. Sci. Mater. Electron. **33**, 21746–21761 (2022). https://doi.org/10.1007/s10854-022-08964-6
28. L. Trupp, M.C. Marchi, B.C. Barja, Lanthanide–based luminescent hybrid silica materials prepared by sol-gel methodologies: a review. J. Sol-Gel Sci. Technol. **102**, 63–85 (2022). https://doi.org/10.1007/s10971-021-05678-9
29. D. Kang, E. Jeon, S. Kim, J. Lee, Lanthanide-doped upconversion nanomaterials: recent advances and applications. BioChip J. **14**, 124–135 (2020). https://doi.org/10.1007/s13206-020-4111-9
30. S. Mohanty, A.M. Kaczmarek, Unravelling the benefits of transition-metal-co-doping in lanthanide upconversion nanoparticles. Chem. Soc. Rev. **51**, 6893–6908 (2022). https://doi.org/10.1039/D2CS00495J
31. B. Mehrdel, A. Nikbakht, A.A. Aziz, M.S. Jameel, M.A. Dheyab, P.M. Khaniabadi, Upconversion lanthanide nanomaterials: basics introduction, synthesis approaches, mechanism and application in photodetector and photovoltaic devices. Nanotechnology **33**, 082001 (2022). https://doi.org/10.1088/1361-6528/ac37e3
32. J. Miranda de Carvalho, C.C.S. Pedroso, M.S. de N. Saula, M.C.F.C. Felinto, H.F. de Brito, Microwave-assisted preparation of luminescent inorganic materials: a fast route to light conversion and storage phosphors. Molecules **26**, 2882 (2021). https://doi.org/10.3390/molecules26102882
33. M. Hasegawa, A. Ishii, Thin-film formation for promoting the potential of luminescent lanthanide coordination complexes. Coord. Chem. Rev. **421**, 213458 (2020). https://doi.org/10.1016/j.ccr.2020.213458
34. F.T.Z. Toma, M.S. Rahman, K.M.A. Hussain, S. Ahmed, Thin film deposition techniques: a comprehensive review. J. Mod. Nanotechnol. **4**, 6 (2024). https://doi.org/10.53964/jmn.2024006
35. R. Balakarthikeyan, A. Santhanam, A. Khan, A.M. El-Toni, A.A. Ansari, A. Imran, M. Shkir, S. AlFaify, Performance analysis of SnS thin films fabricated using thermal evaporation technique for photodetector applications. Optik (Stuttg). **244**, 167460 (2021). https://doi.org/10.1016/j.ijleo.2021.167460

36. D. Dastan, K. Shan, A. Jafari, T. Marszalek, M.K.A. Mohammed, L. Tao, Z. Shi, Y. Chen, X.-T. Yin, N.D. Alharbi, F. Gity, S. Asgary, M. Hatamvand, L. Ansari, Influence of heat treatment on H_2S gas sensing features of NiO thin films deposited via thermal evaporation technique. Mater. Sci. Semicond. Process. **154**, 107232 (2023). https://doi.org/10.1016/j.mssp.2022.107232
37. S. Sahoo, Conduction and switching behavior of e-beam deposited polycrystalline Nb_2O_5 based nano-ionic memristor for non-volatile memory applications. J. Alloys Compd. **866**, 158394 (2021). https://doi.org/10.1016/j.jallcom.2020.158394
38. K. Naveen Kumar, H. Shaik, J. Gupta, S.A. Sattar, R.I. Jafri, A. Pawar, V. Madhavi, A.R. G V, Sputter deposited tungsten oxide thin films and nanopillars: electrochromic perspective. Mater. Chem. Phys. **278**, 125706 (2022). https://doi.org/10.1016/j.matchemphys.2022.125706
39. T.R. Koenig, Z. Rao, E. Chason, G.J. Tucker, G.B. Thompson, The microstructural and stress evolution in sputter deposited Ni thin films. Surf. Coatings Technol. **412**, 126973 (2021). https://doi.org/10.1016/j.surfcoat.2021.126973
40. S.M. Rossnagel, Thin film deposition with physical vapor deposition and related technologies. J. Vac. Sci. Technol. A: Vacuum, Surfaces, Film. **21**(5), S74–S87 (2003). https://doi.org/10.1116/1.1600450
41. M. Curcio, A. De Bonis, S. Brutti, A. Santagata, R. Teghil, Pulsed laser deposition of thin films of TiO_2 for Li-ion batteries. Appl. Surf. Sci. Adv. **4**, 100090 (2021). https://doi.org/10.1016/j.apsadv.2021.100090
42. A.M. Mostafa, E.A. Mwafy, N.S. Awwad, H.A. Ibrahium, Linear and nonlinear optical studies of Ag/Zn/ZnO nanocomposite thin film prepared by pulsed laser deposition technique. Radiat. Phys. Chem. **179**, 109233 (2021). https://doi.org/10.1016/j.radphyschem.2020.109233
43. J. Senthilselvan, R.R. Pillai, R. Kannan, B. Baskar, K. Valleti, J. Manonmani, Cathodic arc deposition of nanocrystalline titanium nitride thin film on fluorine doped tin oxide coated glass substrate: crystal structural, microstructural, mechanical, optical and electrical properties. Ceram. Int. **49**, 37072–37088 (2023). https://doi.org/10.1016/j.ceramint.2023.07.138
44. B.J. May, J.J. Kim, P. Walker, H.R. Moutinho, W.E. McMahon, A.J. Ptak, D.L. Young, Molecular beam epitaxy of GaAs templates on water soluble NaCl thin films. J. Cryst. Growth **586**, 126617 (2022). https://doi.org/10.1016/j.jcrysgro.2022.126617
45. Q. Wan, Z.Y. Wu, Y. Liu, B. Yang, H.D. Liu, F. Ren, P. Wang, Y.Y. Xiao, J. Zhang, G.D. Zhang, Lead-bismuth eutectic (LBE) corrosion mechanism of nano-amorphous composite TiSiN coatings synthesized by cathodic arc ion plating. Corros. Sci. **183**, 109264 (2021). https://doi.org/10.1016/j.corsci.2021.109264
46. Q. Wang, A.T.S. Wee, Photoluminescence upconversion of 2D materials and applications. J. Phys. Condens. Matter **33**, 223001 (2021). https://doi.org/10.1088/1361-648X/abf37f
47. I. Vlassiouk, P. Fulvio, H. Meyer, N. Lavrik, S. Dai, P. Datskos, S. Smirnov, Large scale atmospheric pressure chemical vapor deposition of graphene. Carbon **54**, 58–67 (2013). https://doi.org/10.1016/j.carbon.2012.11.003
48. Y. Zhang, M.R. Karim, Z. Feng, H. Zhao, Ultrafast growth rate and high mobility In2O3 films grown on c-sapphire via low pressure chemical vapor deposition. J. Appl. Phys. **125** (2019). https://doi.org/10.1063/1.5086454
49. F. Tabarkhoon, M. Bazmi, N.A. Welchert, T.T. Tsotsis, M. Gupta, One-pot fabrication of silicon carbide thin films via plasma-enhanced chemical vapor deposition (PECVD) followed by in Situ Pyrolysis. Ind. Eng. Chem. Res. **64**, 5973–5981 (2025). https://doi.org/10.1021/acs.iecr.4c03786
50. C. Manz, S. Leone, L. Kirste, J. Ligl, K. Frei, T. Fuchs, M. Prescher, P. Waltereit, M.A. Verheijen, A. Graff, M. Simon-Najasek, F. Altmann, M. Fiederle, O. Ambacher, Improved AlScN/GaN heterostructures grown by metal-organic chemical vapor deposition. Semicond. Sci. Technol. **36**, 034003 (2021). https://doi.org/10.1088/1361-6641/abd924
51. H.G. Kim, M. Kim, B. Gu, M.R. Khan, B.G. Ko, S. Yasmeen, C.S. Kim, S.-H. Kwon, J. Kim, J. Kwon, K. Jin, B. Cho, J.-S. Chun, B. Shong, H.-B.-R. Lee, Effects of Al precursors on deposition selectivity of atomic layer deposition of Al_2O_3 using ethanethiol inhibitor. Chem. Mater. **32**, 8921–8929 (2020). https://doi.org/10.1021/acs.chemmater.0c02798

52. H.L. Duan, G.A. Zaharias, S.F. Bent, Identification of growth precursors, in *Hot Wire CVD of Amorphous Silicon Films, MRS Online Proceedings Library*, pp. 664 (2001). https://doi.org/10.1557/PROC-664-A3.1
53. F. Teng, M. Li, C. Gao, G. Zhang, P. Zhang, Y. Wang, L. Chen, E. Xie, Preparation of black TiO_2 by hydrogen plasma assisted chemical vapor deposition and its photocatalytic activity. Appl. Catal. B Environ. **148–149**, 339–343 (2014). https://doi.org/10.1016/j.apcatb.2013.11.015
54. Y. Liu, E. Koep, M. Liu, A highly sensitive and fast-responding SnO_2 sensor fabricated by combustion chemical vapor deposition. Chem. Mater. **17**, 3997–4000 (2005). https://doi.org/10.1021/cm050451o

Spectroscopy Techniques for Upconversion Phosphors

Ovica Manners and Vikas Dubey

Abstract Upconversion phosphors, typically doped with rare earth ions, are capable of converting multiple lower energy photons into a single higher energy photon. Owing to their photostability, sharp emission bands, anti-stokes emission, and long luminescence lifetime makes them promising candidates for applications in bio-imaging, opto-electronics, solid-state lighting, and anti-counterfeiting measures. However, optimizing these materials for such applications requires a detailed understanding of structure–property relationships. This chapter provides a comprehensive review on the spectroscopy techniques aimed at characterising distinct properties of the upconversion phosphors. The discussion is structured around six characterisation goals which includes (a) analysis of electronic structure using PL and UV–Visible, (b) investigation of surface ligands bound to the upconversion phosphors using IR and NMR, (c) examination of vibrational modes using FTIR and Raman, (d) determination of elemental composition using XPS, EDXS, and EELS (e) evaluation of local atomic environment around dopant ions using XAS, and (f) assessment of luminescence lifetime dynamics using TRS. By correlating spectroscopic data with optical properties, this chapter highlights the role of spectroscopy as a vital role in the rational design and performance enhancement of upconversion phosphors.

Keywords Upconversion · Spectroscopy techniques · Lanthanides

1 Introduction

Rare earth-doped phosphors capable of converting near-infrared (NIR) long wavelength radiation into shorter-wavelength visible radiation through upconversion (UC) have emerged as a promising class of luminescent materials known as upconversion (UC) phosphors. These materials are a subject of extensive research due to their exceptional photostability, sufficient anti-stokes emission, precise band emission, and adjustable long-emission lifetime [1]. This upconversion process has a larger

O. Manners (✉) · V. Dubey
Department of Physics, North-Eastern Hill University, Shillong, India
e-mail: ovicamanners@gmail.com

V. Dubey et al. (eds.), *Upconversion Luminescence of Phosphors*, Progress in Optical Science and Photonics 39, https://doi.org/10.1007/978-981-95-1762-6_5

probability of occurrence in lanthanide ions (Ln^{3+}), which possesses rich-ladder like energy levels in their 4f inner-electron shells [2–5]. As a result, UC phosphors have found applications in diverse fields like bio-imaging, solid-state lighting, opto-electronics, anti-counterfeiting measures, etc. [1]. Upconversion primarily occurs through a single or multi-photon absorption processes [2]. In the single phonon process, a single photon is absorbed and re-emitted at a higher energy (Anti-Stokes), whereas multi-photon process involves the sequential absorption of multiple lower energy photons, resulting in visible light emission [2]. The use of NIR light rather than UV light in upconversion minimises background autofluorescence and also a reduced photodamage in biological specimens [6]. The luminescence efficiency of UC phosphors is highly sensitive to dopant ion type and their concentration, and surface modifications, all of which influences non-radiative losses and energy transfer dynamics [7]. Therefore, a comprehensive understanding of their optical behaviour requires the use of advanced characterisation techniques capable of probing structure, composition, and energy transfer at different scales [8].

Spectroscopy provides broad range of scientific tools, essential for the exploration of the functional aspects of UC phosphors. By examining their interaction with electromagnetic radiation, these techniques help in elucidating structural, chemical, and optical properties critical to phosphors [2, 6]. This chapter reviews key spectroscopy techniques used to UC phosphors, focusing on their ability to assess specific parameters such as electronic structure (Photoluminescence (PL), UV–Visible), ligands interaction (Infrared, Nuclear Magnetic Resonance (NMR)), vibrational modes (Raman, FTIR), elemental composition (X-ray photoelectron (XPS), Energy dispersive X-ray (EDXS), Electron energy-loss spectroscopy (EELS)), local structure (X-ray absorption spectroscopy (XAS), and its lifetime dynamics (Time resolved (TRS)). Therefore, by grouping these techniques according to their characterisation goal, we aim to provide a practical and application-focused overview for researchers working on the design and optimisation of UC phosphors.

2 Determination of the Electronic Structure

In upconversion phosphors, luminescence arises from electronic transitions between discrete energy levels of rare earth dopant ions. These electronic transitions are responsible for the anti-Stokes emission characteristics seen in UC phosphors, and are highly sensitive to the dopant ion concentration, local crystal field, and surrounding host lattice structure. The incorporation of rare earth dopant ions such as Er^{3+}, Yb^{3+}, and Tm^{3+} into suitable host lattice exhibits well-defined intra-4f transitions, which are largely shielded by the surrounding environment. However, subtle changes such as ion substitution, surface modification, or lattice strain could influence the splitting of energy levels and transition probabilities, consequently modifying the luminescence output. Hence, precise knowledge of the electronic structure is essential for tailoring phosphor performance in targeted applications. To investigate these properties, two spectroscopy techniques including Photoluminescence (PL) and UV–Visible are commonly employed to probe optical band structures in UC phosphors [9, 10]. The two subsections below detail how these two techniques help elucidate electronic structure of UCNPs.

2.1 *Photoluminescence (PL) Spectroscopy*

Photoluminescence (PL) spectroscopy serves as a sensitive, non-destructive technique which is widely used to characterise nanophosphors by probing their electronic structure, luminescent properties, and energy transfer dynamics [11, 12]. In a typical PL measurement, phosphor samples are excited using a specific wavelength, commonly generated from a laser or a xenon lamp, which is selected to match the absorption band of the dopant ions [13]. The fundamental mechanism of PL is illustrated in Fig. 1, showing luminescence emission after absorbing excitation photons.

In UC phosphors, rare earth dopants incorporated into the host material serves as activators (i.e. luminescent centres), significantly affecting the material's optical properties. For enhancing luminescence efficiency, an additional dopant, referred to as a sensitizer is often added, which facilitates energy transfer in phosphors [13]. For instance, Feng et al. investigated the PL properties and photoluminescence excitation spectra (PLE) of two distinct Erbium centres (Er^{3+} (I) and Er^{3+} (II)) observed in $CaSc_2O_4$:Er^{3+}, Yb^{3+} upconverting phosphors [14]. In this UC system, Er^{3+} functions as the activator, while Yb^{3+} serves as the sensitizer. The PL spectra shown in Fig. 2 display emission peaks in the green (510–580 nm) and red (640–700 nm)

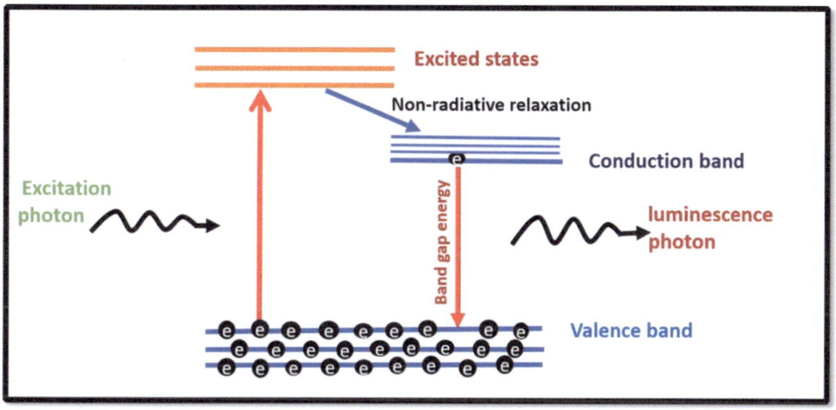

Fig. 1 Photoluminescence mechanism. *Source* Ref Vallikkodi [12] Synthesis, growth, and characterisation of piperazinium P-aminobenzoate and piperazinium P-chlorobenzoate nonlinear optical single crystals. *Alagappa University, Karaikudi, India.* Ph.D. thesis, Alagappa University, 2018

spectral regions, which are associated with the $^2H_{11/2}$, $^4S_{3/2} \rightarrow {}^4I_{15/2}$, and $^4F_{9/2} \rightarrow {}^4I_{15/2}$ transitions from Er^{3+} ions. With increasing Yb^{3+} concentration, it was observed that the intensities of the red emission increase relative to the green emission. This enhancement is attributed to the de-excitation of the $^4S_{3/2} \rightarrow {}^4I_{13/2}$ transition of Er^{3+} by transferring the energy to Yb^{3+}, which is followed by energy transfer from the excited Yb^{3+}, to the same Er^{3+}, in order to promote the excitation from $^4I_{13/2} \rightarrow {}^4F_{9/2}$ transition of Er^{3+}. By systematically varying the dopant concentrations, PL studies reflect how energy transfer between The Yb^{3+} and Er^{3+} ions will affect the emission intensity and spectral distribution. Corresponding PLE spectra is shown in Fig. 3, which determines the preferential excitation wavelength for the phosphor [14]. The spectra exhibit excitation peaks at 487 nm for Er^{3+}(I) and of 488.2 nm for Er^{3+}(II). In another study by Lei et al., $NaCeF_4$: Er/Yb nanophosphor was observed to emit intense near-infrared emission under a 980 nm excitation. This upconversion system with a quantum yield of 32.8% functions as a homogeneous bio probe towards detecting uric acid with a sensitivity range down to 25.6 nM. Furthermore, this bio-probe enables deeper tissue penetration depth detectable up to 10 mm, suitable for in vivo imaging of mouse organs [15]. Therefore, detailed PL analysis allows identification of distinct luminescent centres, and enables to optimise design of phosphors for specific emission wavelengths and enhanced quantum efficiency. Overall, PL spectroscopy is valuable for evaluating dopant interactions, nature of emission centres, efficiency of radiative transitions, and most importantly assists in functional assessment of upconversion nanophosphors beyond structural characterisation.

While PL spectroscopy aids in characterising upconversion phosphors, but certain limitations such as sensitivity to temperature changes can affect measurement accuracy. A small variation in temperature may result in the erroneous results of data [11]. Furthermore, determining quantum efficiency (QE), which is the ratio of the number

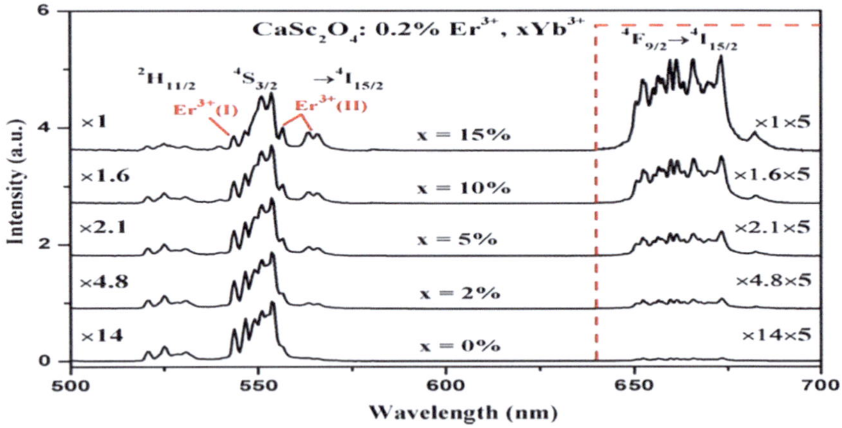

Fig. 2 PL spectra of $CaSc_2O_4$: 0.2% Er^{3+}, xYb^{3+} upon Er^{3+} $^4F_{7/2}$ excitation at 492 nm. Reproduced with permission from ref [14] copyright @Elsevier (5775730947573: Licence number)

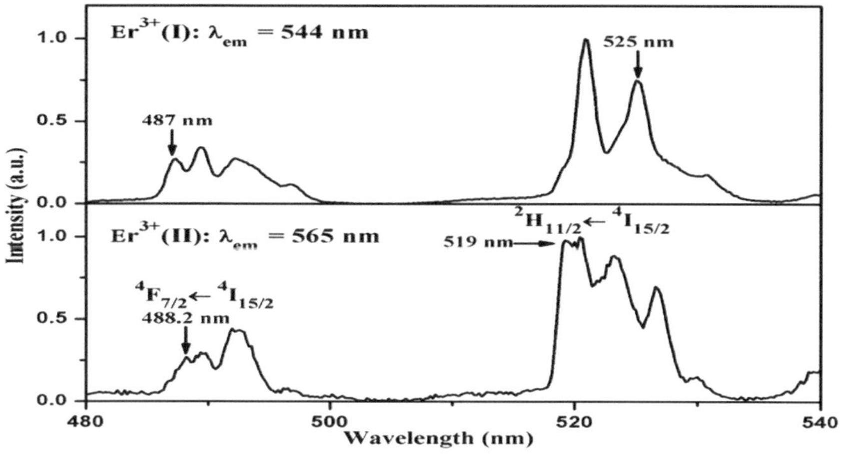

Fig. 3 PLE spectra of $CaSc_2O_4$: 0.2% Er^{3+}, 10% Yb^{3+} monitored at 544 nm of Er^{3+}(I) and 565 nm of Er^{3+}(II). Reproduced with permission from ref [14] copyright @Elsevier (5775730947573: Licence number)

of emitted photons to that of absorbed photons is essential but experimentally difficult to obtain correctly [12]. Despite these challenges, PL spectroscopy remains an indispensable technique for tailoring the properties of the nanophosphors for diverse applications.

2.2 UV–Visible Spectroscopy

This technique is widely employed to characterise optical absorption behaviour in UC phosphors, providing valuable insights into their electronic transitions, charge transfer mechanisms, and host-dopant interactions [12, 16]. A schematic diagram illustrating the analysis process is presented in Fig. 4. In UC phosphors, absorption characteristics arises from the electronic transitions within the rare earth dopant ions [16, 17]. UV–vis analysis can detect spectral shift and intensity changes caused by substitutional doping, offering a non-destructive way to monitor the uniformity and efficiency of dopant incorporation [18, 19]. For instance, Choudhary et al. characterised the red-emitting Yb/Ho–$CaSiO_3$ upconverting nanophosphors using UV–vis spectroscopy. The absorption spectra of the samples So (undoped host) and S3 (20 wt% and 2 wt% Ho) are presented in Fig. 5. The spectra revealed a pronounced absorption edge, indicative of lattice charge transfer, and successful incorporation of the dopant ions. In their study, Yb^{3+} concentration was strictly maintained at 20 wt% and Ho was varied between 1.0, 1.5, 2, 2.5, and 3 wt% leading to their respective designations as S1, S2, S3, S4, and S5 [20]. These findings suggest that UV–Vis spectroscopy allow systematic analysis of dopant-induced spectral changes. Furthermore, UV–Vis spectroscopy is used to estimate the optical band gap, essential for understanding the phosphor's excitation range. Ullah et al. asserted that upconversion phosphors with a higher band gap (Eg ~ 2.93 eV) are more likely to absorb light in the UV region and consequently exhibiting low photoactivity in the visible region [21]. Thus, UV–vis spectroscopy quantitatively supports dopant efficiency, host-dopant interactions, and band gap estimation in upconversion nanophosphors.

Although UV–vis absorption in UCNPs generally follow Beer–Lambert behaviour, deviations can occur in aggregated or concentrated systems, due to dimerization, decomposition, precipitation, and scattering effects [16, 22]. To overcome such limitations, researchers often make use of spectroscopy techniques, such as PL

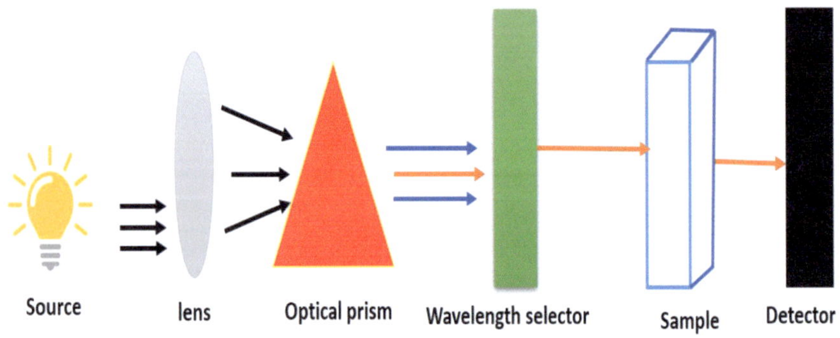

Fig. 4 Schematic diagram of UV–visible spectroscopy. Adapted from Agrawal Ref [17]. Spectroscopy techniques for rare earth-activated phosphors. *Rare Earth-Activated Phosphors*. Elsevier, 2022. 147–175

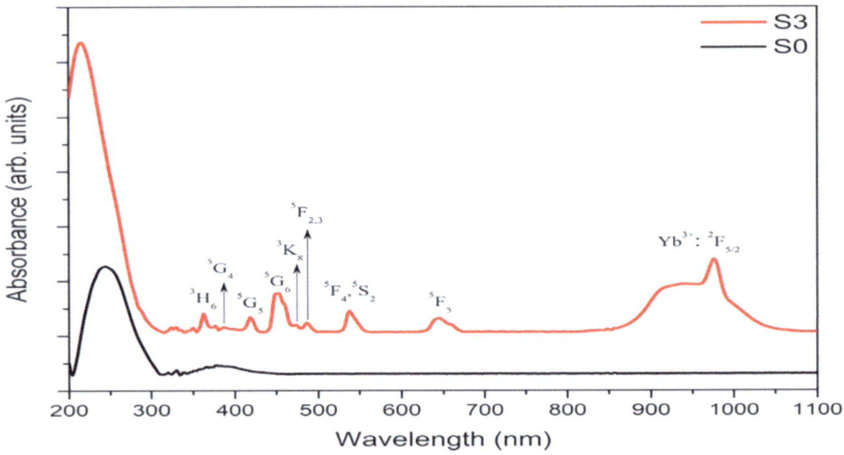

Fig. 5 UV–visible absorption spectra of S0 and S3 samples. Reproduced with permission from ref [20], copyright @Elsevier (5775740257252: Licence number)

spectroscopy in conjunction with UV–visible techniques for enabling comprehensive understanding of behaviour of upconversion phosphors. Thus, UV–vis spectroscopy still remains a valuable component for assessing optical bandgaps, host-dopant interaction, and host-dopant interaction strength in engineered luminescent materials.

3 Characterisation of the Ligands Coated on the Surface of Upconversion Phosphors

The optical performance of UC nanophosphors is immensely affected by the impurities or lattice defects present at its surface. Therefore, in order to preserve the luminescence efficiency, surface modification strategies of bounding ligands to the nanocrystal's surface are essential [23]. UC nanophosphors are then coated with an inert crystalline shell made of undoped material, or using hydrophilic polymers such as polyacrylic acid (PAA), polyethyleneimine (PEI), or polyvinylpyrrolidone (PVP) [24–28]. These surface modifications can improve photostability and also provide increased reactive sites meant for bioconjugation. Thus, this increases the potential utility of UC nanophosphors in bio-imaging and drug delivery [24]. Therefore, probing the presence and distribution of surface ligands is necessary for a better understanding of UC nanophosphors. These properties are widely probed by Infrared spectroscopy and NMR spectroscopy. The following subsection details the use of these techniques to probe surface ligands in UC nanophosphors.

3.1 Infrared (IR) Spectroscopy

Infrared spectroscopy is widely used to analyse surface moieties and ligands on UC nanocrystals, providing qualitative and semi-quantitative information [24]. The schematic representation of IR spectroscopy is given in Fig. 6. IR spectroscopy allows identification of specific functional groups and chemical bonding environment in upconversion phosphors. For example, IR spectroscopy was used to confirm the successful coating of polyethyleneimine (PEI) on the surface of $NaYF_4$: Yb/Er (Tm) upconversion phosphor. Characteristic absorption bands at 3436 cm^{-1} (stretching vibration of N–H or O–H), 1638 cm^{-1} and 750 cm^{-1} (bending vibration of N–H), 2861 cm^{-1} and 2930 cm^{-1} (symmetric and asymmetric stretching vibrations of –CH_2) can be verified through the IR spectroscopy [24, 25]. Other surface ligands such as oleic acid, polyacrylic acid (PAA), polyvinylpyrrolidone (PVP), and Oleylamine, have also been spectroscopically confirmed via their distinct C=O, C–H, or N–H vibrational modes [26–28].

Notably, IR spectroscopy plays a significant role in highlighting the effect of the ligands in the suppression of the UC emission of nanophosphors. This suppression

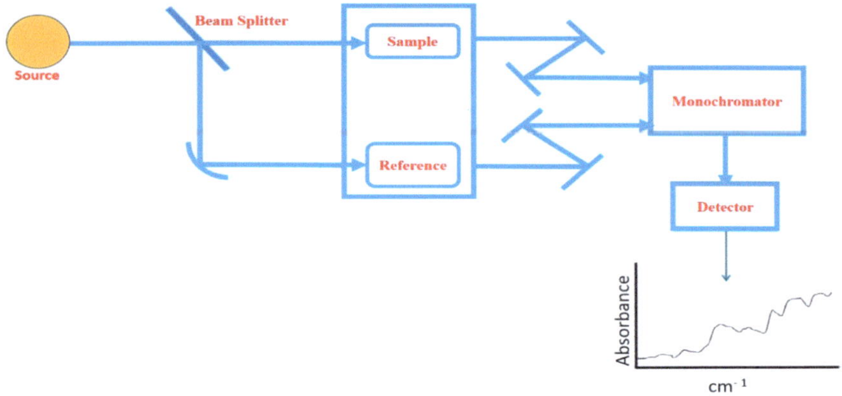

Fig. 6 Schematic representation of a typical conventional dispersive IR instrument in transmittance mode and double-beam operation. Redraw with permission from ref [29], copyright @Elsevier (5775750769384: Licence number)

occurs through a process known as surface quenching, in which non-radiative energy can be released by the surface ligands. Upconversion efficiency of nanophosphors initially coated with surface ligands was observed to gradually increased after the removal of these surface ligands [24]. Additionally, some abnormal optical properties of the UC nanophosphors which is primarily due to the chemical reaction between the ligands can be analysed by utilising IR spectroscopy [24]. Even though IR spectroscopy can assist with the characterisation and analysis of surface ligands, there are certain limitations associated with this technique. Obtaining accurate measurement of intensity is arduous [24]. Also, in order to acquire the IR spectrum, the molecule must exhibit activity in the IR region. Other challenges include establishing the accurate nature of surface ligands, and deciphering the source of the absorption peaks [24].

3.2 *Nuclear Magnetic Resonance Spectroscopy (NMR)*

NMR is a valuable tool for identifying surface ligands bound to the surface of upconversion nanophosphors. A block diagram of NMR spectroscopy is given in Fig. 7, highlighting the probe, magnet stack, and control interface used during analysis. Unlike IR spectroscopy which identifies molecular vibrations, NMR spectroscopy targets the quantification of magnetically distinct nuclei in surface-bound ligands [30]. The presence of these ligands induces characteristic 'Chemical shift' or displacements of NMR signals, particularly important for evaluating ligand coordination in upconversion nanophosphors [24, 30]. In the case of oleic acid-coated $NaYF_4$ nanocrystals, chemical shifts were detected in the range of 0.85–5.45 ppm,

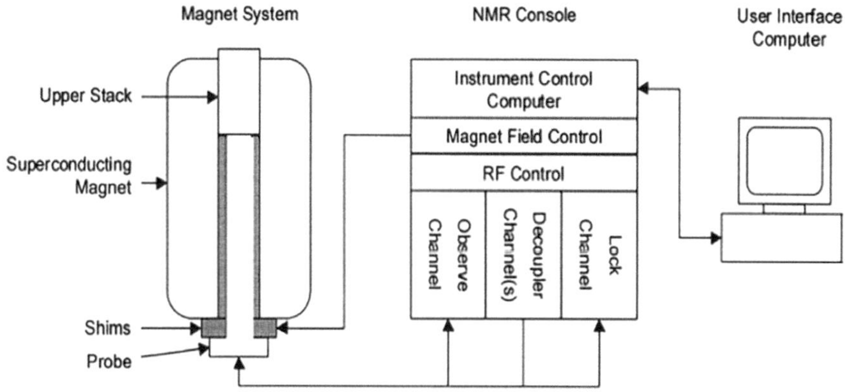

Fig. 7 Block diagram of a typical NMR spectrometer. The sample is introduced through the upper stack of the magnet into the probe. The shims are used to optimise the homogeneity of the magnetic field around the sample. Instrument operation is carried out from the NMR console through the user interface computer. Reproduced with permission from ref [31], copyright @Elsevier (5775931354012: Licence number)

while Oleylamine-coated nanocrystals show chemical shift in the range of 2.64–2.68 ppm. This shifting of NMR signals represents the presence of the ligands bound to the surface of upconversion nanocrystals [32–34].

Liu et al. asserted that the larger particle size is causing the NMR spectrum's signals to overlap, making it difficult to obtain the optimum resolution in NMR. NMR is a sensitive technology that can identify even the smallest changes in ligands, but it has some limitations that can limit its application. This includes substantial line broadening which makes it difficult to precisely characterise UC nanophosphors. The widening of the spectral lines seen in the NMR spectrum is due to irregularities in the electronic and chemical environments of the surface ligands [24]. NMR can be used to monitor inter-ligand the reaction of multiple ligands in upconversion phosphors. For instance, if the NMR signals of one ligand is well-matched with that of the other ligand, then it is evident of the presence of the inter-ligand reaction during the synthesis process [24]. Furthermore, a condition known as "Saturation" occurs when the applied radio frequency wave energy is strong enough, causing a transition that is induced both downwardly and upwards [30].

4 Understanding the Vibrational Modes

Vibrational properties of UC nanophosphors, such as bending and stretching vibrations arising from the lattice dynamics of atoms, offer valuable insights into the bonding characteristics and local environment of dopant ions [35]. In UC systems, where the optical performance is closely linked to the lattice symmetry and defect states, the analyses of vibrational states enable the identification dopant incorporation, crystallinity, and phase purity. Moreover, vibrational bands can reveal subtle distortions caused by the impurities present in UC nanocrystals, which often lead to non-radiative losses that compromises UC efficiency [35–38]. Therefore, to investigate these features, Fourier-transform infrared (FTIR) and Raman spectroscopy are widely employed due to their sensitivity towards lattice and molecular vibration. The details of these techniques are provided in the following sub section.

4.1 Fourier-Transform Infrared Spectroscopy (FTIR)

This technique is extensively applied to probe vibrational modes and surface modifications in both doped and undoped upconversion phosphors [39, 40]. A simplified line diagram of FTIR is presented in Fig. 8. The FTIR spectra of undoped and doped $CaSiO_3$: Yb/Ho samples, reported by Choudary et al. are presented in Fig. 9. Absorption bands linked to the vibration of SiO_4 tetrahedra were observed, with stretching vibrations of O–Si–O and Si–O–Si occurring at 677 and 713 cm^{-1}, respectively. A distinct band at 1469 cm^{-1} attributed to the Ca–O stretching bond experiences a notable decrease in intensity following Yb and Ho doping. In another study [20], Dubey et al. analysed the presence of distinct functional groups of $NaYF_4$:Er^{3+}/Yb^{3+}/Li^{+} upconversion phosphor. FTIR spectra revealed characteristic vibration bands of –CH, –OH, and –CO, with –OH stretching vibrations being dominant as a result of the hygroscopic nature of the host [41, 42].

FTIR is also used to study surface modifications in upconversion phosphors. FTIR analysis conducted on $NaYF_4$: Yb, Er phosphor by Meng wang et al. is as shown in Fig. 10, The FTIR spectra uncovered the polymers such as PEI, PVP, PAA, and PEG, which are coated on the surface of the nanophosphor [43–47]. These polymers exhibit great solvency, thereby the efficiency of the UC nanoparticles is significantly affected by them. Additionally, the presence of these water-soluble

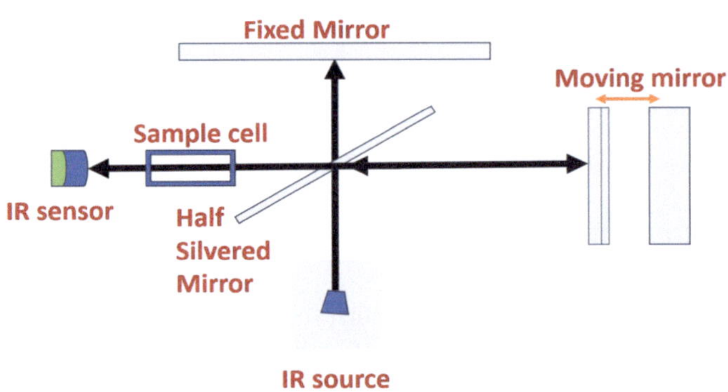

Fig. 8 Line diagram of a Fourier-transform infrared instrument. IR, infrared. Redraw with permission from ref [39], copyright @Elsevier (5775950598996: Licence number)

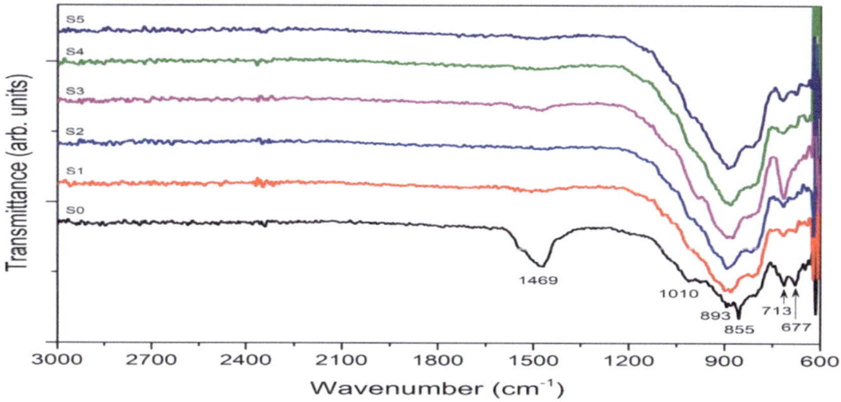

Fig. 9 FTIR spectra of the undoped and Yb/Ho-doped $CaSiO_3$ samples. Reproduced with permission from ref [20], copyright @Elsevier (5775941336678: Licence number)

polymers such as the PAA and PEI, and having functional groups such as COOH and NH_2, capped on the surface can promote the affinity between the biomolecules and the nanophosphor [43]. While FTIR helps in understanding surface modification of the phosphors, the characteristics of the upconversion phosphor as a whole will not be easily depicted to us. This technique also falls short in exploring the upconversion process mainly because it placed emphasis on the vibrational modes rather than the electronic transitions.

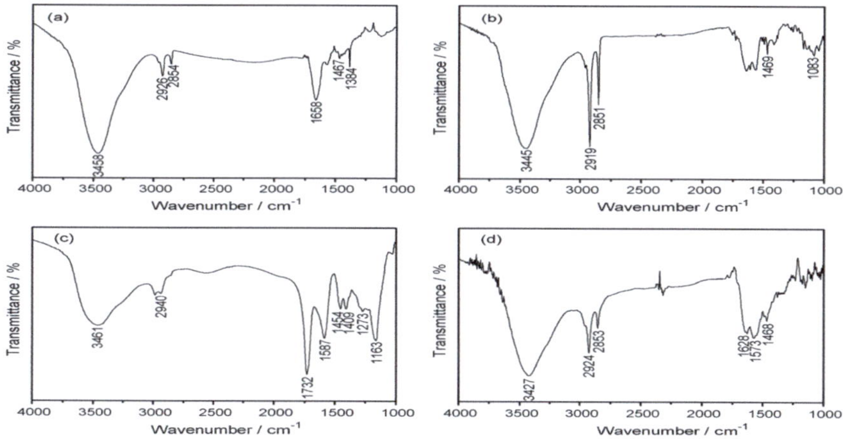

Fig. 10 FTIR spectra of PVP (**a**), PEG (**b**), PAA (**c**) and PEI (**d**) capped $NaYF_4$: Yb, Er UCNPs. Reproduced with permission from ref [43], copyright @Elsevier (5777510984067: Licence number)

4.2 Raman Spectroscopy

Raman spectroscopy is employed to investigate the lattice vibrational modes in UC phosphors providing insights into structural dynamics and bonding environment. The scattering phenomenon inherent to this technique provides unique information concerning the molecular bonds and structure of the compounds, making it akin to a molecular fingerprint [48–52]. Figure 11 illustrates Raman scattering mechanism for conceptual clarity. This technique offers complementary information by detecting changes in the molecular polarizability. Molecules which undergo changes in polarizability will be considered as Raman active. Polarizability changes arise from the interaction of the laser beam with the molecular bonds in a sample, inducing electric dipole moment that leads to vibrational transitions in a molecule [53–56].

In the context of UC luminescence, Sales et al. utilised Raman spectroscopy to characterise the vibrational properties of $BiNbO_4$: Yb^{3+}/Er^{3+} phosphors, revealing important structure–property relationships. The Raman spectra of this co-doped upconversion system are presented in Fig. 12 [57]. The cause of these vibrational bands was attributed to the internal vibrations of NbO_6 and also because of the external vibrations that spring up at lower frequencies. Bond stretching modes were observed at 537 and 624 cm^{-1}, while angle-bending vibrations appeared at 200 and 274 cm^{-1}. The peak at 886 cm^{-1} is due to the stretching vibrations of the Bi^{3+} ions in their highly distorted surroundings, and the highest phonon energy for the $BiNbO_4$ nanocrystals is observed at this peak which indicates that this host lattice is suitable for upconversion luminescence. Additional peaks at 270, 393, 426, 447, and 510 cm^{-1} appears in the region of F_{1u} mode, and peaks at 531, 551, 577, and 603 cm^{-1}, appears in the region of E_g mode. However, one limitation of Raman spectroscopy in upconversion systems is the potential interference from Raman signal

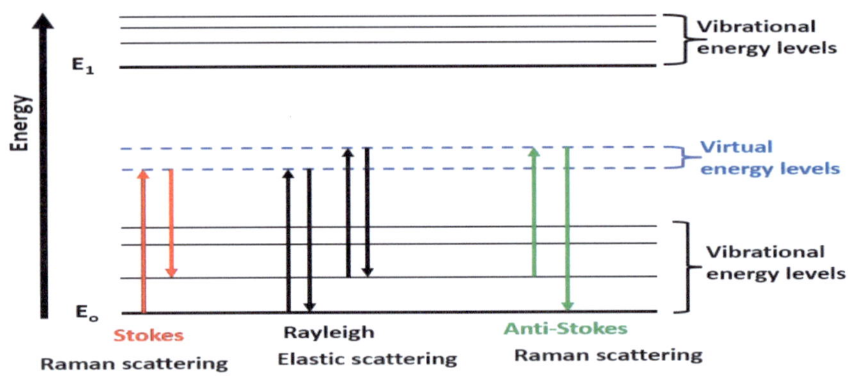

Fig. 11 "Jablonski" style diagram of energetic transitions involved in Raman scattering. Rayleigh scattering is elastic; the incident photon is of the same energy as the scattered photon. Raman scattering is inelastic; in Stokes scattering, the incident photon is of greater energy than the scattered photon, while in anti-Stokes scattering, the incident photon is of lower energy. From ref [53]: Open access source

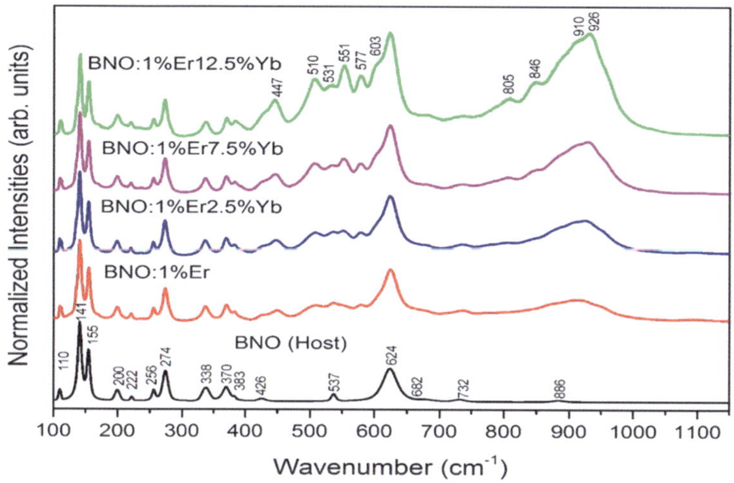

Fig. 12 Raman spectra recorded for undoped BNO as well as RE^{3+} doped BNO samples. Reproduced with permission from ref [57], copyright @Elsevevier (5777030190989: Licence number)

from strong fluorescence backgrounds, which can obscure weak Raman signals and reduce sensitivity.

5 Identification of the Elemental Composition

UC efficiency in lanthanide-doped phosphors is highly dependent on the type and concentration of dopant ions used, as well as the composition of the host lattice. Lanthanide ions, with their rich electronic configuration, play a crucial role in determining the optical characteristics and energy transfer mechanisms in UC nanophosphors. Therefore, it is essential to determine precise elemental composition, which not only confirms the successful incorporation of the dopant ions, but also identify the impurities which could quench luminescence efficiency [1, 24]. Therefore, several advanced micro-analytical techniques such as X-ray photoelectron (XPS), Energy dispersive X-ray (EDXS), and Electron energy-loss spectroscopy (EELS) are extensively used to detect elemental presence and distribution, making them highly valuable for structural and compositional characterisation of UC nanophosphors. The following subsections elaborate on how each technique contributes to elemental analysis in UC nanophosphors.

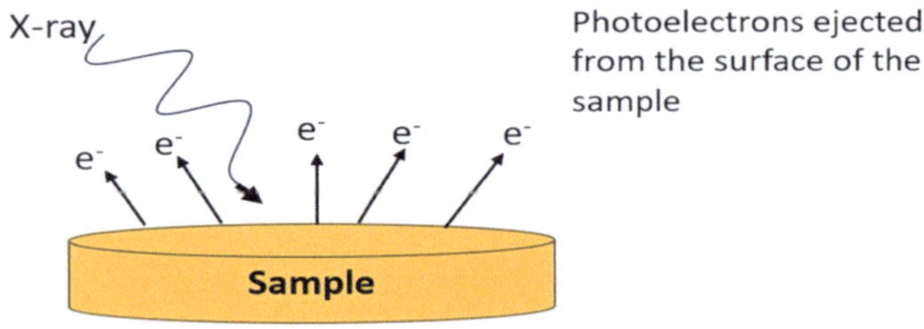

Fig. 13 Mechanism of X-ray Photoelectron Spectroscopy. Adapted from Evans Analytical Group (n.d.). https://www.eag.com/techniques/spectroscopy/x-ray-photoelectron-spectroscopy-xps-esca/

5.1 X-Ray Photoelectron Spectroscopy (XPS)

XPS, initially known as Electron Spectroscopy for Chemical Analysis (ESCA), is a powerful surface-sensitive technique primarily focused on chemical state identification and elemental composition of a sample. A schematic diagram illustrating the photoelectron emission process is presented in Fig. 13, showcasing how the monochromatic incident light induces the emission of core electrons, which are then analysed to determine their binding energies. Chemical shift, which in the context of XPS, refers to the variation in the elemental binding energies helps in identifying the chemical state of the sample being analyzed [58]. Since the binding energy is element specific and is sensitive to surface modifications, thereby making XPS a critical tool for evaluating dopants incorporation, surface chemistry, and the local chemical environments [58]. In the context of upconversion phosphors, XPS is notably helpful in determining the presence and oxidation states of the rare earth dopants and the host lattice [24].

In the case of upconversion phosphors such as $NaYF_4$/$NaGdF_4$ doped with Gd^{3+}, XPS confirms the successful incorporation of Gd^{3+} on the surface of the nanocrystal [59]. In a co-doped upconversion phosphor material such as $NaYF_4$: Eu, Yb, Trilok et al. reported the detection of elements which includes Y, Na Eu, C, F and Yb using XPS [60]. In their study, the binding energy of the C signal (284.6 eV) is used as a reference point to compare the binding energy of the various elements detected [60]. While XPS spectroscopy is very helpful regarding the surface chemistry and elemental composition, this method cannot provide us in-depth knowledge about the luminescent behaviour and emission dynamics in upconversion phosphors. Furthermore, XPS is limited to only the surface modification of the nano-crystals in view of the weak escape capability of the photo electrons [24].

5.2 Energy Dispersive X-Ray Spectroscopy (EDXS)

EDXS is a non-destructive analytical technique for generating X-ray maps showcasing the spatial distribution of elements existing within the material [61–63].

EDXS is typically in conjunction with electron microscopy techniques such as SEM or TEM, allowing correlation of the spectral components with its composition and microstructure [64–69]. In the context of upconversion phosphors, EDXS is predominantly utilised for elemental mapping to confirm the presence and spatial distribution of rare earth dopants. It also plays a critical part in the investigation of core–shell structure of the nanocrystal through a line scanning mode [24, 70–72]. In this line scan approach, a defined line or path is chosen across the region for finding out the elemental composition. This line scan is particularly valuable for visualising the spatial distribution of the lanthanide dopants within the upconversion nanoparticle. Variations in the intensity of the signal of the lanthanides along the scanned line will indicate the existence of features such as hollow nanostructures or core–shell formation [73, 74]. Additionally, EDXS also facilitates the characterisation of the UC nanophosphors-based heterostructures and also understanding the growth mechanism of the crystal [24]. Yadav et al. analysed the Y_2O_3: Yb^{3+}, Pr^{3+} co-doped nano-phosphor and found the presence of Pr, Yb, O, and Y elements within the phosphor material as shown in Fig. 14 [75]. This elemental verification demonstrates the successful incorporation of the dopants. However, EDXS does have certain limitations. It can also analyse the impurities present in the sample, as in the case of the presence of C in Fig. 14. The presence of these impurities could very well interfere with the spectral pattern and generate overlapped spectra [61]. Thus, interpretation of this overlapped spectra remains an arduous task. Furthermore, analysis is also dependent on the inhomogeneity and roughness of the sample, which could affect the result of the EDXS spectroscopy [76].

5.3 *Electron Energy-Loss Spectroscopy (EELS)*

EELS is a high-resolution technique which offers superior elemental mapping of the upconversion phosphors compared to EDXS. This enhanced resolution is due to the use of an angle–limiting aperture, which reduces the size of the volume element responsible for production of the energy-loss signal [24, 77]. Figure 15 illustrates a typical EELS spectrum which provides valuable insights into the valuable gradients and interfacial chemistry of upconversion phosphors [78].

In UC nanophosphors, the spatial distribution of dopants influences its electronic, optical, and structural properties [79]. Therefore, by elucidating the local environments of these dopants through EELS analysis, researchers can confirm the uniform incorporation of the dopants into the upconversion host matrix [61, 80, 81]. In addition to compositional information, this technique can also look into the core–shell structure of the upconversion phosphor. For example, EELS has been applied to $NaYF_4$/$NaGdF_4$ upconversion nanocrystal, whereby it showed Y occupying the core position and Gd located in the shell. Abel et al. further demonstrated the non-uniform distribution of the shell among all the nanoparticles. Even though most of the particles were core/shell, however EELS data showed that the nanocrystal tends to have an anisotropic growth, with some of the particles having thicker/thinner shells than

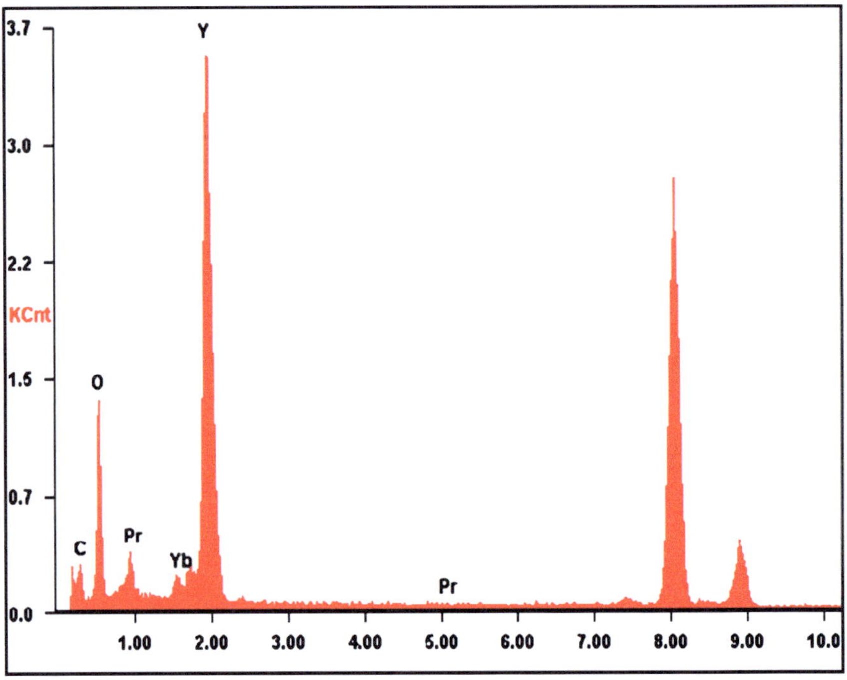

Fig. 14 EDS spectrum of (3 mol%) Yb^{3+}, (0.08 mol%) Pr^{3+} co-doped Y_2O_3 nanophosphor annealed at 1200 oC/5 h. Reproduced with permission from ref [75], copyright @Elsevier (5775981445460-Licence number)

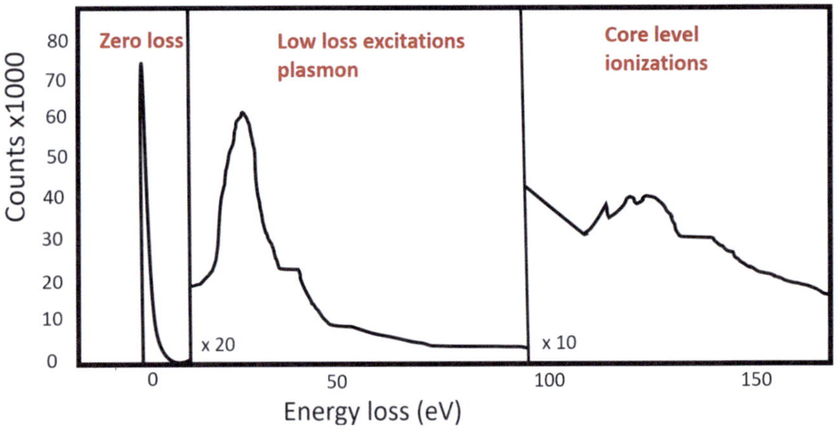

Fig. 15 An example of EELS spectrum with the different regions (zero-loss, low-loss, core loss) labelled. Redraw with permission from ref [78], copyright @Elsevier (5775310393644-Licence number)

the others [82, 83]. Using this technique can lower the damage caused to the crystal in view of the technique's fast acquisition time and uncomplicated operation [81]. However, spectral interpretation of dense sample can be inherently challenging since EELS is sensitive to thick samples. This is due to the multiple scattering of the inelastic signal in thick samples [24].

6 Characterisation of the Local Structure Around the Dopant Atom

In UC nano phosphors, the local atomic structure around dopant ions is intricately linked to their optical behaviour, which includes coordination number, bond length, and site symmetry. Each of these features could significantly influenced the splitting of energy levels and transition probabilities. Small distortions in the local structures could either enhanced or decrease the upconversion efficiency depending on their impact on non-radiative losses and energy transfer [84–86]. To investigate these local structure features, X-ray Absorption spectroscopy (XAS) or a specific type of XAS called as Extended X-ray absorption fine structure (EXAFS) is used as a probing instrument. The following subsection details on how these techniques is used to study the impact of the local structure around the dopant ions.

6.1 X-Ray Absorption Spectroscopy (XAS)

XAS is a powerful synchrotron-based technique use to probing the local atomic structure and chemical environment of specific elements within upconversion phosphors. This technique is particularly used for understanding site symmetry, coordination geometry, and host-dopant interactions, critical for governing upconversion [87–90]. Figure 16 represents an XAS spectrum, which is composed of two components: X-ray Absorption Near Edge Structure (XANES) and Extended X-ray Absorption Fine Structure (EXAFS) [88, 91]. XANES provides information on the oxidation state and coordination chemistry of the absorbing atom, while EXAFS offers details on the bond length, coordination numbers, and local symmetry around the dopant ions [24].

In the context of UC nanophosphors, XAS has proven effective in correlating the dopant positioning with luminescence efficiency. For instance, Bai et al. employed EXAFS to analyse the impact of Li^+ on the local structure around Er^{3+} in ZnO nanocrystals. The results revealed slight changes in the Er–O bond length and Er–Er bond, indicating that distortions in local symmetry could influenced the intra-4f transitions governing Er^{3+} ions [92]. Similarly, Liu et al. demonstrated that incorporating alkali earth ions (Li^+, Na^+, and K^+) into Er^{3+}-co-doped Y_2O_3 phosphors induces site asymmetry, thereby enhancing upconversion efficiency [93]. Although

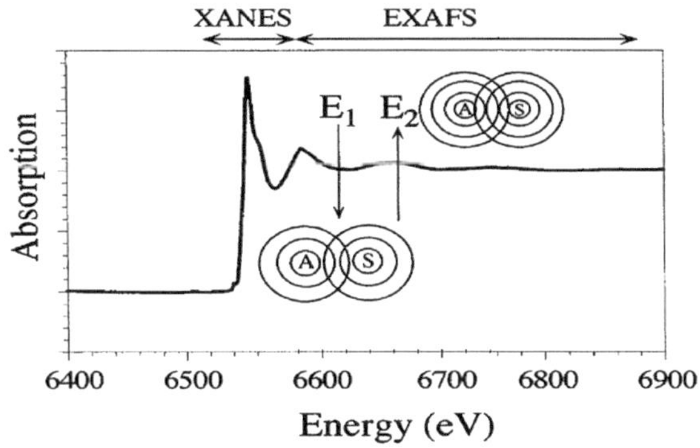

Fig. 16 Schematic illustration of an X-ray absorption spectrum at the Mn K edge. The insets show the interference phenomena that give rise to the modulations in absorption cross-section. In the inset, **A** and **S** refer to the absorbing and scattering atoms, respectively. The concentric circles around **A** and **S** represent the maxima in the photoelectron wave that describes the propagation of the X-ray excited photoelectron. The 1 s.3d transition is the very small peak at ca. 6550 eV, before the rising edge. Reproduced with permission from ref [91], copyright @Elsevier (5775991312561-Licence number)

XAS is highly sensitive and can even detect lower concentration of dopants, it is limited to studying the local structures of a single rare earth element at a time. Therefore, it may not be able to possible to capture multi-elemental interaction in complex upconversion systems. Nevertheless, XAS is still a valuable tool for correlating local structures to the functional performance of UC nanophosphors.

7 Understanding the Lifetime Dynamics

The lifetime dynamics of UC nanophosphors offers critical insights into dopant-relaxation behaviour, energy transfer efficiency, and the impact of particle size, dopant concentration, and core–shell structure [24, 93, 94]. To analyse these dynamics, time-resolved spectroscopy is used to monitor the emission decay curve following pulse excitation. From this decay curve, the luminescence lifetime (τ) reflects on the rate of radiative versus non-radiative emission. τ can be estimated using the exponential decay Eq. (1) [24]:

$$I(t) = I_o e^{\left(-\frac{t}{\tau}\right)} \tag{1}$$

where I_o represents the initial luminescence intensity and $I(t)$ refers to the luminescence intensity at a specific time t. If the decay deviates from the ideal exponential behaviour, then we can determine the effective lifetime using Eq. (2):

$$\tau_{\text{eff}} = \frac{1}{I_o} \int_0^{\infty} I(t)\mathrm{d}t \tag{2}$$

Generally, longer lifetime of UC nanophosphors indicates minimised non-radiative losses and increased UC efficiency. Additionally, lifetime dynamics correlates with the physical attributes of UC nanophosphors. For increasing dopant concentrations, the lifetime shortens due to cross-relaxation effects. Therefore, measuring lifetime dynamics is ideal for identifying quenching pathways, evaluate energy transfer mechanism, and optimise UC nanophosphors for targeted applications.

7.1 Time-Resolved Spectroscopy (TRS)

Time-resolved spectroscopy is a powerful technique for probing dynamical processes in nano UC nanophosphors, especially under pulse excitation conditions [93, 94]. By monitoring the dynamical changes of an excited species following excitation, it enables the direct determination of lifetimes of excited state of a material under operational conditions. In case of up conversion phosphors, 'Transient Absorption spectroscopy', which is a key implementation of time-resolved spectroscopy, allows the observation of changes in a material for a very short interval of time ranging from femtoseconds to milliseconds [95]. This capability makes this technique particularly useful for studying energy and charge transfer processes, vibrational relaxation, isomerisation, as well as the coupling of electronic and vibrational degrees of freedom, and other related phenomena [96, 97].

In the context of UC nanophosphors, the decay curve could insinuate on the time scale of process of excitation and their return to the ground state, and the lifetime determined from the decay curve basically ranges from nanoseconds to milliseconds [96]. For instance, Kumar et al. measured the decay times of ($^5F_4$, $^4S_2$), $^5F_5$ levels of Ho^{3+} in YVO_4 host following with excitation under a 266 nm UV light. The respective decay times were measured to be 1.8 and 2μS as illustrated in accompanying Fig. 17. Interestingly, these decay times appear to be comparatively shorter in YVO_4 host materials when compared to other material [98]. In another study, Andresen et al. described the qualitative and quantitative characterisation of the ligands-tethered upconversion phosphors by employing time-resolved luminescence method. The surface-coated UC nanophosphors might behave differently when dissolved in a particular solvent, so we need to have a better understanding of their dissolution dynamics. Additionally, the disintegration kinetics of the UC phosphors is also dependent on the particle size. It was found that the disintegration kinetics of the larger particles proceeds at a slower rate as compared to their smaller counterparts [97]. The paper also concluded that the dissolution of the upconversion nanocrystal

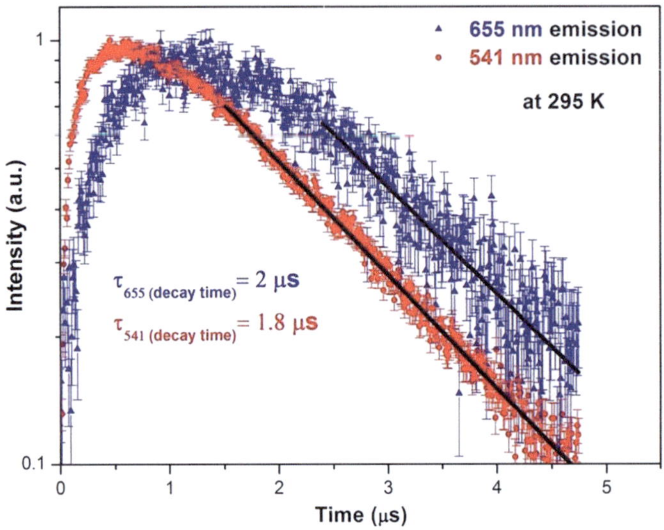

Fig. 17 Measured decay times of ($^5F_4$, $^4S_2$), $^5F_5$ levels of Ho^{3+} in YVO_4 host. Reproduced with permission from ref [98], copyright @Elsevier (5776000304871: Licence number)

is improved for smaller particles. Moreover, by virtue of the time-resolved spectroscopy, the exceptional optical and electrochemical properties of the UC luminescent properties were discovered, thereby affirming the potentiality of this technique in monitoring the stability and dissolution of the UC nanophosphors.

Time-resolved spectroscopy offers an efficient approach for analysis of the excited kinetics of the UC nanophosphors without the need for isolation of the particles from the solution. However, UC is a complex process and involves multiple simultaneous events, which makes it challenging to interpret accurate results. Additionally, this technique faces difficulty when analysing UC nanophosphors in complex environment. Therefore, it is anticipated that future advancements would enable this to analyse the toxicity concerning the different sizes and surface coating of the nanocrystal [97].

8 Conclusion

The last decade has brought about an expansion concerning the study and investigation of UC phosphors to satisfy the requirement for newly advanced technique. Thus, since the small variation in the dopant ion, and their concentrations can alter the spectroscopic properties of upconversion phosphors therefore, different spectroscopy technique are required for better understanding. Awareness of these techniques is imperative in considering how the application of these techniques can be enhanced in various industries. As in any other case, understanding of the strength or potential is equally important coupled with an understanding of their limitation so as to achieve a preeminent performance of UC phosphors. The complexity that is involved is mainly brought about by the nature of the upconversion crystal and the various reactions that can be used to modify the upconversion process when dopants are incorporated. Also, the ability of the upconversion materials to absorb several photons in the process of upconversion could also make their characterisation quite challenging. As a result, in their further attempts to obtain superior results, researchers instead of using single technique, have been making efforts to combine multiple methods. An in-depth understanding of the functioning of the rare earth elements as dopants is needed for a potentially successful integration of these techniques.

References

1. H. Dong, L.D. Sun, C.H. Yan, Upconversion emission studies of single particles. Nano Today **35**, 100956 (2020)
2. J.F. Suijver, Upconversion phosphors, in *Luminescence: From Theory to Applications*, pp. 133–177 (2007)
3. J.C.G. Bünzli, C. Piguet, Taking advantage of luminescent lanthanide ions. Chem. Soc. Rev. **34**, 1048–1077 (2005)
4. J. Shen, L.D. Sun, C.H. Yan, Luminescent rare earth nanomaterials for bioprobe applications. Dalton Trans. **42**, 5687–5697 (2008)
5. W. Zheng, P. Huang, D. Tu, E. Ma, H. Zhu, X. Chen, Lanthanide-doped upconversion nanobioprobes: electronic structures, optical properties, and biodetection. Chem. Soc. Rev. **44**, 1379–1415 (2015)
6. F. Wang, X. Liu, Recent advances in the chemistry of lanthanide-doped upconversion nanocrystals. Chem. Soc. Rev. **38**, 976–989 (2009)
7. A. Sedlmeier, H.H. Gorris, Surface modification and characterization of photon-upconverting nanoparticles for bioanalytical applications. Chem. Soc. Rev. **44**, 1526–1560 (2015)
8. A. Hofmann, Spectroscopic techniques: I Spectrophotometric techniques, in *Principles and Techniques of Biochemistry and Molecular Biology*, pp. 477–521 (2010)
9. H. Zhu et al., Eu^{3+}-doped β-Ga_2O_3 nanophosphors: Annealing effect, electronic structure and optical spectroscopy. Phys. Chem. Chem. Phys. **13**(10), 4411–4419 (2011)
10. Z. Li, H. Yuan, W. Yuan, Q. Su, F. Li, Upconversion nanoprobes for biodetections. Coord. Chem. Rev. **354**, 155–168 (2018)
11. T. Aoki, Photoluminescence spectroscopy, in *Characterization of Materials*, pp. 1–12 (2002)
12. M. Vallikkodi, Synthesis, growth, and characterization of piperazinium P-aminobenzoate and piperazinium P-chlorobenzoate nonlinear optical single crystals. Ph.D. thesis, Alagappa University, Karaikudi, India, 2018
13. M. Anpo, S. Dzwigaj, M. Che, Applications of photoluminescence spectroscopy to the investigation of oxide-containing catalysts in the working state. Adv. Catal. **52**, 1–42 (2009)
14. L. Feng et al., Observation and photoluminescence properties of two Er^{3+} centers in $CaSc_2O_4$: Er^{3+}, Yb^{3+} upconverting phosphor. J. Alloy. Compd. **708**, 827–833 (2017)
15. X. Lei, R. Li, D. Tu, X. Shang, Y. Liu, W. You, C. Sun, F. Zhang, X. Chen, Intense near-infrared-II luminescence from $NaCeF_4$: Er/Yb nanoprobes for in vitro bioassay and in vivo bioimaging. Chem. Sci. **9**(20), 4682–4688 (2018)
16. F.S. Rocha et al., Experimental methods in chemical engineering: Ultraviolet visible spectroscopy—UV-Vis. Can. J. Chem. Eng. **96**(12), 2512–2517 (2018)

17. S. Agrawal, Spectroscopy techniques for rare-earth-activated phosphors, in *Rare-Earth-Activated Phosphors* (Elsevier, Amsterdam, 2022), pp. 147–175
18. G.S. Patience, *Experimental Methods and Instrumentation for Chemical Engineers* (Elsevier, Amsterdam, 2017)
19. M. Lewis, *Advanced Chemistry Through Diagrams*, 2nd edn. (Oxford University Press, Oxford, 2001)
20. B. Chaudhary et al., Synthesis and characterization of red-emitting Yb/Ho-$CaSiO_3$ upconversion phosphors. Progr. Nat. Sci. Mater. Int. **32**(5), 594–601 (2022)
21. S. Ullah, E.P. Ferreira-Neto, C. Hazra, R. Parveen, H.D. Rojas-Mantilla, M.L. Calegaro, Y.E. Serge-Correales, U.P. Rodrigues-Filho, S.J.L. Ribeiro, Broad spectrum photocatalytic system based on BiVO4 and $NaYbF_4$: Tm^{3+} upconversion particles for environmental remediation under UV–vis-NIR illumination. Appl. Catal. B **243**, 121–135 (2019)
22. M. Sauer, J. Hofkens, J. Enderlein, Basic principles of fluorescence spectroscopy, in *Handbook of Fluorescence Spectroscopy and Imaging*, vol. 1, pp. 30 (2011)
23. Z. Chen, H. Chen, H. Hu, M. Yu, F. Li, Q. Zhang, Z. Zhou, T. Yi, C. Huang, J. Am. Chem. Soc. **130**, 3023–3029 (2008)
24. X. Liu et al., Probing the nature of upconversion nanocrystals: instrumentation matters. Chem. Soc. Rev. **44**(6), 1479–1508 (2015)
25. N.J. Johnson, N.M. Sangeetha, J.C. Boyer, F.C. van Veggel, Facile ligand-exchange with polyvinylpyrrolidone and subsequent silica coating of hydrophobic upconverting β-$NaYF_4$: Yb^{3+}/Er^{3+} nanoparticles. Nanoscale **2**(5), 771–777 (2010)
26. H. Qiu, G. Chen, L. Sun, S. Hao, G. Han, C. Yang, Ethylenediaminetetraacetic acid (EDTA)-controlled synthesis of multicolor lanthanide doped $BaYF_5$ upconversion nanocrystals. J. Mater. Chem. **21**(43), 17202–17208 (2011)
27. B. Stuart, Infrared spectroscopy, in *Kirk-Othmer Encyclopedia of Chemical Technology* (2000)
28. L.M. Ng, R. Simmons, Infrared spectroscopy. Anal. Chem. **71**(12), 343–350 (1999)
29. E. Capuano, S.M. Van Ruth, Infrared spectroscopy: applications, in *Encyclopedia of Food and Health* (Elsevier, Amsterdam, 2015), pp. 424–431
30. D.L. Pavia, G.M. Lampman, G.S. Kriz, J.R. Vyvyan, in *Introduction to Spectroscopy* (2015)
31. E.A. Williams, Nuclear magnetic resonance spectrometry, in *Encyclopedia of Materials: Science and Technology*, pp. 6319–6324 (2001)
32. J.C. Boyer, F. Vetrone, L.A. Cuccia, J.A. Capobianco, Synthesis of colloidal upconverting $NaYF_4$ nanocrystals doped with Er^{3+}, Yb^{3+} and Tm^{3+}, Yb^{3+} via thermal decomposition of lanthanide trifluoroacetate precursors. J. Am. Chem. Soc. **128**(23), 7444–7445 (2006)
33. J.C. Boyer, L.A. Cuccia, J.A. Capobianco, Synthesis of colloidal upconverting $NaYF_4$: Er^{3+}/Yb^{3+} and Tm^{3+}/Yb^{3+} monodisperse nanocrystals. Nano Lett. **7**(3), 847–852 (2007)
34. Y. Liu, K. Ai, L. Lu, Designing lanthanide-doped nanocrystals with both up-and down-conversion luminescence for anti-counterfeiting. Nanoscale **3**(11), 4804–4810 (2011)
35. A. Rohman, A. Windarsih, E. Lukitaningsih, M. Rafi, K. Betania, N.A. Fadzillah, The use of FTIR and Raman spectroscopy in combination with chemometrics for analysis of biomolecules in biomedical fluids: a review. Biomed. Spectrosc. Imaging **8**(3–4), 55–71 (2019)
36. M.J. Baker, S.R. Hussain, L. Lovergne, V. Untereiner, C. Hughes, R.A. Lukaszewski, G. Thiéfin, G.D. Sockalingum, Developing and understanding biofluid vibrational spectroscopy: a critical review. Chem. Soc. Rev. **45**(7), 1803–1818 (2016)
37. A.L. Mitchell, K.B. Gajjar, G. Theophilou, F.L. Martin, P.L. Martin-Hirsch, Vibrational spectroscopy of biofluids for disease screening or diagnosis: translation from the laboratory to a clinical setting. J. Biophotonics **7**(3–4), 153–165 (2014)
38. Z. Konkoli, D. Cremer, A new way of analyzing vibrational spectra. III. Characterization of normal vibrational modes in terms of internal vibrational modes. Int. J. Quant. Chem.**67**(1), 29–40 (1998)
39. A. Dutta, Fourier transform infrared spectroscopy, in *Spectroscopic Methods for Nanomaterials Characterization*, pp. 73–93 (2017)
40. B.C. Smith, *Fundamentals of Fourier Transform Infrared Spectroscopy*, 2nd edn. (CRC Press, 2011)

41. A. Dubey, A.K. Soni, A. Kumari, R. Dey, V.K. Rai, Enhanced green upconversion emission in $NaYF_4$: $Er^{3+}/Yb^{3+}/Li^+$ phosphors for optical thermometry. J. Alloy. Compd. **693**, 194–200 (2017)
42. W.D. Perkins, Fourier transform-infrared spectroscopy: Part l Instrumentation. J. Chem. Educ. **63**(1), A5 (1986)
43. M. Wang et al., One-step synthesis and characterization of water-soluble $NaYF_4$: Yb, Er/Polymer nanoparticles with efficient up-conversion fluorescence. J. Alloy. Compd. **485**(1–2), L24–L27 (2009)
44. Z. Li, Y. Zhang, Monodisperse silica-coated polyvinyl-pyrrolidone/$NaYF_4$ nanocrystals with multicolor upconversion fluorescence emission (2006)
45. B. Yan, X. Chen, J. Wu, Induced assembly and photoluminescence of lanthanum (Tb, Eu, Dy) complexes/ZnO/polyethylene glycol hybrid phosphors. Appl. Surf. Sci. **253**(21), 8575–8580 (2007)
46. G.S. Yi, G.M. Chow, Water-soluble $NaYF_4$: Yb, Er (Tm)/$NaYF_4$/polymer core/shell/shell nanoparticles with significant enhancement of upconversion fluorescence. Chem. Mater. **19**(3), 341–343 (2007)
47. F. Wang, X. Fan, M. Wang, Y. Zhang, Multicolour PEI/$NaGdF_4$: Ce^{3+}, Ln^{3+} nanocrystals by single-wavelength excitation. Nanotechnology **18**(2), 025701 (2006)
48. A. Kudelski, Analytical applications of Raman spectroscopy. Talanta **76**(1), 1–8 (2008)
49. A. Ceja Fernández, *Biological application of ceramic and metallic nanoparticles by excitation in the near infra-red*. Ph.D. thesis, 2016
50. F.A. Miller, G.B. Kauffman, CV Raman and the discovery of the Raman effect. J. Chem. Educ. **66**(10), 795 (1989)
51. G.S. Bumbrah, R.M. Sharma, Raman spectroscopy–Basic principle, instrumentation and selected applications for the characterization of drugs of abuse. Egypt. J. Forensic Sci. **6**(3), 209–215 (2016)
52. I. Mukerji, Resonance Raman spectroscopy, in *eLS* (Wiley, USA, 2000)
53. K.J. Ember, M.A. Hoeve, S.L. McAughtrie, M.S. Bergholt, B.J. Dwyer, M.M. Stevens, K. Faulds, S.J. Forbes, C.J. Campbell, Raman spectroscopy and regenerative medicine: a review. NPJ Regenerative Med. **2**(1), 12 (2017)
54. S. Reitzenstein, P. Rösch, M.A. Strehle, D. Berg, M. Baranska, H. Schulz, E. Rudloff, J. Popp, Nondestructive analysis of single rapeseeds by means of Raman spectroscopy. J Raman Spectrosc: An International Journal for Original Work in all Aspects of Raman Spectroscopy, Including Higher Order Processes, and also Brillouin and Rayleigh Scattering, **38**(3), 301–308 (2007)
55. B. Schrader, H. Schulz, M. Baranska, G.N. Andreev, C. Lehner, J. Sawatzki, Non-destructive Raman analyses–polyacetylenes in plants. Spectrochim. Acta Part A Mol. Biomol. Spectrosc. **61**(7), 1395–1401 (2005)
56. J.M. Antosiewicz, D. Shugar, UV–Vis spectroscopy of tyrosine side-groups in studies of protein structure. Part 1: Basic principles and properties of tyrosine chromophore. Biophys. Rev. **8**(2), 151–161 (2016)
57. A.J.M. Sales et al., Power dependent upconversion in Er^{3+}/Yb^{3+} co-doped $BiNbO_4$ phosphors. Ceram. Int. **42**(6), 6899–6905 (2016)
58. J. Chastain, R.C. King Jr., Handbook of X-ray photoelectron spectroscopy. Perkin-Elmer Corporation **40**(221), 25 (1992)
59. K.A. Abel, J.C. Boyer, F.C.V. Veggel, Hard proof of the $NaYF_4$/$NaGdF_4$ nanocrystal core/shell structure. J. Am. Chem. Soc. **131**(41), 14644–14645 (2009)
60. T.K. Pathak, A. Kumar, L.J.B. Erasmus, A. Pandey, E. Coetsee, H.C. Swart, R.E. Kroon, Highly efficient infrared to visible up-conversion emission tuning from red to white in Eu/Yb co-doped $NaYF_4$ phosphor. Spectrochim. Acta Part A Mol. Biomol. Spectrosc. **207**, 23–30 (2019)
61. R.K. Mishra, A.K. Zachariah, S. Thomas, Energy-dispersive X-ray spectroscopy techniques for nanomaterial, in *Microscopy Methods in Nanomaterials Characterization* (Elsevier, Amsterdam, 2017), pp. 383–405

62. F. Hofer, F.P. Schmidt, W. Grogger, G. Kothleitner, Fundamentals of electron energy-loss spectroscopy, in *IOP Conference Series: Materials Science and Engineering*, vol. 109, no. 1. (IOP Publishing, 2016), p. 012007
63. M. Felderhoff, K. Klementiev, W. Grünert, B. Spliethoff, B. Tesche, J.M.B. von Colbe, B. Bogdanović, M. Härtel, A. Pommerin, F. Schüth, C. Weidenthaler, Combined TEM-EDX and XAFS studies of Ti-doped sodium alanate. Phys. Chem. Chem. Phys. **6**(17), 4369–4374 (2004)
64. S.A. Katsyuba, N.I. Monakhova, L.K. Ashrafullina, R.R. Shagidullin, S.A. Terent'eva, Spectroscopic study of compounds with P-H bonds: Vibrational spectra and rotational isomers of ethyl ethylphosphite. J. Appl. Spectrosc. **54**(6), 540–544 (1991)
65. S. Auzoux-Bordenave, C. Brahmi, A. Badou, M. de Rafélis, S. Huchette, Shell growth, microstructure and composition over the development cycle of the European abalone Haliotis tuberculata. Mar. Biol. **162**, 687–697 (2015)
66. R. Hillel, M. Maline, F. Gourbilleau, G. Nouet, R. Carles, A. Mlayah, Microstructure of chemically vapour codeposited SiC TiC C nanocomposites. Mater. Sci. Eng. A **168**(2), 183–187 (1993)
67. W.U. Anake, G.R. Ana, N.U. Benson, Study of surface morphology, elemental composition and sources of airborne fine particulate matter in Agbara industrial estate Nigeria. Int. J. Appl. Environ. Sci **11**(4), 881–890 (2016)
68. M. Abd Mutalib, M.A. Rahman, M.H.D. Othman, A.F. Ismail, J. Jaafar, Scanning electron microscopy (SEM) and energy-dispersive X-ray (EDX) spectroscopy, in *Membrane Characterization* (Elsevier, Amsterdam, 2017), pp. 161–179
69. M. Scimeca, S. Bischetti, H.K. Lamsira, R. Bonfiglio, E. Bonanno, Energy Dispersive X-ray (EDX) microanalysis: a powerful tool in biomedical research and diagnosis. Eur. J. Histochem. EJH **62**(1), 2841 (2018)
70. D.H. Loughridge, The direction of ejection of photo-electrons produced by X-rays. Phys. Rev. **26**(6), 697 (1925)
71. Y. Shinohara, R. Imai, H. Kishimoto, N. Yagi, Y. Amemiya, Indirectly illuminated X-ray area detector for X-ray photon correlation spectroscopy. Synchrotron Radiat. **17**(6), 737–742 (2010)
72. H. Jho, B. Lee, Y. Ji, S. Ha, Discussion for the enhanced understanding of the photoelectric effect. Eur. J. Phys. **44**(2), 025301 (2023)
73. F. Zhang, Y. Shi, X. Sun, D. Zhao, G.D. Stucky, Formation of hollow upconversion rare-earth fluoride nanospheres: nanoscale kirkendall effect during ion exchange. Chem. Mater. **21**(21), 5237–5243 (2009)
74. F. Wang, L.D. Sun, J. Gu, Y.F. Wang, W. Feng, Y. Yang, J. Wang, C.H. Yan, Selective heteroepitaxial nanocrystal growth of rare earth fluorides on sodium chloride: synthesis and density functional calculations. Angew. Chem. **124**(35), 8926–8929 (2012)
75. R.S. Yadav, R.K. Verma, A. Bahadur, S.B. Rai, Structural characterizations and intense green upconversion emission in Yb^{3+}, Pr^{3+} co-doped Y_2O_3 nano-phosphor. Spectrochim. Acta Part A Mol. Biomol. Spectrosc. **137**, 357–362 (2015)
76. J.T. Grant, T.W. Haas, Combined low-energy electron diffraction and auger electron spectroscopy studies of Si, Ge, Gaas, and Insb surfaces. J. Vac. Sci. Technol. **8**(1), 94–97 (1971)
77. J. Verbeeck et al., in *Electron Energy Loss Spectrometry* (Elsevier, Amsterdam, 2005), pp. 324–331
78. S. Porter, A.K. Datye, in *Imaging of Single Atom Catalysts* (Elsevier, Amsterdam, 2023), p. 222243
79. R.D. Leapman, M.A. Aronova, EELS and EFTEM analysis of biological materials. Microsc. Microanal. **20**(S3), 582–583 (2014)
80. J. Geiger, Inelastic electron scattering in thin films at oblique incidence. Physica Status Solidi (B) **24**(2), 457–460 (1967)
81. R.F. Egerton, R.F. Egerton, An introduction to EELS, in *Electron Energy-Loss Spectroscopy in the Electron Microscope*, pp. 1–28 (2011)
82. K.A. Abel, J.C. Boyer, C.M. Andrei, F.C. Van Veggel, Analysis of the shell thickness distribution on $NaYF_4$/$NaGdF_4$ core/shell nanocrystals by EELS and EDS. J. Phys. Chem. Lett. **2**(3), 185–189 (2011)

83. H. Wen, H. Zhu, X. Chen, T.F. Hung, B. Wang, G. Zhu, S.F. Yu, F. Wang, Upconverting near-infrared light through energy management in core-shell-shell nanoparticles. Angewandte Chemie **125**(50) (2013)
84. M. Novita et al., Study on Local-Structure Symmetrization of K2XF6 Crystals Doped with Mn^{4+} Ions by First-Principles Calculations. Materials **16**(11), 4046 (2023)
85. N.C. George, K.A. Denault, R. Seshadri, Phosphors for solid-state white lighting. Annu. Rev. Mater. Res. **43**(1), 481–501 (2013)
86. S.E. Brinkley et al., Robust thermal performance of Sr2Si5N8: Eu^{2+}: an efficient red emitting phosphor for light emitting diode based white lighting. Appl. Phys. Lett. **99**(24) (2011)
87. F. De Groot, High-resolution X-ray emission and X-ray absorption spectroscopy. Chem. Rev. **101**(6), 1779–1808 (2001)
88. P.T. Andrews, *Unoccupied Electronic States: Fundamentals for XANES, EELS, IPS and BIS*, vol. 69, pp. 348–351 (1992)
89. J. Yano, V.K. Yachandra, X-ray absorption spectroscopy. Photosynth. Res. **102**, 241–254 (2009)
90. M. Jain et al., Ce doping induced trapping states and local electronic structure modifications in $SrZnO_2$ nanophosphors. J. Mater. Chem. C **10**(31), 11379–11387 (2022)
91. J.E. Penner-Hahn, X-ray absorption spectroscopy in coordination chemistry. Coord. Chem. Rev. **190**, 1101–1123 (1999)
92. Y. Bai, Y. Wang, K. Yang, X. Zhang, G. Peng, Y. Song, Z. Pan, C.H. Wang, The effect of Li on the spectrum of Er^{3+} in Li-and Er-co doped ZnO nanocrystals. J. Phys. Chem. C **112**(32), 12259–12263 (2008)
93. L. Liu, Y. Wang, Y. Bai, X. Zhang, K. Yang, Y. Song, Effects of alkali metal ions on upconversion photoluminescence intensity of Er^{3+}-doped Y_2O_3 nanocrystals. Appl. Phys. B **110**, 111–115 (2013)
94. C. Grieco, Time-resolved optical spectroscopy: A versatile, complementary tool for advancing cutting-edge materials technologies. MRS Bull. **44**(7), 519–520 (2019)
95. I.H. Van Stokkum, D.S. Larsen, R. Van Grondelle, Global and target analysis of time-resolved spectra. Biochimica et Biophysica Acta (BBA)-Bioenergetics **1657**(2–3), 82–104 (2004)
96. H. Chander, S. Chawla, Time resolved spectroscopic studies on some nanophosphors. Bull. Mater. Sci. **31**, 401–407 (2008)
97. E. Andresen et al., Time-resolved luminescence spectroscopy for monitoring the stability and dissolution behaviour of upconverting nanocrystals with different surface coatings. Nanoscale **12**(23), 12589–12601 (2020)
98. M.K. Mahata, T. Koppe, H. Hofsäss, K. Kumar, U. Vetter, Host sensitized luminescence and time-resolved spectroscopy of YVO_4: Ho^{3+} nanocrystals. Phys. Proc. **76**, 125–131 (2015)

Upconversion Phosphors for Forensic Applications

Kanchan Tiwari, Nameeta Brahme, D. P. Bisen, and Akesh Kumar

Abstract Like a coin, every crime has two sides—the truth and the lie. However, just like in coin toss games, you cannot use probability mathematics in crime scene investigation to play flip games without solid proof. When examining the evidence in a real-world criminal case, one must make sure that unfairness is completely ruled out. The fight for justice is based on a rigorous assessment of evidence, which necessitates technologies that reveal hidden details and view the unknown. In recent years, luminous materials (phosphors) known as up-conversion phosphors have emerged as viable tools for forensic investigations, providing distinct benefits in scene interpretation and evidence collecting. This chapter digs into the fascinating potential of upconversion phosphors for forensic applications, investigating their principles, functions, and promising future developments. Unlike traditional phosphors, which absorb high-energy light and release lower-energy photons (down conversion), upconversion phosphors use elegant physics (anti stokes shifting) to convert low-energy near-infrared (NIR) radiation into higher-energy visible or ultraviolet (UV) light. This extraordinary phenomenon results from consecutive photon absorption in specially developed lanthanide-doped materials. As NIR light stimulates the lanthanide ions, they undergo several energy level transitions, eventually emitting higher-energy light. In present chapter, details of upconversion phosphor for various forensic based applications, current research and potential phosphor for the same is discussed in detail and some of the valuable applications are given: (1) Latent fingerprint detection (2) Tracing Evidences (3) Security markers (4) Anti-counterfeiting. While upconversion phosphors have intriguing promise, some problems remain, which developers can overcome. These issues include: sensitivity to various surfaces, reduced toxicity, improved biocompatibility, selectivity to specific analytes, economic cost, availability, and practicality. The chapter also covers all of these obstacles and provides scientific papers that offer ways to mitigate them. Along with-it futuristic application is also discussed.

K. Tiwari · N. Brahme · D. P. Bisen · A. Kumar
S.o.S. in Physics and Astrophysics, Pandit Ravishankar Shukla University, Raipur, India

N. Brahme (✉)
Center for Nanoscience and Nanotechnology, Pandit Ravishankar Shukla University, Raipur, India
e-mail: namitabrahme@gmail.com

V. Dubey et al. (eds.), *Upconversion Luminescence of Phosphors*, Progress in Optical Science and Photonics 39, https://doi.org/10.1007/978-981-95-1762-6_6

Keywords Upconversion phosphors · Forensic applications · Fingerprint detection · Security markers · Evidence tracing

1 Introduction

The public's knowledge and interest in this field of criminal investigation have grown in recent years thanks to the forensic science-focused television shows, online series, films, and dramas as well as the extensive coverage of several criminal trials. Although not brand-new, the field of using upconversion (UC) for forensic applications is still developing. UC luminescence is an optical phenomenon that arises from anti-Stokes luminescence and involves the absorption of two or more low-energy photons to produce high-energy photons. It has attracted a lot of attention. Compared to organic dyes and quantum dots, UC nanomaterials are preferable due to their unique properties. They are promising luminophores for forensic-based applications. This chapter provides a thorough introduction to lanthanide-doped UC phosphors, covering the basic idea, the synthesis and design of desired properties, a variety of applications in forensic-based applications, applied techniques and applied phosphors, reported research works, the benefits and drawbacks of current methods and anticipated future results.

2 Overview of Lanthanide Ions and UC Process

2.1 Overview of Lanthanide Ions

Sir William Crookes (February 16, 1887) once stated an amazing thought about rare earth (RE) elements, "These elements perplex us in our reaches [sic], baffle us in our speculations and haunt us in our very dreams. They stretch like an unknown sea before us – mocking, mystifying and murmuring strange revelations and possibilities." [1]. This statement poetically reflects the weirdness, misunderstood concepts, mystical physics as well as multiple applications of RE elements. Because of their comparable electron configurations ($[Xe]\,4f^{[0-14]}\,5d^{[0-1]}\,6s^2$), the fifteen elements that make up the lanthanide (Ln) series (from lanthanum to lutetium) exhibit strong chemical similarities. The Ln elements usually form trivalent cations with the electronic configuration of $[Xe]\,4f^n$, where n varies from 0 to 14. Certain Ln elements have oxidation states of $+\,2$ or $+\,4$, in addition to the trivalent state, such as Sm^{2+}, Eu^{2+}, Yb^{2+}, and Tb^{4+}, Ce^{4+}, and Pr^{4+}. The common ions of Ln elements and their electron configurations are compiled in Table 1. Except for lanthanum and lutetium (d-block), the Ln family is classified as f-block elements. The huge number of energy levels in the 4f electron configuration causes numerous intra-configurational changes. The trivalent Ln ions exhibit distinctive spectroscopic features. Because the closed 5 s and 5p shells effectively cover the 4f shell, the chemical environments of connected ligands have only a minor effect on electronic transitions within the 4f orbital. Although weak, this perturbation is responsible for the tuning of spectral fine structures. The absorption spectra of Ln ions doped into single crystals are composed of several tiny lines that resemble an atomic spectrum. In contrast, line transitions have been found to be expanded in liquids, ceramics, inorganic matrix, and glasses. The chosen matrix affects the widening, but the linewidth of this band is still substantially lower than in the absorption spectra of transition metal ions. Many distinct optical and magnetic features, extended lifetimes of the excited states, massive anti-Stokes shifts and narrow emission bands are all caused by the transitions of the 4f electrons in Ln ions. These characteristics, which make them appropriate for a variety of applications, are summed up in Table 1. The promotion of 4f electrons into the 5d subshell and the intra-f-f transitions of the Ln ions are the sources of the luminescence emission colors (4f–5d transitions). Ln ions are therefore thought to be promising luminous centers. The electric dipole (ED), magnetic dipole (MD), induced ED and electric quadrupole transitions all permit any transition in RE. Even though there are very few permitted and observed MD transitions for trivalent Ln ions, these transitions are crucial for comprehending the matrix environment. Since MD transitions typically do not depend on the ligand environment, they can be utilized as benchmarks for intensity. Furthermore, by employing theoretical models such energy matrix formulation, wavefunction knowledge, density functional theory (DFT)-based research and Judd–Ofelt theory, the intensity of an MD transition may be precisely computed. Compared to MD transitions, ED transitions happen far more frequently. Though ED transitions, in contrast to MD transitions, cannot be analytically assessed, they are nonetheless sensitive to the host environment, which is why they are frequently referred to as hypersensitive transitions. Although induced electric dipole and electric quadrupole transitions follow the same selection rules, they are occasionally prohibited and are consequently referred to as pseudo transitions [2–5].

2.2 Upconversion Process

UC is a nonlinear optical process that is defined by the emission of high-energy photons after two or more successive absorptions of low-energy photons. In other words, UC occurs when lower electronic levels of low-energy (longer wavelength) radiation (NIR light) excite the generation of higher energy (shorter wavelength) emission (visible or UV light) at higher electronic levels as a result of an anti-Stokes mechanism. Bloembergen's 1959 study of the UC phenomena with an infrared quantum counter device paved the way for subsequent research and Auzel's 1966 observation of the process confirmed it. Since then, a lot of work has gone into expanding the range of UC materials [6]. The latest research is based on modifying UC process and as a consequence, many efficient, tunable and advanced luminescence can be obtained by a continuous-wave (CW) lasers, low power density laser source, pulse duration lasers, infrared laser beams, typical xenon or halogen lamps

Table 1 A summary of different properties of RE ions like atomic number (AN), Atomic weight (AW) along with their uncertainty (U), electronic configuration (EC), ionic radius (IR), Valancies (V), number of isotopes (I), and nature of isotopes (N), Luminescence property (L), circular dichroism (CD), nuclear magnetic resonance (NMR), and magnetic resonance imaging property (MRI) [1]

Element	AN	AW (U)	EC	IR (Å) (RE^{3+})	Valency	I-N	L	CD	NMR	MRI
La	57	138.905(2)	{Xe}$5d^1 6s^2$	1.045	+ 3	3-g		*	*	
Ce	58	140.116(1)	{Xe}$4f^2 6s^2$	1.01	+ 3, + 4	4-g		*		
Pr	59	140.90765(2)	{Xe}$4f^3 6s^2$	0.997	+ 3, + 4	1		*		
Nd	60	144.24(3)	{Xe}$4f^4 6s^2$	0.983	+ 3	7-g	**	*	*	
Pm	61	145	{Xe}$4f^5 6s^2$	0.97	+ 3, + 2	–		*		
Sm	62	150.36(3)	{Xe}$4f^6 6s^2$	0.958	+ 3, + 2	7-g		*	*	
Eu	63	151.964(1)	{Xe}$4f^7 6s^2$	0.947	+ 3, + 2	2-g	**	**	**	
Gd	64	157.25(3)	{Xe}$4f^7 5d^1 6s2$	0.938	+ 3	7-g		*	*	**
Tb	65	158.92534(2)	{Xe}$4f^9 6s^2$	0.923	+ 3	1	**	**		
Dy	66	162.50(3)	{Xe}$4f^{10} 6s^2$	0.912	+ 3	7-g		*	**	
Ho	67	164.93032(2)	{Xe}$4f^{11} 6s^2$	0.901	+ 3	1		*	**	
Er	68	167.26(3)	{Xe}$4f^{12} 6s^2$	0.89	+ 3	6-g	**	*	**	
Tm	69	168.93421(2)	{Xe}$4f^{13} 6s^2$	0.88	+ 3, + 2	1		*		
Yb	70	173.04(3)	{Xe}$4f^{14} 6s^2$	0.868	+ 3, + 2	7-g	**	**	**	
Lu	71	174.967(1)	{Xe}$4f^{14} 5d^1 6s^2$	0.85	+ 3	2-g		*		

In column 7 number of isotopes (I)—their nature (N) is demonstrated. Here, alphabet "g" is indicating "Geological Isotopic Composition", these elements have isotopes having AW exceeded to the uncertainty [1]
Here, * refers to important and ** refers to very important property of that specific RE element

or uniform focused sunlight [6, 7]. The UC process can be explained quantitatively in various ways. The most widely used quantitative word to characterize the efficiency of the UC process is quantum yield. The photoluminescence and quantum yield are highly reliant on the excitation optical density due to the nonlinear character of the UC process. The ratio of the absorbed power to the emitted power is known as conversion efficiency. The more or less intricate photophysical conversion pathways cannot be handled by quantum yield and conversion efficiency. Usually, rate equations are the most effective way to characterize them since they enable a quantitative description of the processes [6, 8]. The precise UC efficiency value of lanthanide-doped UC nanoparticles (UCNPs) is hard to measure. At the moment, the most often used parameter in lanthanide UCNP investigations is the UC emission intensity. Comparing the UC efficiency of lanthanide UCNPs can be done using the UC emission intensity provided that uniform measurement circumstances are maintained across several samples [7].

Many mechanisms have now been identified to explain the process of UC, such as energy migration UC, photon avalanche, cooperative sensitization UC, excited-state absorption, Förster resonance energy transfer (FRET), and energy transfer UC (Fig. 1) [6, 7]. Essentially, though, UC processes fall into three primary categories: photon avalanche, excited-state absorption, and energy-transfer UC. Though photon avalanche is infrequently observed in nanoscale lanthanide materials, all of which have been extensively covered in numerous prior published works. A suitable pumping mechanism for singly doped UC materials is excited-state absorption. The overall energy plan associated with the subsequent absorption of two photons is displayed in Fig. 1a. The ground state 1 transitions to the excited metastable level 2 under appropriate excitation and the excited state 2 then absorbs another photon with energy that is resonant with the state 2 transition to a higher excited state 3. Higher-energy photon emission is the outcome of a radiative transition from this excited state 3 back to the ground state or another lower-energy state. To enhance the gain in the excited-state absorption process and prevent transfer losses due to cross-relaxation between the luminous centers, the doped particles must have a low concentration of active ions. Materials doped with rare earth elements (RE) (such as Yb^{3+}, Pr^{3+}, Nd^{3+}, Er^{3+}, Ho^{3+}, and Tm^{3+}) have this kind of process, known as energy excited state absorption since their excitation states are well matched with the excitation energy of commercial lasers (~ 800–1000 nm).

Transfer of energy UC is the most effective UC procedure by far. The majority of RE-doped phosphors use an energy-transfer UC technique to produce efficient UC emissions. Schematic illustrations of several energy-transfer UC processes can be viewed in Fig. 1b–h. Two conditions must be met by the sensitizer and activator in the energy transfer UC process: first, their excited energies must be nearly identical and second, their distances must be sufficiently close. For the benefit of

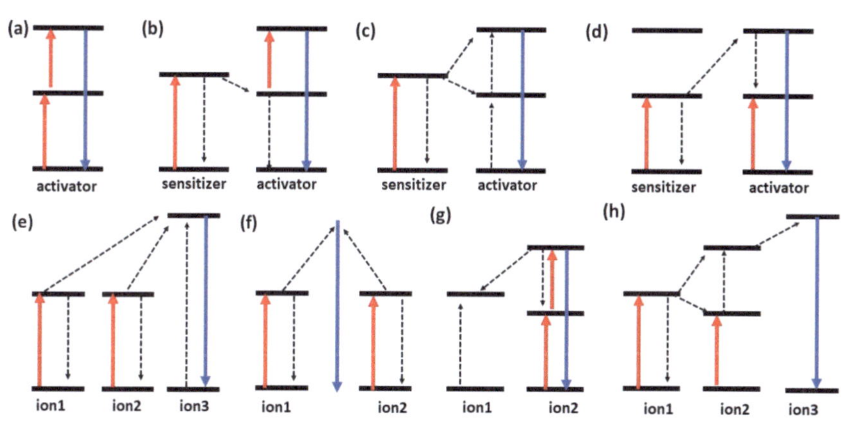

Fig. 1 **a** General energy transfer process of excited-state absorption process; **b–h** general energy transfer process of energy-transfer UC processes; **b** energy-transfer by excited-state absorption; **c** successive energy-transfers; **d** cross-relaxation UC; **e** cooperative sensitization; **f** cooperative luminescence; **g** photon avalanche; **h** energy migration UC

readers' comprehension, the remaining requirements are listed in Sect. 5. As a result of the sensitizer's ability to transmit energy to the activator prior to photon emission, the activator has a tendency to transition between the ground and excited states. Well-researched combinations of sensitizers and activators are: Er^{3+}/Yb^{3+}, Yb^{3+}/Tb^{3+}, Yb^{3+}/Eu^{3+}, Yb^{3+}/Pr^{3+}, Nd^{3+}/Yb^{3+}, Yb^{3+}/Tm^{3+}, etc. Compared to excited-state absorption, energy-transfer UC has the benefit of requiring only one pump source. One important outcome of ion–ion interaction is the cross-relaxation process, whose effectiveness is directly correlated with the dopant concentration. Although it is the primary source of the well-known "concentration quenching mechanism" of emission, it can be intentionally exploited to modify the color output in UC emission [8].

An alternative method for attaining ultraviolet–visible light emission is the triplet triplet annihilation (TTA) technique. Parker and associates published the initial report on this procedure in the 1960s. Since it has been demonstrated that TTA-based UC luminescence emission spans the UV to the near-infrared (NIR) spectrum and can even be seen with the unaided eye when excited by a commercial green laser at low levels. Too far, TTA has been used to study a wide range of organic dyes and metal complexes organic solutions, solid-state polymer matrices, and hydrophilic nanoparticles for use in bioimaging, lighting, and solar energy systems [9]. This technique has distinct benefits and drawbacks despite having comparable spectroscopic features and application fields. Here, long-wavelength excitation of the sensitizer leads to its singlet excited state; intersystem crossing then produces the triplet state. Afterwards, the sensitizer regenerates its ground state by transferring its triplet energy to the annihilator. In order to generate a high number of annihilator long-lived triplets and facilitate an efficient TTA process, this step is repeated. Ultimately, the annihilator's excited state undergoes a radiative transition process to return to the ground state. The UC emission is the result of this. The emitter is another name for the annihilator sometimes. The sensitizers and annihilators in the TTA-based UC process are molecule-based chromophores with their triplet states serving as the intermediate excited states, in contrast to the lanthanide ions utilized in the UC process. In order to sustain a high population in the intermediate excited state before the second excitation energy, it is anticipated that this metastable intermediate excited state will have a comparatively long lifetime [6, 8].

Nevertheless, there are additional processes in which one higher energy photon is emitted after two or more photons are successively absorbed and their aggregate energy is absorbed. We refer to this procedure as sequential absorption. The occurrence of actual metastable states in the material is thought to be the cause of the sequential absorption phenomenon. For shorter wavelength production, this results in successive discrete absorption and luminescence stages. Under some circumstances, these several phases result in a nonlinear dependence of emission on excitation light intensity, where the conversion efficiency increases steadily as the excitation wavelength increases. Compared to the typical UC emission process, this emission mechanism is quite simple, although it is difficult to see [8]. There are three energy levels where two photons can be absorbed; however, the second photon needs to be absorbed by the electron at the metastable level, which is why the population is typically very low. This population is influenced by variables such the metastable state's lifespan, the power of the excitation source and the absorption cross-section. By incorporating the energy transfer process from other ions, where energy transfer occurs between the sensitizer and the activator, the absorption cross-section of ions can be efficiently increased. This enables the use of lower power excitation sources by enabling the collection of numerous photons over a longer time scale.

2.3 Properties of Upconverting Rare Earth Ions

Large anti-Stokes shifts of several hundred nanometers (even $>$ 600 nm, about 2 eV), sharp and narrow emission lines, long UC luminescence lifetimes ($\sim$ms), reduced photobleaching, minimal autofluorescence background interference, low toxicity, high conversion efficiency, outstanding chemical stability, and superior photostability are just a few of the unique UC luminescence properties exhibited by RE-doped materials. Since some of the above qualities are innate, they can be used to create lasers, optical devices, anti-counterfeiting appliances, bioimaging of living cells and tiny animals, biosensors, chemosensors, and other optical disciplines. However, they can be applied to forensic investigations after being synthesized.

3 General Overview of UC Process in Forensic Investigations

The word "forensic" is derived from the Latin forum, which was the city's hub and the public forum where arguments and conflicts were discussed and resolved, ultimately establishing local law. The term "forensic science" refers to the application of several scientific fields, including chemistry, biology, pathology, anthropology, toxicology, and engineering, to legal concerns. The notion of "individualization," which entails identifying an item's unique source, classifying it or identifying it and accounting for the varied contribution of all forms of evidence, was applied by forensic science [10]. The trace is the fundamental unit of forensic science. In essence, traces are the unprocessed material found at a crime scene that is used to create forensic information or intelligence. Traces are indicators or signals for the objects or events that give rise to them, meaning that they signify more than just themselves. A trace can be fiber, saliva, semen, vaginal fluids, blood stain, fingerprint, soil, gun powder, shoe print, etc. The goals of evidence tracing are to: ensuring the integrity of evidence, preventing contamination or tampering, establishing a clear chain of custody, providing a clear and transparent record of evidence handling, supporting the admissibility of evidence in court [11].

Conventional approaches rely on chemical amplification procedures, which have limited sensitivity and can contaminate evidence. Because of their special capacity to transform low-energy radiation into higher-energy emission, UC phosphors provide a non-destructive and incredibly sensitive substitute that is useful for identifying and evaluating evidence. The capacity of UC phosphors to find minute amounts of evidence is a key benefit. They may detect even the smallest bits of evidence by boosting weak signals by transforming low-energy radiation into higher-energy emission. This is particularly important in forensic science, where every detail counts. UC phosphors are also used in forensic imaging techniques, such as UC luminescence imaging. This method uses a specialized camera to capture the upconverted emission, allowing for the visualization of evidence that would otherwise be invisible. UC process has been used to detect fingerprints, shoe prints, and other traces. Another significant application of UC phosphors in forensic science is in the detection of biological fluids. These materials can detect trace amounts of blood, saliva, and other bodily fluids, which is crucial in criminal investigations. UC phosphors can also detect drugs and explosives, making them valuable in narcotics and explosives detection. In addition to their sensitivity and versatility, UC phosphors offer several other advantages in forensic applications. The non-toxic, non-radioactive, and non-destructive nature make them safe to handle and use. They also have a long shelf life, reducing waste and costs. The upcoming sections of this chapter discuss the detail about use of UC materials for forensic investigations, their drawbacks, latest inventions, etc. [12, 13].

3.1 In Latent Fingerprint Detection

For many years and in numerous texts, the convoluted tale of the unavoidable fingerprint imprint of identity has been repeated and retold. The friction ridge skin, which leaves behind impressions of its shapes when it comes into touch with an object, is present on the palm side of each person's hands and the soles of each person's feet. Fingerprints are the imprints left by the final joint of the finger. It is an interesting fact that no two people have the same fingerprints Eccrine glands, which generate salt, a mixture of water and sweat, are typically responsible for the natural secretion that occurs on the feet and hands. When a person contacts any surface with hands, this chemical phenomenon creates a particular pattern on it [14].

The LFPs that are observed at the crime scenes are exceptional but undetectable to the naked eye, making them difficult for investigators to see. There are various methods for creating LFPs, including powder dusting, iodine fuming, cyanoacrylate fuming, ninhydrin spraying, silver nitrate soaking, optical detection, vacuum metal deposition, laser detection, etc. However, these conventional methods have a number of drawbacks, including diminished optical contrast, weak discrimination, extraordinary background intervention, non-biocompatible and toxic, which limits their application in forensic sciences. The images of FPs produced by iodine fuming quickly disappear when exposed to air. Of all the methods, the powder dusting method is one of the most popular, simple, and useful for identifying LFPs. To detect these fingerprints, a regular, metallic and luminous powder is used, which contains an adhesive resinous polymer, a contrast agent and heavy metals like gold, lead, and silver for longer shelf lives. But, in many situations, it results in poor adherence, reduced contrast and occasionally toxicity. In order to prevent misidentification, a more affordable phosphor that can provide in-depth analyses of the ridge, furrow, sweat pores, and bifurcation angles is still required. For deciphering an LFP, it is typically divided into three categories. Basic pattern recognition skills including the delta, loop, and whorl are shown at level 1 (basic level). In level 2 (macro level), small details like bifurcations, eyes, islands, lakes, hooks, short ridges, bridges, and scars are mentioned. In level 3 (micro level), ridge and edge route deviations, as well as very definite ridges comprising sweat pores, are watched as particularly crucial measurable data to classify people (Fig. 2). The advanced study of the LFP friction ridge part may be additionally sectioned into ridgeology, edgeoscopy, and poroscopy, which are the study of ridge details, contours of edge and pores, respectively. These three level tests are mostly used by forensic professionals to identify fingerprints; however, they disclose only fundamental information [15–17].

To further differentiate between two distinct fingerprints, the colorimetric images of fingerprints are viewed using level 3 ridge features, such as the width and gap

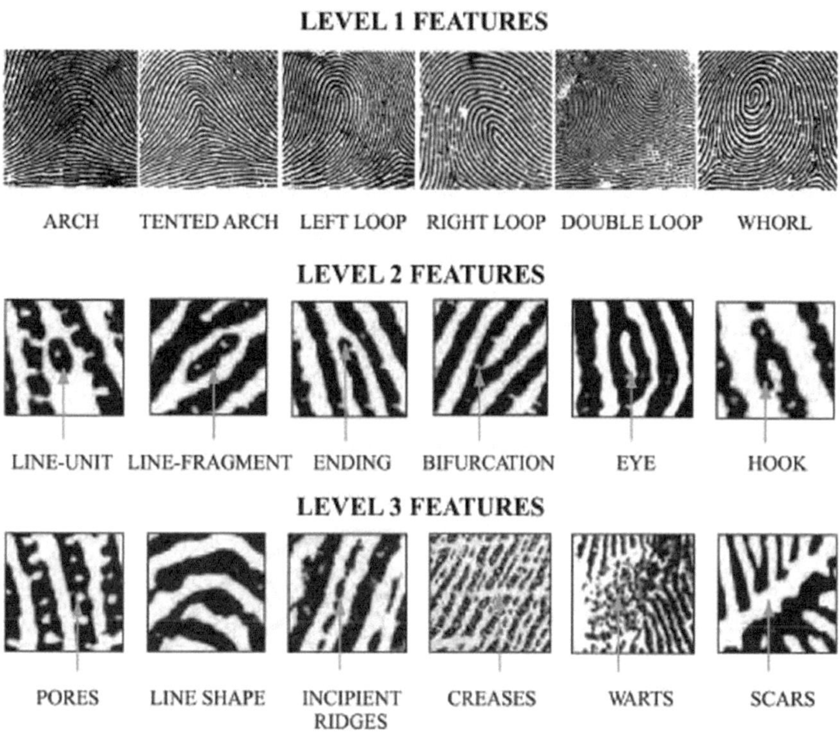

Fig. 2 General demonstration of fingerprint patterns and corresponding specific minutiae, which make the fingerprints individual to each person (Reprinted with permission [17]. Copyright 2015 Elsevier)

between ridges, the bifurcation angle and the successive distance between sweat pores, among others (Fig. 3). It is often known as the"12-point test" because sources state that the precise number of details needed for a formal identification range from 8 to 18. However, "12 points" is a widely accepted informal benchmark. However, demanding a minimum number of matching ridge characteristics to reach an identity conclusion lacks scientific support [14, 16].

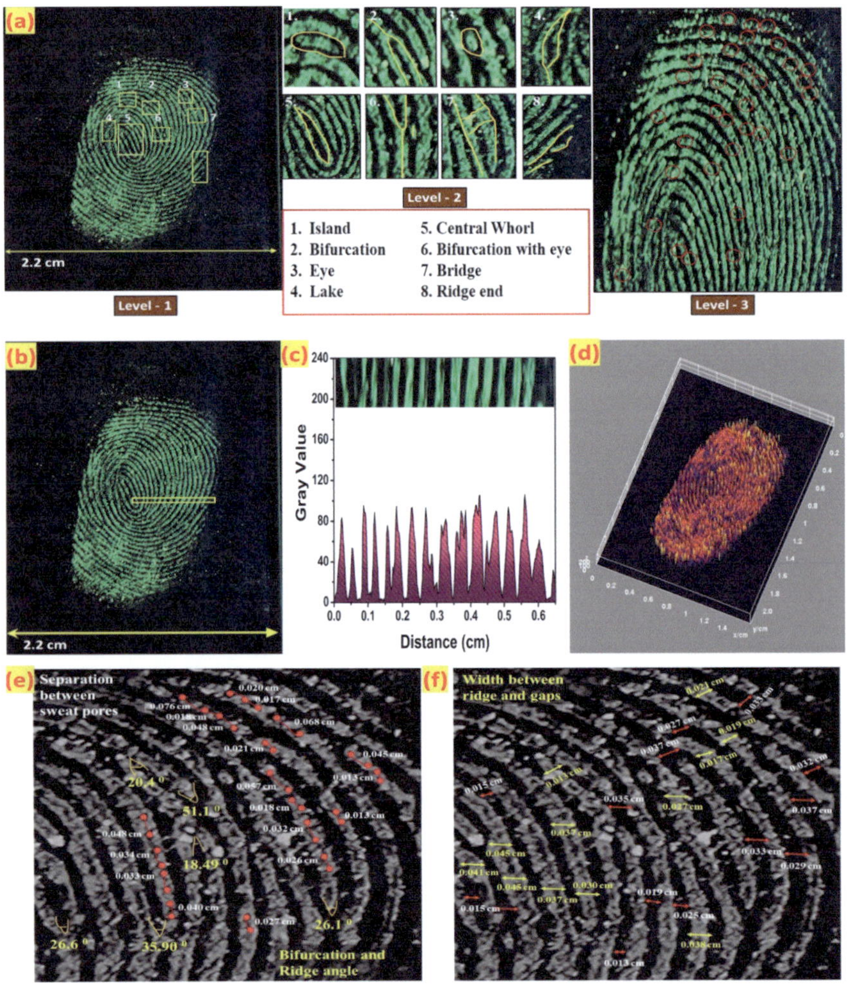

Fig. 3 **a** 3-Level testing of the LFP using Ln doped inorganic phosphor. **b** Selected region for the pixel profile plot from the RGB image of the LFP of the donor (yellow line) and **c** pixel plot profile of the selected region from the (**b**, **d**) interactive 3D plot of the LFP to see the uniformity of the spread phosphor. **e** Microscopic detailed level-3 analysis of the bifurcation angle (yellow) and sweat pore distance (red) and **f** microscopic detailed level-3 analysis of the ridge length (yellow) and furrow length (white) (Reprinted with permission [14] Copyright 2024 American Chemical Society)

Average distance between the sweat pores and ridge and furrow are recorded to be between several micrometers (100–300 μm). To study 12-point test, a phosphor morphology should be homogeneous, well distributed, and nanosized. Figure 4 demonstrates UC luminescence of some of the reported works for LFP detection under suitable excitation wavelength along with their SEM/TEM images to understand the reported morphology and particle size. Some of the UCNPs are able to detect blood stained LFP under NIR radiation via dust cleaning tools. Since hemoglobin is primarily responsible for the absorption behavior of blood, it is difficult to detect blood fingerprints using visible-light-emissive luminescent materials because of the significant quenching effect, though it is possible in NIR-irradiated UPNPs. Hemoglobin is known for its remarkable absorbance in the visible light region and low absorbance in the NIR region [12, 23]. Another benefit that UP materials provide is that NIR radiation, which is safe for human health and living tissues and has exceptional sensitivity, minimal background, and good efficacy for LFP detection, is safer for DNA in fingerprint residuals when used in fingerprint development. It also poses less of a risk to the skin and eyes of operators than UV light.

Recently numerous inorganic phosphors have been created over time by using various synthesis techniques for controlled morphology and improved detection. For a luminescent material to be utilized in forensic applications, it must maintain its emission intensity and sensitivity, even after extended storage. Hence, it must not lose its luminescence due to light exposure over time, which would compromise its usefulness in gathering and preserving evidence. Since UC materials display exceptional characteristics (Sect. 2.3), they demonstrate unique emission properties which offer noteworthy rewards for the identification of LFP images owing to their good images, good imagining, better ridge minutiae, higher sensitivity, and selectivity with improved background difference as they display contrasting visible emission on NIR excitation. Some of the noteworthy inorganic phosphors doped with UC rare earth metals are $NaGdF_4$:Yb,Er, $NaYF_4$:Yb, Er, Gd, $NaYbF_4$:Ho, LaOF:Eu/Pr,$NaGdF_4$:Nd,Yb, NaYbF4:Tm, $NaYF_4$:Yb [25], $NaYF_4$:Yb,Er (green emission 980 nm NIR excitation [20], YVO4:Yb,Er NaYF4:Yb,Er/Ce (green emission 980 nm NIR excitation [26], $YAlO_3$:Nd (violet under 365 nm excitation) [24], $YAlO_3$:Tm (indigo under 365 nm excitation) [21], Ho^{3+}/Yb^{3+}-doped KYF_4 (green emission 980 nm NIR excitation) [18], $NaYF_4$:2%Tm, 48%Yb (red or green depending on irradiation) [19] AF_3: Tm^{3+}/Yb^{3+} (A = La, Y, and Gd) (purple under 980 nm NIR excitation) [22], $CaYF_5$:Ho^{3+}/Yb^{3+} (green emission 980 nm NIR excitation) [27], NaGdF4:Yb,Er@ NaGdF4:Nd,Yb core–shell nanoparticles, $NaYbF_4$:Tm (green emission 808 nm NIR excitation [28]. Yb and Er doping combination usually give green emission but this emission intensity is also dependent on the host structure. Yb^{3+} ion is experimented to captivate the NIR excitation (usually at 980 nm) and it handover the energy to the suitable emitter ion such as trivalent Er, Tm, or Ho. Though the application of Yb^{3+} ion sensitizer for Ln-UCNPs has been partial limited to low conversion efficiency because of the thermal effect induced by a 980 nm laser and low absorption cross-section of the Yb^{3+} ion. In many papers, Nd^{3+} ion has been described as an appropriate sensitizer to intensify the UC production owing to its

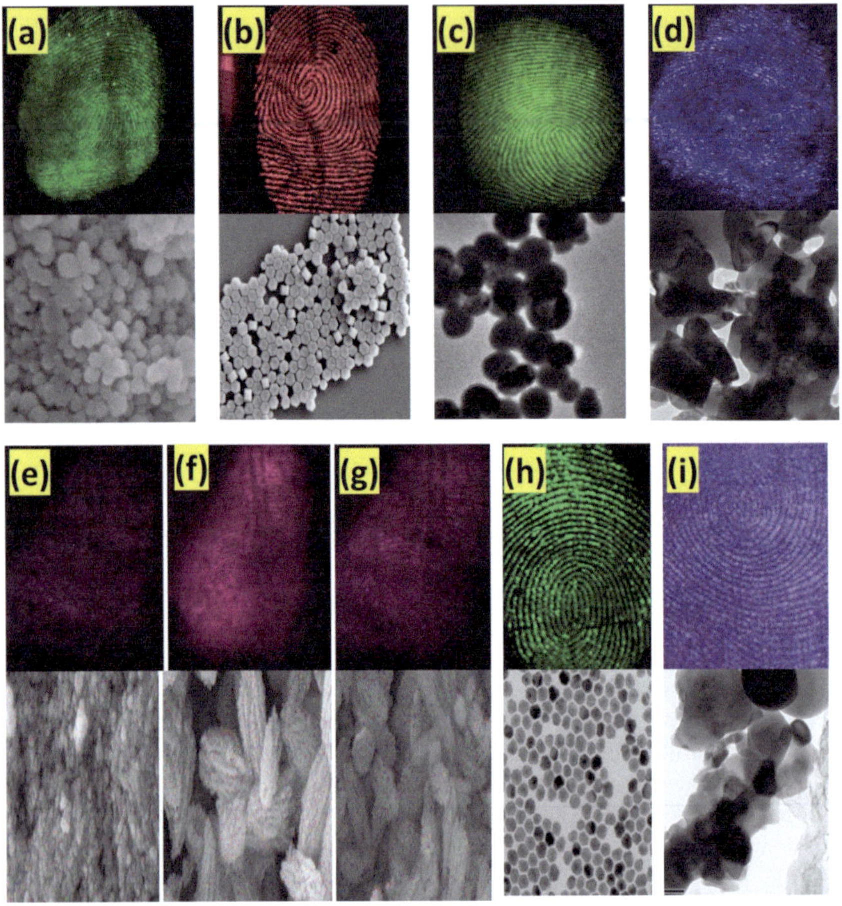

Fig. 4 Demonstration of UC-doped phosphors for LFP detection under suitable excitation wavelength along with their SEM/TEM images, **a** $KYF_4:Ho^{3+}/Yb^{3+}$ nanoparticles under 980 nm excitation, corresponding SEM image under 10 nm magnification (Reprinted with permission [18]. Copyright 2023 Elsevier), **b** β-$NaYF_4$:2%Tm, 48%Yb nanoparticles under 978 nm excitation, corresponding SEM image under 100 nm magnification (Reprinted with permission [19]. Copyright 2019 American Chemical Society), **c** $NaYF_4$:Yb,Er nanoparticles under 980 nm excitation, corresponding TEM image under 30 nm magnification (Reprinted with permission [20]. Copyright 2015 American Chemical Society), **d** $YAlO_3:Tm^{3+}$ nanoparticles under 254 nm excitation, corresponding TEM image under 200 nm magnification (Reprinted with permission [21]. Copyright 2016 Elsevier), **e**, **f**, **g** $LaF_3:Tm^{3+}/Yb^{3+}$, YF_3: Tm^{3+}/Yb^{3+} and GdF_3: Tm^{3+}/Yb^{3+} nanoparticles under 980 nm excitation, corresponding SEM image under 20 nm magnification (Reprinted with permission [22]. Copyright 2023 Elsevier), **h** $NaYbF_4$:Tm@$NaYF_4$:Yb@$NaYbF_4$:Nd nanoparticles under 808 nm excitation, corresponding TEM image under 100 nm magnification (Reprinted with permission [23]. Copyright 2016 American Chemical Society), **i** $YAlO_3:Nd^{3+}$ nanoparticles under 365 nm ex citation, corresponding TEM image under 30 nm magnification (Reprinted with permission [24]. Copyright 2016 Elsevier)

great absorption cross-section at ~ 800 nm and efficient energy transfer from Nd^{3+} to Yb^{3+} ions.

3.2 In Lip Print Detection

Along with LFP, lip print detection is another fascinating and new area of forensic science for personal identification. In forensic investigations, the authorizations of human presences based on their lip patterns is known as "cheiloscopy". Everybody has a unique lip print, which is rarely altered, much like a fingerprint. Since it may be easily lifted from cigarette butts, glassware and coffee cups used for drinking, letters, windows, and doors, it can also be used for the identification of personal traits at crime scenes like murder, rapes or robbery. Cheiloscopy gives a reliable assessment of the association between gender and different lip segments and also, it is as trustable as LFP detection in courts [29]. A lip print is classified into six zones: upper left corner, upper middle corner, upper right corner, lower left corner, lower middle corner, and lower right corner. These zones help in representing the location of lip lines with ease [30]. K. Suzuki and Y. Tsuchihashi studied and categorized lip prints into six major lip features (Fig. 5) for reference model. Differentiation leads towards features such as: a clear linear groove that runs vertically across the lips (Type I), a straight groove that is similar to type I but disappears at midway (Type I'), branched grooves or Y shaped grooves (Type II), intersecting or cross grooves (Type III), reticular grooves (Type IV) and some miscellaneous features that do not fall under any category i. e. undefined point holes (Type V) [29, 31].

The latest work was reported on $La_2Zr_2O_7$ nanoparticle doped with Er ions for lip print detection, LFP detection, and anticounterfeiting applications (Fig. 6). To illustrate the significance of cheiloscopy, Fig. 6 shows the potential lip print of two different donors (male and female) detected using powder dusting method under 375 nm illumination. Primary observation led towards the fact that both prints were quite distinguishable in the terms of shape, size, sections, length and width from each other. For first donor, the most dominating pattern was Type I (complete vertical lines). Lip prints of donor 2 was more rigorous than first one due to existence of intersecting cross patterns [32]. There are very few works that are reported on lip print detection using a UP-based materials. However, it is seen that a good adhesive material that can detect LFP could be used for it also such as Er^{3+}-doped β-Ca_2SiO_4, $La_2Zr_2O_7:Er^{3+}$, La_2O_3:Yb,Er, Gd_2O_3:Yb,Er, CuS@SiO_2@Y_2O_3:Er,Yb, $NaYbF_4$, $NaGdF_4$, YF_3, and $LiLuF_4$ doped with Yb^{3+}, Er^{3+} [25, 32].

3.3 In Anticounterfeiting

The phenomenon of counterfeiting is pervasive in all spheres of economic and social life and poses a threat to national security from the perspective of both individuals

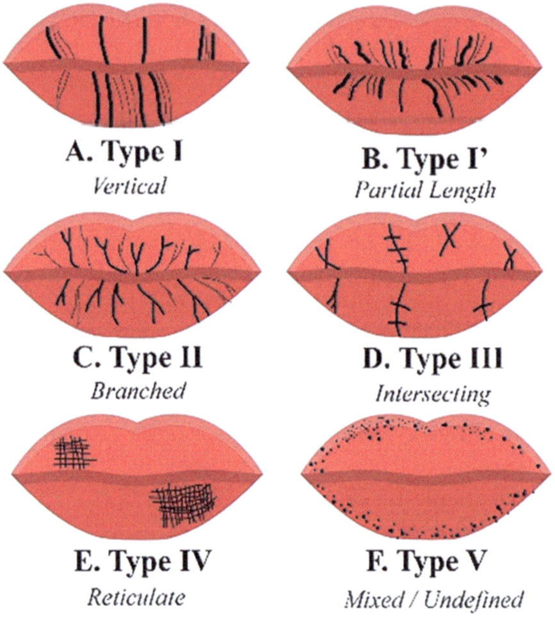

Fig. 5 Six basic types of lip prints (Reprinted with permission [14]. Copyright 2024 American Chemical Society)

and society at large. In terms of recognizing money, passports, identity cards, bank accounts, important documents, high-end goods, and branded products, it poses the greatest threat to the world economy. Anti-counterfeiting technologies, while not directly involved in traditional forensic analysis like fingerprint or DNA identification, can act as a hidden weapon in the investigator's arsenal. Criminals might attempt to mislead investigations by introducing counterfeit evidence—fake documents, IDs, or even physical objects designed to look genuine. Because of this, every nation uses the finest security measures to safeguard official papers, currency, and goods; yet, the ease of counterfeiting has been facilitated by the availability of sophisticated technologies and current, high-tech equipment. Luminescence anticounterfeiting has gained increasing attention due to the attractive properties of luminescent materials used, such as colorful and visually identifiable emissions, diverse emission lifetimes, and distinct emission modes in terms of different excitation channels. As a result, advanced anticounterfeiting strategies and technologies are in high demand nowadays. Special inks visible only under specific light, microscopic printing invisible to the naked eye or even hidden codes detectable through specialized equipment can help forensic examiners expose these forgeries. These measures go beyond simple detection. Many anti-counterfeiting strategies like track and trace systems or unique identifiers embedded in products offer a path to follow. By tracing the origin of counterfeit goods, investigators can potentially reach production facilities or distribution networks involved in the illegal activity. This information becomes crucial

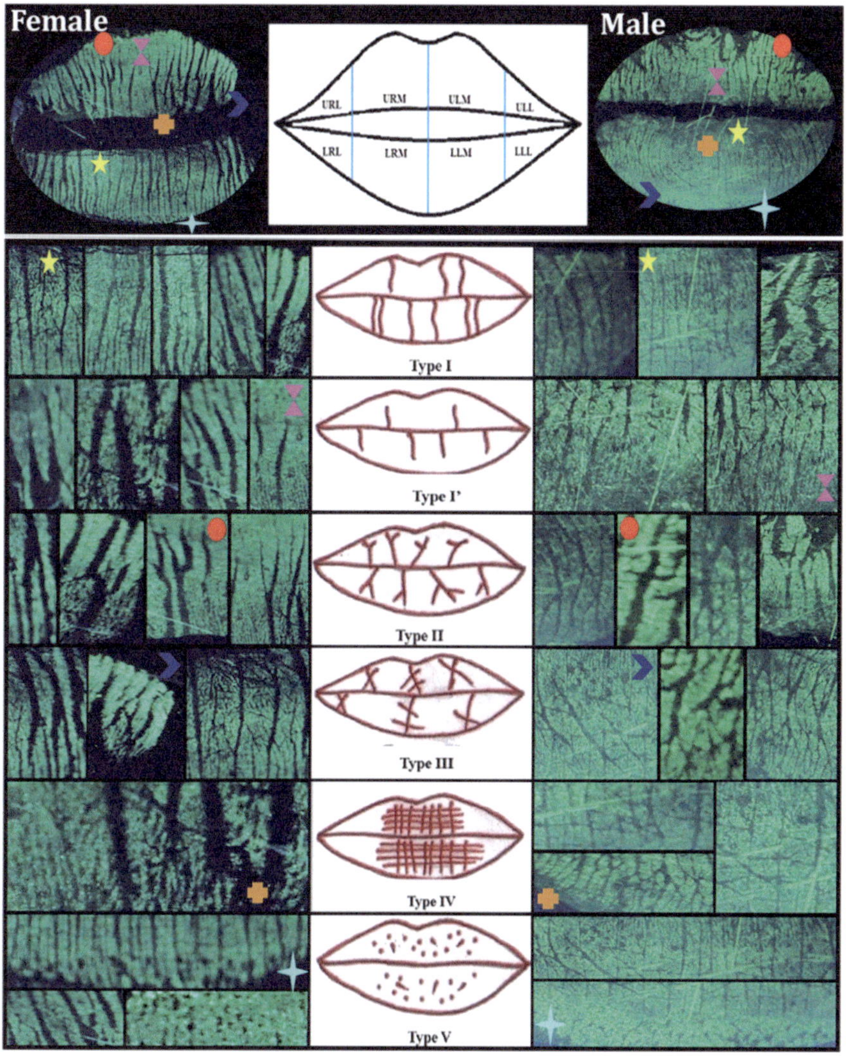

Fig. 6 Lip print obtained from $La_2Zr_2O_7$:Er nanoparticles under 375 nm excitation (Reprinted with permission [32]. Copyright 2024 American Chemical Society)

in dismantling counterfeiting operations and apprehending those responsible. Anti-counterfeiting measures like registered trademarks or patents can serve as strong evidence in court [33].

In order to prevent counterfeiting, luminescent materials that generate visible light when excited by UV or infrared radiation have recently been employed as security labels, holograms, basic luminous markers, and invisible ink for printing barcodes or security codes. UV-to Vis fluorescent security inks are generally used in conjunction with various printing techniques, such as stamp, screen and inkjet printing, to create highly intricate and high-resolution luminescence patterns for anti-counterfeiting applications. These techniques are used to identify currency, identity cards, bank accounts, valuable documents, high-value merchandise, and branded products. But advancement in technologies and information have made it is easy to mimic the specific color of any RE doped materials. For example, if any ink displays red color under UV irradiation due to its specific dopant, one fraud can easily mimic any material that has potential to display that special color under UV light. As a result, unique materials, nanoparticles, and phosphors are being created nowadays that can be used in a variety of anticounterfeiting schemes that include multichannel encoding, direction-based irradiation, time-resolution displays and multi-stimulus responsiveness. Today, scientists can create UC-based anti-counterfeit labels that remain hidden when installed on smooth surfaces and prevent information from being seen from any viewpoint, even when exposed to UV or W-LED light. Due to the persistence of the RE elements, even after a lengthy irradiation, the label only shows minute features, making the decryption still incomplete. The actual data can now be decrypted by a second NIR laser that is continuously scanning the preirradiated film. The emission color shifts throughout scanning, going from one color to another and so on, delivering precise information piece by piece [33–35]. Some of the reported blended phosphors are mixtures of $NaYF_4$:Eu@$NaYF_4$:Yb,Er nanophosphors in a polyvinyl alcohol (PVA) solution, which emit strong red ($NaYF_4$:Eu) and green ($NaYF_4$:Yb, Er) colors at irradiation wavelengths of 254 nm and 980 nm, respectively [13].

There are a lot of outstanding UC-based materials that are available and largely reported but the problem of dissolving those materials in any organic solvent is tough. Most of the inorganic host do not get soluble in any organic/inorganic solvents, they get precipitated, display unprofessional whiteness, uneven distribution on surface, non-adhesive nature, damaging nature (damaging the surface or printer nozzle), and diminished intensity on solution form remains. In addition to luminescent materials, selecting the right ink medium is crucial to creating luminescent ink. This is because the ink must have the right viscosity, which is necessary for the ink to be sticky enough to print on a variety of surfaces and the luminescent nanomaterials must be evenly distributed to form colloidal suspensions without clumping or sinking to the bottom [36]. The nanoparticles (NP) based on UC can be used to tackle problems associated to phosphor powder. Because of their tiny size, excellent colloidal stabilities, and strong luminescence, UC-based NPs are advantageous for mitigating the interference of background fluorescence in information encryption. They can also be put into an inkjet printer. They can therefore be swiftly loaded into expert inkjet printers to generate multilevel printing in large quantities. In particular, the UCNPs with variable irradiation or emission wavelengths can be easily created by modifying the host matrix and adding ions [13]. These UCNPs can be used to create security patterns, pictures, bar codes and luminous sheets and stickers that can be printed on critical documents that are authentic and branded goods. Among the instances are: Ho^{3+}/Yb^{3+} doped Hexagonal host, Yb^{3+}/Er^{3+} doped KY_3F_{10}, $NaLaF_4$:Pr^{3+},Yb^{3+} and Pr^{3+}-doped YF_3, $NaYF_4$:Er,Yb and $NaYF_4$:Tm,Yb, β-$NaYF_4$:Er,Yb [36, 37]. Along with it some of the most utilized solvents are: Xylene, cyclohexanone, methyl benzoate, polymethyl methacrylate (PMMA), toluene, polyvinyl alcohol (PVA), aqueous solution of sodium hexametaphosphate (SHMP), polyvinyl chloride (PVC), etc. [36].

Another progressive choice of creating security is multilayer covers/fibers. For instance, when excited at 980 nm, Er^{3+} doped NPs show NIR-Vis UC luminescence, whereas Tm^{3+}-doped NPs show NIR-to-NIR UC luminescence at precisely the same wavelength. These kinds of multilayer films can be used for color-coded sequencing, where different hues of UCL are produced by the distinct composition of each surface. While the NIR light does not activate the other chemicals in the paper [13]. It can be utilized to make security barcodes. N. Jaroch and coworkers synthesized and patented $GdVO_4$: 20% Yb, 0.5% Tm, 5% Eu material for fiber processing for application against counterfeiting. The color and intensity of the resulting luminescence are independent of the medium, but they are reliant on the excitation sources [38]. The study of "nano barcodes" is the upcoming treasure. They are helpful in healthcare settings for quick, easy and precise target identification in addition to other things from grocery stores, stationery stores, packages, and letters. These barcodes are designed to be used in a variety of applications, from bio-detection to anti-counterfeiting and feature much lower scales, with micron- and nano-sizes. Unlike the parallel line bars of standard macroscopic barcoding, micron- and nano-sized structures can be encoded with different elements to generate micro- and nano-barcodes with varied features including different sizes, forms, colors, etc., so extending beyond the bars as codes. For some applications, such cell and biomolecule labeling and imaging, the size of micro-barcodes becomes a constraint. In this sense, a variety of uses for nano-barcodes have been reported, including theragnostic, imaging, security, medication delivery and the detection of biomolecules (nucleic acids and proteins) (Fig. 7) [11, 34, 35].

The use of color-tunable UP materials on heat variation is another extensively studied strategy. More authentication features can be provided by temperature-responsive color changes and different emission colors in hybrid materials, which opens up the possibility of creating anticounterfeiting patterns that are more secure. It will not be as practicable to use the heating procedure to cause color variations in the emission. Alternatively, the intrinsic photothermal conversion of UC-based NP combined with laser irradiation can accomplish the color shifts. One reported grouping, consisting of $NaYF_4$:Yb/Tm and $NaGdF_4$:Yb/Er@$NaYbF_4$ NCs, displays color changes from blue to green with growing temperature. With temperature increasing from 30 to 150^{0}C, the temperature-affected multicolor productions are experimented. As well as, $NaYF_4$:Yb/Ce/Ho + $NaGdF_4$:Yb/Er@$NaYbF_4$ NCs and hybrid $NaYF_4$:Yb/Er MRs + $NaGdF_4$:Yb/Ce/Ho@$NaYbF_4$ NCs display color shifts from red → green and green → red respectively. Also $NaGdF_4$:Yb/Er@$NaGdF_4$:xYb ($x = 5\%, 20\%$) core/shell NPs shows variation on Yb concentration changes [34, 35, 39].

Another most misleading thing that happens in the field of document misleading is the use of false signature or copying the signature of any other individuals. In the beginning Albert S. Osborn published a book "Questioned Documents" containing the straightforward values that text inspectors still use. Let us take an example, if

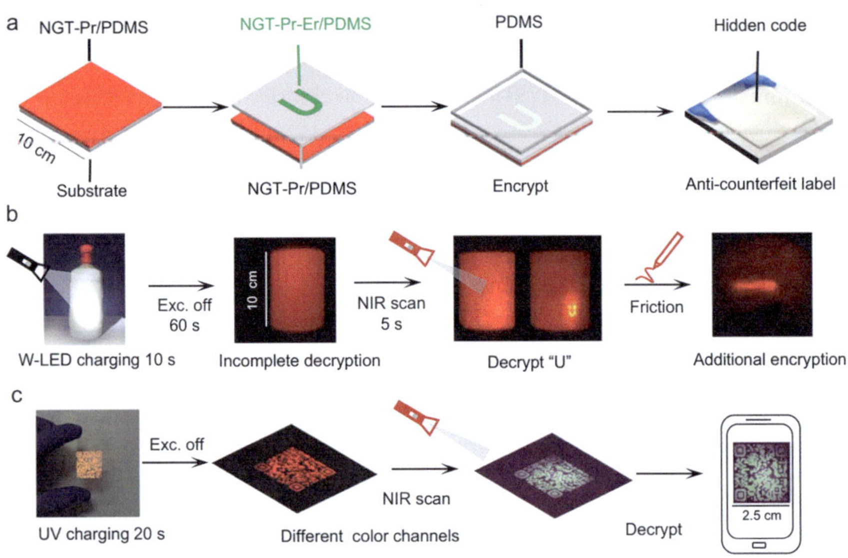

Fig. 7 Universal design of the concealed two-dimension patterns completed with $NaGdTi_2O_6$ (NGT) phosphor doped with Pr^{3+}, Er^{3+} of diverse composition. **a** The three-layered design was put on a glass substrate (10 × 10 cm). The as ready multimode luminescent particles and PDMS were mixed and used for different functions: (from left to right) NGT:0.1%Pr^{3+}@PDMS as background layer, NGT:0.1%Pr^{3+}, 0.1%Er^{3+} @ PDMS as the info layer, see-through PDMS as smoothing surface. The anti-counterfeit tag with concealed information "U" (1.5 × 0.8 cm) was lastly fabricated. **b** The as-prepared elastic anti-counterfeit label is functional to a bottle. Emission outlines of the patterns chronicled under dissimilar irradiation or incentives conditions. The power density of the laser used in NIR scan is ca. 0.82 W/cm^2. **c** Anti-counterfeiting barcode invented via screen printing. Barcode (made with NGT: 0.1%Pr^{3+}, 0.5%Er^{3+}@PDMS) photographs under and after 365 nm UV excitation as well as 980 nm NIR laser photo-stimulation, showing various colors constructing different coding channels (Reprinted with permission [33]. Copyright 2022 Nature Communication)

I am signing my documents, obviously the pressure by my pen, steps during my signatures and flow of alphabets while writing would have a specific style. But if any other individual tries to copy my signature style, he would definitely lack some minute details. The ML-based sensitive sensors can provide 3D analysis of pressure inserted by a writer to distinguish the writing styles. It is commonly recognized that the most straightforward application of ML phosphors is in stress or pressure sensing. The combination of elastomers and ML phosphor allows the stress distribution to be observed in real time. One of the main uses for the stress sensing technology is the gathering of electronic signature recordings or handwriting analysis (Fig. 8). While it is feasible to deceitfully replicate someone else's signature or handwriting, no human is in the habit of writing with the same amount of pressure on every curve. When the ML-based system precisely records both the signature graphics and the signing habits aligned with digital cameras and other sensors, it generates a 3D image of pressure applied to each character by the signees, signing speed, and force, with high

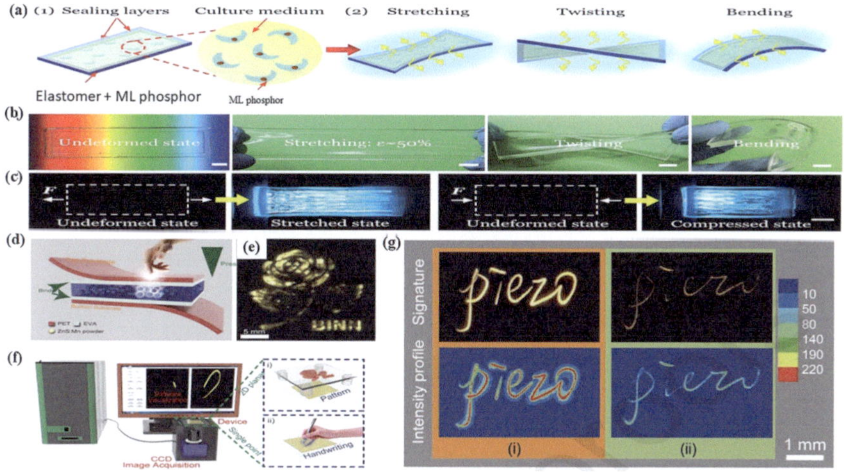

Fig. 8 **a** The usual synthesis method of a mechanoluminescent sensor, **b**, **c** The stretchable transparent film under normal condition and under after long, **d** An example of how touch sensors are used for stress sensing **e** A rose pattern made by stamping pressure onto a tool, The two-dimensional planar and single-point dynamic pressure image acquisition and processing system's schematic representation is shown in (**f**). The signing habits and intensity profile of the various signees are shown in (**g**) (Reprinted with permission from [40]. Copyright 2024 Elsevier)

resolution and quick response time. ML-based sensing systems require less device association and are sufficiently flexible to be adaptive for stress detection, in contrast to platforms based on electric or fiber-type strain gauges [40, 41]. Some machine learning materials are so good that they can be used to hide messages or fingerprints that are hidden under normal, stress-free settings. Then, using ultrasonic waves, the materials may be decoded into readable form. Furthermore, in reaction to a range of stimuli, such as heat, light, and stress, they are capable of emitting multiple patterns of light. These qualities make it promise for state-of-the-art optical anticounterfeiting [10].

3.4 In Encryption Restoring

In addition to stress and damage sensing for document falsification verification, machine learning approaches are also showing promise for recovering the intended nano-barcodes. The capacity of machine learning (ML)-based sensors (barcodes) to exhibit persistent luminescence (PersL), which can be reactivated by friction or heat, gives additional characteristics for the creation of photonic barcodes for information encryption [33]. ML offers several intriguing possibilities for forensic investigations [40, 42]:

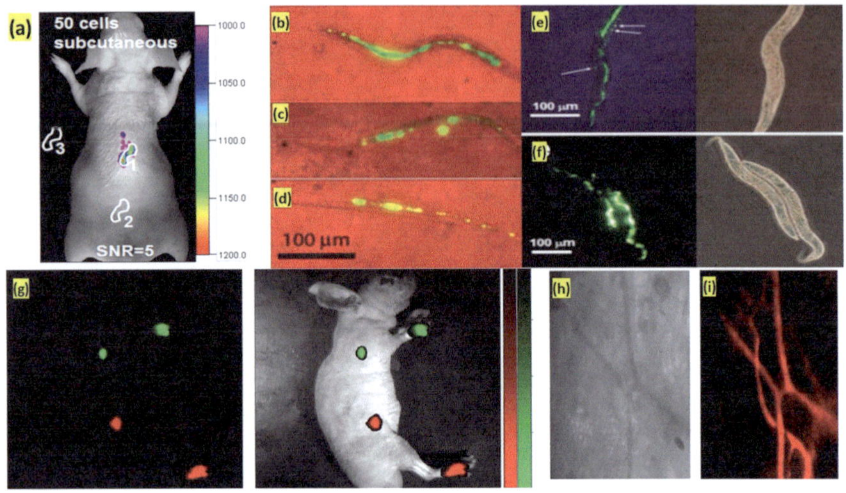

Fig. 9 **a** In vivo UC bioimaging of mice after subcutaneous injection 50 lanthanide $NaLu(Gd)F_4$:Yb,Tm-treated KB cells, **b** – **d** Two-photon images of *C. elegans* worms under 980 nm excitation. Here, the red color is representing the bright field and green color is for the phosphor emission under excitation. **e**, **f** UC images of *C. elegans* worms incubated with 5 mg mL^{-1} of PEI-modified $NaYF_4$:Yb,Er nanoparticles for 6 h and 20 h respectively, **g** In vivo TTA-based UC lymphatic imaging after 30 min of postinjection of 20 μL of UCNP under excitation wavelength 635 nm, **h** Photoluminescent imaging of blood vessels in the mouse ear following tail vein injection of $NaYF_4$:Yb,Er nanoparticles (10 mg) imaged with a blue light filter, **i** UC image with excitation at 980 nm (Reprinted with permission [7]. Copyright 2014 American Chemical Society)

- **Hidden Trace Detection**: Imagine a scenario where a thief attempts to open a locked safe using a knife. Traditional methods like fingerprint dusting might prove ineffective due to the smooth metal surface. However, ML can potentially reveal the hidden imprint of the knife. The act of damaging lock could leave faint traces on the safe that, under controlled mechanical stimulation, might emit a faint glow. This ML emission, captured by a specialized camera, could reveal the location and shape of the knife, aiding investigators in identifying the type of tool used.
- **Time Since Last Activity**: Studies suggest that the intensity of ML emission might diminish over time. This phenomenon holds the potential for estimating the timeframe since a mechanical interaction occurred. For instance, a shoeprint left at a crime scene might exhibit a stronger ML signal compared to a shoeprint left days or weeks earlier. Although further research is needed to refine this application, the ability to estimate the timeframe of activity based on ML emission intensity could be invaluable in reconstructing the timeline of events at a crime scene. This kind of studies might be helpful in the hit-and-run accident where the driver fled the scene, investigators might find faint tire marks on the road. Traditional methods might not yield much information about the time of the accident.
- **Material Identification**: The spectral signature of the emitted light in ML can be characteristic of the material generating it. This fingerprint-like quality could

be used to identify the origin of unknown materials found at a crime scene. For example, fragments of glass left at the scene of a hit-and-run accident might emit a specific ML signature that can be compared to a database of glass compositions from different car manufacturers. This comparison could potentially identify the make and model of the vehicle involved in the accident.

3.5 In Bioimaging

Another forensic science technology that deals with the real-time viewing of biological processes is called bioimaging. This process involves the imaging techniques to study anatomical structures with applied electrical and magnetic fields, moving functional materials, mechanical movements, etc. in forensic science it could provide some vital minute details regarding cell damages, internal injuries, hidden diseases, pathological damages. To resolve criminal and civil matters, modern forensic investigation makes use of cutting-edge technologies and innovative instruments. As an addition to or substitute for the conventional invasive autopsy, forensic bio-imaging makes use of X-ray, computed tomography (CT), magnetic resonance imaging (MRI), multi-slice CT, enhanced minimally invasive techniques via CT and MRI, such as angiography and biopsy and three-dimensional (3D) surface scanning. Since radiation exposure can be disregarded for postmortem evaluation, a high-resolution CT scan can be utilized to produce the highest quality images in forensic imaging. Since traditional autopsies have the potential to disseminate infectious illnesses, postmortem imaging is a preferable substitute. In order to precisely identify traumas and pathological changes in the corpse, it is frequently employed in pathology in place of a typical autopsy when the case is delicate or a traditional autopsy is refused by family members for other reasons, such as religious conviction. Since forensic imaging has been employed in forensic investigations, its primary focus has been on forensic pathology research and applications [11, 43]. There are benefits and drawbacks to each traditional diagnostic imaging method, such as MRI, CT, nuclear and optical imaging. One benefit of both MRI and CT is that they are noninvasive methods for 3D tomography and in vivo imaging. In vivo, nuclear imaging modalities such as PET and SPECT show exceptionally high sensitivity. But MRI's low sensitivity places limitations on it. Additionally, individuals undergoing nuclear imaging and CT scans are at risk for radioactivity and X-ray radiation, respectively. UC imaging has a relatively modest tissue penetration depth ($\sim$cm), which is a downside despite its comparatively strong subcellular sensitivity. When it comes to providing more comprehensive or accurate information, multimodality imaging can make up for the shortcomings of single imaging modalities. Several studies have detailed the integration of UC imaging with traditional diagnostic imaging methods to create such multifunctional imaging systems [7].

Aspects of UC luminescence microscopy, in vivo imaging, and three-dimensional (3D) imaging have all been successfully developed so far for the purpose of seeing living cells and small animals by making use of the unique UC process. While they use different excitation sources and mirror arrangements, the laser scanning UC luminescence confocal microscope (LSUCLM), small-animal UC luminescence imaging system and TTA-based microscopy UC system all operate on a similar premise. TTA-based UC bioimaging often requires lower laser excitation power than LSUCLM. For TTA-based UC systems, the small-animal UC luminescence imaging system configuration can also be changed by substituting CW lasers at 543 or 633 nm for the 980 nm laser as the excitation source. Currently, a few commercial devices with comparable modifications are offered for UC luminescence imaging in vivo. The ability of a small-animal UC luminescence imaging system to function efficiently in the presence of ambient illumination light is crucial for future uses like image-guided surgery. By obtaining the border fluence utilizing several source-detector pairs, Fluorescence Diffuse Optical Tomography (FDOT) is a 3D imaging technique that is used to rebuild the internal fluorophore distribution inside a highly scattering material. A recently designed optical tomography system uses CCD cameras to record a picture for each scanned position under excitation at 980 nm in order to reconstruct the UC emission image. Since the ultralow autofluorescence background produces an exceptional UC signal-to-noise ratio (SNR), the tomographic reconstruction images from lanthanide UCNPs are of higher quality than those from ordinary organic dyes [11, 44].

It is commonly recognized that the bio-safety and clearance pathway of inorganic nanoparticles from the body provide a significant hurdle and cause the most concern when it comes to their therapeutic application. Generally speaking, nanoparticles (NPs) must have a hydrodynamic size of less than 10 nm. This is because nanoparticles are easier to enter intracellular spaces, undergo rapid kidney filtration and cause less toxicity. Additionally, nanoparticle stability in a physiological environment can be extended through appropriate surface modification and design. The last ten years have seen a rapid growth of UC activated NPs for biological applications due to their distinct properties. These include achieving multifunctionality with ease and obtaining biocompatibility through varied surface modification, along with producing adequate signal output. Because of these characteristics, UC-doped NPs are a perfect tool for next-generation medicines and sophisticated bioimaging. Among the benefits for the bio labeling applications are a minimal auto-fluorescence background to enhance the signal-to-noise ratio. Moreover, UCNPs exhibit non-blinking behavior, reduced light scattering, and enhanced light penetration depth into tissues as a result of stimulating the near-infrared region situated within the optically transparent window. Because of their anti-Stokes shift property, UCNPs are able to effectively convert non-visible infrared light (NIR) instead of ultraviolet radiation, resulting in significantly less photodamage to biological samples and a significant increase in the excitation light's penetration depth. These outstanding properties make UCNPs of tremendous potential for sensing and imaging in biomedical applications. Moreover, UCNPs may be driven by low-power NIR lasers, generating emissions under moderate light intensities, which is usually a basic prerequisite in biological inquiries [45, 46].

Recently, a high spatial resolution and great sensitivity type of UC luminescent NP has been created. Additionally, luminous nanoparticles (NPs), peptides, antibodies, and small molecules can be conjugated with specific targeting ligands to enable high-sensitivity and selectivity NPs to be employed as probes to identify and target certain cells of interest. These days, therapeutic medicines and genes, including photosensitizers, RNA, DNA, and doxorubicin, can also be incorporated into these luminous nanoparticles for therapies guided by optical imaging [45]. It has been claimed that UCNPs coated with a variety of ligands, including silica, small compounds and polymers, are helpful for bioimaging living cells in order to control their propensity to interact with water (Sect. 8). Up till now, various small molecules have been employed, such as glutamic acid, diethylene triamine pentacetic acid, citrate, β-cyclodextrin, α-cyclodextrin, 6-aminohexanoic acid, azelaic acid and 3-mercaptopropionic acid. 3-mercaptopropionic acid-functionalized $NaYF_4$:Yb,Tm with 980 to 800 nm UC emission can be castoff to label cells. Amphiphilic LaF_3:Yb,Ho NPs bearing polyethylene glycol monomethyl ether can willingly infuse through cell membranes. Silica coatings deliver a harmless barrier, with the consequence that SiO_2- coated UCNPs can be used for imaging live cells. UCNPs coated with polymers can also be rummage-sale as cell imaging labels. Tuning the external charge is predicted to change the acceptance of lanthanide UCNPs by existing cells significantly. UC luminescence bioimaging of living cells can be understood even at the single-particle level. For multiple studies $NaYF_4$:Yb,Er nanocrystals (~ 27 nm to 15 nm), $NaYF_4$:Yb,Tm hexa-plates (~ 180 nm), $NaYF_4$:Yb,Er@SiO_2 (~ 100 nm), Yb^{3+}, Tm^{3+} and Sm^{3+} co-doped $NaLuF_4$, citrate-modified $NaYF_4$:Gd,Yb,Er, α-CD-modified NaYF4:Yb,Tm and PAA-modified $NaYF_4$:Yb,Er NPs (~ 100 nm) have also been reported. In order to obtain in vivo anatomical and physiological details, UC bioimaging has been successfully investigated for different small animals [7, 47].

3.6 In the Identification of Biological Fluids

We are interested in technologies like DNA analysis because of how distinct they are and how certain it is that a person can be identified. In most cases, a clinical serologist works with fresh bodily fluid samples. Other bodily fluids, such as saliva, semen, and urine, are sampled in stain form and frequently deteriorated or damaged, which makes accurate analysis more challenging. The forensic analyst has no control over a variety of factors that may affect the amount, state of degradation, and condition of evidential stains. A stain's biochemical information can be largely destroyed by uncontrolled exposure to heat and humidity, as this accelerates the breakdown of chemical compounds that are significant to the analyzer. This may happen prior to or following a laboratory receiving the evidence. The quantity, location, quality, and adhesiveness of the samples at a crime scene determine how the crime was committed [10, 11].

Sexual assault crimes entail physical contact between the offender and the victim, which leads to the transfer of materials from one person to another, including hairs, fibers, and bodily fluids, especially seminal fluid and saliva. Many times, the victim does not know who the attacker is and witnesses are not present when the assault occurs. With today's laboratory techniques, forensic experts can analyze DNA and find and identify minuscule amounts of bodily fluids. When it comes to incidents of sexual assault, the detection of saliva on human skins, glassware, stamps and envelopes, and cigarette butts can help identify the offender through DNA analysis. Because of the insensitivity of the tests and the poor success rate of DNA profiling, forensic laboratories do less chemical detection of urine, even though it may be relevant in situations of sexual assault, harassment, and mischief. Urine detection, like many other analyses, depends on visually examining stains for distinctive appearances. While stains can shine in different light sources, diluted stains are more difficult to find. Another feature of urine is odor. A forensic investigation begins with a visual inspection of the goods under a bright light source to look for any obvious stains. A UC-based laser with the required wavelength of emission is employed for this. However, long-wave (320–400 nm) ultraviolet lights are the most popular and commonly utilized wavelength for detecting blood, saliva, and semen—even at diluted quantities. These UV rays are extremely dangerous to people because they can burn exposed skin, blur vision without protective goggles, and damage DNA by forming thymine–thymine dimers. In reality, in laboratories, UV light is frequently employed as a sterilization/decontamination technique to inactivate extraneous DNA. It is evident that UV light is not always the ideal option for a backup light source with the development of modern DNA technology. The other options include high-output UC laser light sources combined with filters that efficiently reduce the generated light to a single wavelength, much as longwave UV light [10, 11].

Blood is one bodily fluid that is examined and analyzed in serology. Blood is tested in a medical context to determine a patient's condition. Blood analysis is used in forensics to identify the source of blood on an item of evidence or at a crime scene. Bloodstains' direction can be used to determine the sort of crime, the weapon used, the perpetrators' hand movements and the location of the incident. In many scenarios, blood is presumed to be there based on witness reports or predicted in a certain area; nevertheless, minimal blood is visible in typical lighting conditions. Typical instances could be a washed-out splatter pattern on a wall or a drag pattern across a floor that was cleaned or washed away. Latent fingerprints in blood or on certain unusual types of surfaces require special techniques, enhancement modalities, or approaches [12]. By definition, bloodstain fingerprints are not latent. Blood appears to have some sort of pattern and while investigators may determine that the pattern is most likely a fingerprint (or palm or foot print), the ridge characteristics are not distinct enough in the print to allow for comparison. Using a bloody fingerprint enhancement reagent could be an option in these situations. Many people view crystal testing as confirmatory of presumptive tests. Hemoglobin's nonprotein heme group is used in these tests. Blood is detected by its well-coordinated structure, whether the blood is moist or dry. Using luminol and fluorescein spray, even a minute amount of blood stain can be found when the offender attempts to remove it. Stain patterns will typically diffuse as a result of excessive spraying. Fluorescein was first identified as a useful tool for blood detection in 1910. Prepared for usage similarly to phenolphthalein is fluorescein. Applying fluorescein to the probable bloodstained region involves reducing fluorescein in an alkaline solution over zinc. The heme's catalytic activity speeds up the process of hydrogen peroxide's oxidation of fluorescin to fluorescein, which fluoresces when exposed to ultraviolet light. Although luminol and fluorescein both generate light to signal the presence of blood, they differ in practical use [11]:

1. Fluorescein and luminol both produce light, although they do so in distinct ways. It should be recognized that the chemical processes behind each of these tests produce two completely distinct phenomena: fluorescence and chemiluminescence. The mechanism by which light is released as a byproduct of a chemical reaction is known as chemiluminescence. The reaction can occur without the need for extra light. Chemiluminescence is necessary for luminol. When a chemical substance is exposed to a specific wavelength of light—typically a short wavelength, like the ultraviolet—it fluoresces, releasing light energy at longer wavelengths. This irradiation—usually by UV light in the 425–485 nm range—produces the light that fluorescein emits.
2. To efficiently cause the combination to stick to a surface containing probable bloodstain residues, an aqueous solution of luminol reactants is sprayed on it. The fluorescein system comprises a commercial thickener. Fluorescein works better on vertical surfaces as a result. After the mixture is sprayed, a different light source that is usually tuned at 450 nm is needed to visualize the fluorescence.

Numerous metabolic activities that are highly particular to the kind of substrate and its catalytic function require the presence of enzymes. Enzyme dynamics in biological systems may be better tracked and revealed by employing activatable probes in in situ real-time fluorescence imaging experiments. Immunodetection is another program that is frequently utilized in forensic investigations. It explains how structural and physico-chemical interactions enable antibodies to specifically recognize molecular targets. The molecular makeup of a donor's sweat, such as its proteins, lipids, and amino acids, along with other molecular components, can now be detected as possible antigens in a forensic context thanks to the use of cutting-edge procedures and tools. By focusing on a small range of substances found in some donors' perspiration or impurities connected to odd behaviors (such as handling chemicals, explosives, or narcotics), the number of marks found will be restricted to a small number of donors. The only method used on any item would be the combination of chemical imaging and nanoparticles for antigen detection. Chemical imaging has seen a transition from detection to security/drug testing, much like the usage of nanoparticles [11, 48]. $NaGdF_4$: Yb, Er/AuNPs@Pt satellite assemblies, $NaYF_4$: Yb, Er-DNA H_1, $NaYF_4$: Yb, Er-NH_2/$NaYF_4$: Yb, Er-COOH/ dye UC, $NaGdF_4$: Yb, Er@NaYF4-DNA/AuNPs, $NaYF_4$: Yb, Tm, Er-Ti_3C_2, CaF_2: Yb, Ho@MSNs@SiO_2-ssDNA/ Polyurethane fibers@GO, FexCuySe@$NaYF_4$: Yb^{3+}, Tm^{3+} are used in fluorescence imaging and MRI for the detection of mRNAs and miRNAs [25].

3.7 In Forensic Odontology

Sometimes at crime scenes, criminals attempt to harm the victim by biting him or her. The human body, chewable foods, ropes, etc., can all leave these kinds of scars. The study of a deceased person's dental data in order to identify the unidentified person is known as forensic odontology. Forensic odontology relies heavily on dental radiology. The use of teeth and other oral structures for identification in a legal setting is the main focus of this study. A range of forensic odontology techniques are employed to assist in the identification of human remains in various occurrences, including but not limited to terrorist attacks, road, train, and airline accidents, fires, mass murders, and natural catastrophes like floods, earthquakes, and tsunamis. The body's toughest and most protective tissues are those found in the teeth. These buildings are among the last to crumble after death because they are resistant to both high temperatures and decomposition. The main source of dental identity is the individuality of each person's teeth and the fact that no two mouth cavities are same. For identification, the ante-mortem records are compared with the dental evidence of the dead that was found at the crime site. Every person has a distinct identity based on their dental characteristics, which include tooth morphology, differences in shape and size, restorations, pathologies, missing teeth, wear patterns, crowding of the teeth, color, and position of the tooth, rotations and other unusual dental anomalies. The teeth can be used to determine an individual's age, sex, race or ethnicity, habits, occupation, and other details that may provide further hints about who they are, even in the absence of ante-mortem dental records for comparison. Even though determining a person's sex from their teeth is not definitive, teeth can provide an indication if there is no other supporting information. The forensic scientist can further corroborate and validate the results with other facts and data [49, 50]. The ability to see human dentition for forensic reasons can be achieved by the effective use of ceramic-based modalities, pressure-sensitive analysis and 3D volumetric analysis. Dental scan records are frequently available at forensic labs and they can be a useful extra resource for identifying a person's teeth. Human dental anatomy and chewing patterns can be determined with the use of UC-based mechanoluminescent materials. A remarkable phenomenon known as mechanoluminescence occurs when a substance responds to mechanical stress by emitting light. Some helpful ML resources based in UC are: Pr^{3+}/Er^{3+} coactivated $Ba_2Ga_2GeO_7$, $NaYF_4:Pr^{3+}$, $KAlF_4$:Yb/Er, Er-Doped CaZnOS, $Sr_3Sn_2O_7$: Nd, Yb, and $NaGdTi_2O_6$:Pr,Er [40].

3.8 In Forensic Ballistics

Rather of destroying the original evidence through surgery, the non-destructive collection of forensic imaging findings might disclose the original injuries and pathological abnormalities. Forensic imaging is undoubtedly a great way to look at the initial trajectory rather than destructively dissect the courses. Forensic ballistics, for instance, requires knowing the bullet path in victims. The primary applications of forensic imaging in forensic ballistics include the examination of gunshot residues, shooting distance, the placement of foreign bodies and paths in both the living and the dead. Numerous pathologists and forensic specialists have been studying gunshot wounds for forensic assessment purposes since the year 2000. This research has the potential to provide information on bullet size, caliber, shot distance and angle, firing ranges and other relevant factors. A laser is typically used to examine evidence linked to ballistics. Light Amplification by Stimulated Emission of Radiation is what laser stands for. It is a gadget that emits a very focused, bright beam of light. This little beam can cover a large distance to assess the trajectory and angle at which a criminal is firing a bullet [11, 42]. Imagine a scenario where a bullet pierces a window and kills the person sitting next to that window. Now this place becomes the crime scene. Traditionally, investigators might rely on measuring the entry and exit points to reconstruct the bullet's trajectory. However, lasers can be used to project a highly focused beam along the bullet path. This beam can illuminate trajectory of bullet or create a linear glow, making the path more visible. This visual aid allows investigators to trace the bullet's trajectory more accurately, potentially revealing exact building location from where the criminal has shot that bullet (Fig. 10). Along with it, Laser scanners can capture intricate 3D models of crime scenes, including bullet holes and surrounding areas. This detailed data allows for precise reconstruction of the bullet's path in three dimensions. By analyzing the entry and exit angles within the 3D model, investigators can gain valuable insights into the shooter's position and potential weapon characteristics. Another use of laser is in detecting gunshot residues. While holding a gun, microscopic remnants may leave behind. Specific laser wavelengths can excite certain elements present in these residues, causing them to emit characteristic light signatures. These signatures can be analyzed by specialized detectors to identify the type of residues. Specificity, Sensitivity, Non-Destructive Analysis, low cost are some of the major reasons behind increasing UC based research and development. $NaYF_4$, BaY_2F_8 hosts are some of the most studied lattices for UC RE doped laser developments.

3.9 In Toxicology

The traditional methods of forensic toxicology often rely on identifying signs like visible wounds, discoloration around the mouth or external traces of toxins. However, these methods have limitations. In cases of poisoning or certain diseases, a lack of external signs can hinder determining the cause of death. This is where advancements in cellular-level ion sensing offer a revolutionary approach. Imagine a scenario where an individual has passed away with no visible injuries or cyanosis (blueness) on the lips or nails, suggesting an unnatural death. By employing highly sensitive ion-sensing technologies, investigators could potentially analyze the deceased's tissue or bodily fluids at the cellular level. These sensors could detect the presence and concentration of specific ions, including heavy metals organic toxins or even abnormal

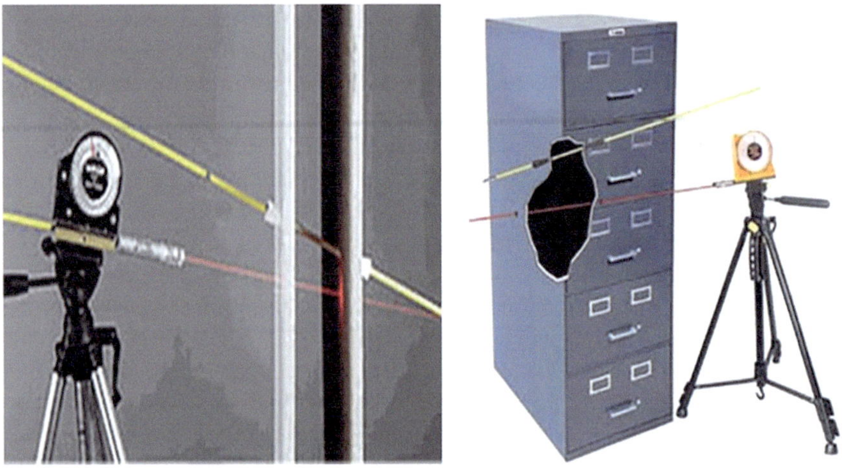

Fig. 10 Bullet Trajectories determination using LASER at Crime Scenes. *Source* https://www.crime-scene-investigator.net

levels of natural elements within the cells. This information could provide crucial clues about the cause of death. For instance, an elevated level of potassium ions might indicate a heart attack, while the presence of specific heavy metal ions could point towards poisoning. Through the examination of the ionic fingerprint present in cells, forensic experts might acquire important information that conventional techniques would overlook. This method of investigating mortality at the cellular level holds great potential to transform forensic examinations, especially when the cause of death is not immediately apparent. In actuality, additional study is required in the area of UCNP for toxicity [6, 8, 51].

3.9.1 pH Detection

An essential biological process that controls numerous organelles' functions as well as cellular homeostasis and metabolism is the equilibrium of intracellular pH. It is essential for biological systems that subcellular compartments maintain the proper pH levels. In various organelles and physiological settings, such as the cytoplasm (pH ~ 7.2), mitochondria (pH ~ 8.0), and lysosomes (pH 4.0–5.5), the internal pH values range from slightly basic to severely acidic. Numerous illnesses, such as cancer, metabolic disorders, neurological diseases, and inflammation are strongly linked to abnormal pH fluctuations. Consequently, it is thought that precise monitoring of pH variations in living cells is a useful technique for investigating cellular processes and learning about cellular internalization mechanisms. The UCNPs pH sensor has the following notable benefits [6, 8]:

(i) Can be manufactured very small;

(ii) Can perform non-contact sensing and imaging;
(iii) Operational even in high electromagnetic field;
(iv) Does not require reference elements.

Experimental research studies have suggested UCNPs matrix better with addition of any hydrogel matrix, embedded pH indicator, pH-sensitive dyes organic dyes as energy acceptors, polymeric hydrogel matrix etc. for this, NIR ratiometric fluorescent pH sensor by assembling hemicyanine dyes on PAA-modified $NaGdF_4$@$NaYF_4$:Yb,Tm @$NaYF_4$ core–shell–shell UCNPs have been reported for showing high biocompatibility for pH detection in living cells. Experimentally, many researchers developed fluorophore with one or two core and shells structures for pH detection. The pH-responsive ratiometric dual-component sensor devices based on UCNPs are shown in Fig. 11a, where they exhibit distinct emission colors (depending on pH) at a comparable excitation wavelength of 980 nm. When an environment has an acidic pH (less than 7), absorbance rises and the UC process produces considerable conventional fluorescence, which results in green emission. These probes continued to emit red light in a non-fluorescent, non-basic state. These probes demonstrated exceptional low cytotoxicity, outstanding photostability, great cell permeability, and sensitive detection of NIR conventional fluorescence. They also effectively monitored pH changes in cells [6, 8, 25, 48].

3.9.2 Ion Detection

Because they help cells and organs operate properly, ions like cations and anions are also essential to biological processes. A wide range of biological events will be strongly impacted by the disorder of particular ion concentrations caused by any external source. The monarchs of cells, K^+ ions, function as tiny pumps. They control electrical signals in muscles and nerves and help maintain fluid balance, keeping your body functioning smoothly. K^+ can be detected using NIR nano-sensor based on $NaYF_4$:Yb,Tm@$NaYF_4$:Yb,Nd UCNPs and a profitable K^+ pointer in SiO_2 nanoparticles. Ubiquitous Ca^{2+} ions orchestrate a symphony of vital functions within the human body. They serve as the structural foundation for bones and teeth, facilitate muscle contraction for movement and act as signaling messengers for nerve impulse transmission and blood clotting. This essential mineral regulates a multitude of physiological processes, ensuring optimal cellular function and overall health. Ca^{2+} ions can be noticed using nanoprobe using sandwich structured UCNPs ($NaYF_4$@$NaYF_4$:Yb,Tm@$NaYF_4$) with confined emitters and bare surface as the luminophore. Fe^{3+}, ferric iron, is crucial for oxygen transport throughout the body. It binds to hemoglobin in red blood cells, carrying oxygen from the lungs to tissues. Fe^{3+} also plays a vital role in enzyme function, DNA synthesis and energy production, ensuring cellular health and proper metabolism. Fe^{3+} ions could be probed using $NaYF_4$:Yb,Er,Tm@$NaGdF_4$ UCNPs. Cu^{2+} act as vital co-factors in numerous human enzymes, aiding in iron metabolism, oxygen transport, and energy production. They also contribute to connective tissue health, neurotransmitter function, and

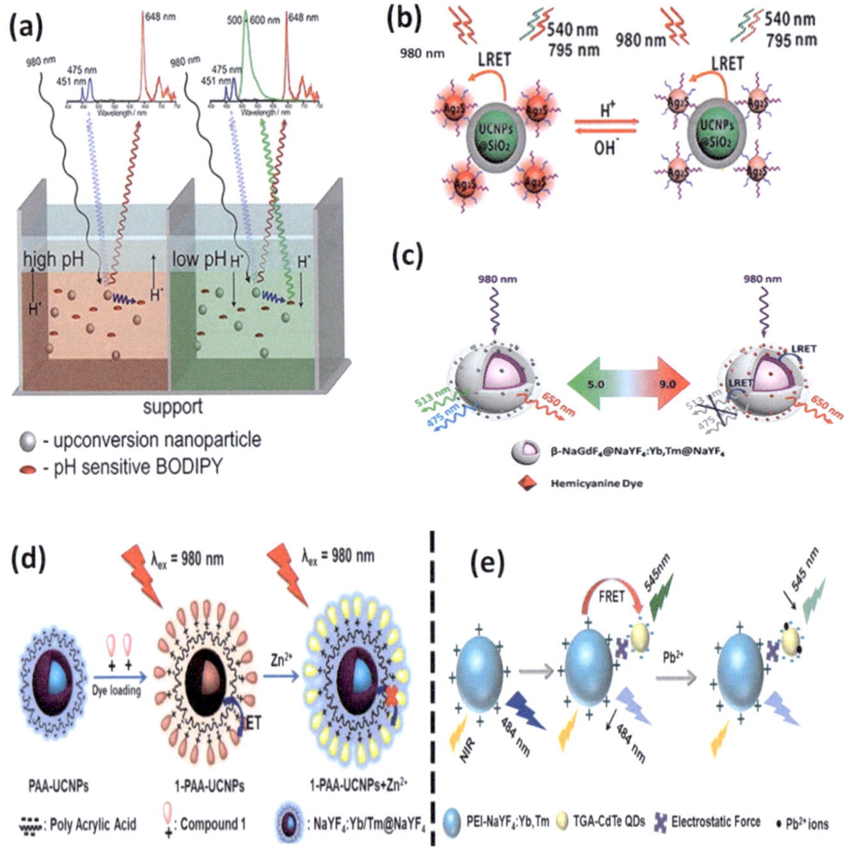

Fig. 11 **a** Creation of UCNP-based, reversibly pH-responsive ratiometric dual-component sensor devices that exhibit distinct emission colors (pH-dependent) at comparable excitation wavelengths of 980 nm, **b** Diagram showing how core/shell/shell UCNPs modified with hemicyanine dyes are used to sense pH, **c** Schematic illustration of core shell structured UCNPs@SiO_2–Ag_2S ratiometric nanoprobe for pH sensing in tumor cells. **d** Schematic illustration for the synthesis of chromophore-assembled UCNPs and their response to Zn^{2+}. **e** Schematic illustration for the synthesis of UCNPs as the donor and CdTe QDs as the acceptor and their response to Pb^{2+} (Reprinted with permission [6]. Copyright 2021 Elsevier)

the body's antioxidant defenses. However, maintaining a delicate balance is crucial, as excess Cu^{2+} can be toxic. In-depth detection of Cu^{2+} could be done by incorporating and covalent conjugation of RBH to $mSiO_2$ coated $NaYF_4$:Yb,Er UCNPs. Zn^{2+} ions, the body's stealthy guardians, play a critical role in immunity, wound healing and cellular growth. They function as enzymatic co-factors in over 300 enzymes, aiding metabolism, DNA synthesis and protein production. Zn^{2+} presence ensures a robust immune system, promotes healthy cell division and accelerates wound closure, keeping human functioning at its best and this ion can be detected using $NaYF_4$:Yb,Tm@$NaYF_4$ CNPs with Zn^{2+} -responsive chromophores (Fig. 11d).

While not directly involved in human metabolism, Cr^{3+} acts as a supportive player. It enhances the action of insulin, influencing blood sugar control and potentially aiding in fat, sugars, lipids, proteins, and cholesterol metabolism. However, excessive Cr^{3+} intake can be detrimental, highlighting the importance of balanced dietary chromium levels. It could be detected using Ce^{3+} codoped $LiYF_4$:Yb,Ho@$LiYF_4$ UCNPs, followed by assembling of Cr^{3+} -CRD onto their surface. Most of the toxic items consist of Pb^{2+}, Cd^{2+}, and Hg^{2+} ions. These are the most harmful heavy metal contaminants that can enter the human body through the food chain and seriously harm the body's organ systems. Chronic low-level exposure to Pb^{2+} (Fig. 11e) and Hg^{2+} will result in cardiovascular problems in the kidney, brain and central nervous system, as well as dysfunction of the nervous system. Utilizing quantum dots based on UC, $NaYF_4$:Yb,Er,Tm UCNPs, $NaYF_4$:Yb,Er@$NaYF_4$:Yb UCNPs, they can be identified [6, 8, 25, 48].

3.9.3 Small Inorganic Molecules

Some of the smaller inorganic molecules have their own fatal disadvantages. CO is a highly toxic substance. Carbon monoxide (CO) gas is invisible, odorless, and deadly. It binds to hemoglobin in blood, blocking oxygen delivery to vital organs. This can cause headaches, dizziness and even unconsciousness or death in high concentrations. CO exposure is a serious risk from faulty appliances, car exhaust and fires. CO can be detected using $NaYF_4$:Yb,Er,Mn UCNPs. Another hazardous gas is H_2S but unlike CO, it is smelly toxic gas. Its Low levels irritate eyes and cause respiratory problems. Inhaling high concentrations can quickly lead to unconsciousness and even death due to oxygen deprivation. Its flammable nature adds another layer of danger, with the potential for explosions in confined spaces. By adsorbing merocyanine into the mesopores of $NaYF_4$:Yb,Er,Tm@$mSiO_2$, sensing of H_2S can be done [6, 8, 25, 48].

3.9.4 Small Organic Molecules

Organic compounds include glucose, energy acceptors, cancer cells, stem cells, dopamine, color molecules and metal–organic complexes. Numerous independent studies have demonstrated the detection of ATP, cysteine, and amino acids such as arginine, histidine, and tyrosine utilizing UCNPs. These are not co- or singly-doped NPs, though. They are helped by particular nanoprobes or sensors for improved detection (10–100 ppm). Among the notable pieces are: Dopamine biosensor featuring a core–shell–shell architecture $NaYF_4$: Yb@ $NaYF_4$:Er@ $NaYF_4$:Yb UCNPs as the energy donor and energy acceptor, nano sensor for glucose determination based on LRET from protein-labeled UCNPs to chitosan-labeled GO, UC-based nanodevice for the detection of ATP using UV light-activatable DNA aptamer probe modified with $NaGdF_4$:Yb,Tm@$NaGdF_4$ UCNPs, hybrid organic dye modified yolk-shell structured UCNPs, PEI assisted $NaYF_4$:Yb,Tm UCNPs as the energy donor, GSH sensor by MnO_2 nanosheets modified $NaYF_4$:Yb,Tm@$NaYF_4$ UCNPs, CoOOH-modified $NaYF_4$:Yb, Tm@$NaYF_4$ UC Nano system for fluorescence sensing of ascorbic acid [6, 8, 25, 48].

3.9.5 Macromolecules and Bacterial Activities

The study of DNA and RNA is crucial to molecular biology, genetics, and medicine. The design of DNA/RNA sensors based on UCNPs has been the subject of numerous published studies. For instance, ssDNA-functionalized $NaYF_4$:Yb,Er UCNPs based on DNA biosensors. Furthermore, UC-based fluorescent probes have proven successful in identifying the DNA of bacteria and viruses. UCNP helped with energy acceptors such as noble metals, carbon compounds, and organic dyes, such as Au-UCNP pyramids. Utilizing UC-based platforms, the activity of other enzymes such as thrombin, trypsin, hyaluronidase, etc., has been identified. Detecting enzymatic reactions, particular biomarkers, proteins, protein–protein interactions, microorganisms, reactive species, radicals, etc. is another use for UCNPs [6, 8, 25, 48].

3.10 In UV Light Generation

Even if the UV light can damage biological samples because of their high energy and its origins harm to the eyes of examiners but it plays a critical role in forensic investigations due to its portability, instant results, and simplicity. Unlike white light, UV light interacts with certain materials in a unique way, causing them to fluoresce (emit visible light) or otherwise react. Because of this property, UV light can be used to both find hidden evidence and improve the analysis of already-found evidence, such as body fluids like saliva, blood, or semen. Sometimes simple oblique illumination can reveal a latent fingerprint. Given that it does not require any chemicals or specialized light sources, this method could be categorized as physical. Another example of how the concept of a latent fingerprint is largely operational is a latent print that may be seen in oblique lighting. There appeared to be some sort of pattern that encouraged the observer to "enhance" the surface by illuminating it in an oblique manner. Another extremely high-intensity light source is a laser. Fibers discovered at a crime scene may contain vital information. By examining them under UV light, investigators can sometimes identify the source of the fiber (e.g., clothing, carpet) based on its fluorescent properties. Along with it, UV light can reveal alterations or hidden markings on documents. For instance, erased ink or forgery attempts might become visible under UV light due to the use of different chemicals in the fraudulent ink. The paper used in documents can degrade over time, sometimes exhibiting a characteristic fluorescence under UV light. This information can be helpful in estimating the age of a document. Certain materials, like plastics or explosives, can fluoresce under UV light in distinct ways. This characteristic can be used to differentiate materials or identify specific substances present at a crime scene [11, 52]. Materials scientists commonly use garnets, silicates, phosphates, and perovskites-based host crystals or poly crystals to manufacture UV emitting materials. They also use these hosts to dope Pr^{3+}, Gd^{3+}, and Tb^{3+} are used. The down converting rare earth Ce^{3+} ion gives emission in the UV-region (200–400 nm). The ones with such capability are Ce^{3+}, Pr^{3+}, Pm^{3+} (also radioactive), Gd^{3+}, and Tb^{3+}. Lu^{3+} doping is also promising though the emission is in the vacuum ultra-violet region UVC radiation are reported to be generated from Pr^{3+}-doped $LaPO_4$, Lu_5SiO_5, Cs_2NaYF_6 [53].

4 Synthesis of the UCNPs

The selected synthesis procedure directly affects the regulated dimensions, morphologies, and stoichiometries of UCNPs, which in turn affects their surface properties, quantum efficiencies and light output. Consequently, considerable study has been devoted to optimizing UCNP synthesis techniques. But this time, the emphasis is on more recent developments made especially for forensic use. Various strategies are available to researchers based on the desired features of UCNP. It takes some trial and error, but getting a repeatable synthesis process is essential. The several methods for creating UCNPs with desired properties for a variety of applications and approaches are compiled in the subsections below.

4.1 High-Temperature Pyrolysis Method

One of the most popular techniques for creating rare-earth UCNPs is high-temperature pyrolysis. To create rare-earth nanoparticles with a consistent shape and regulated size, the rare-earth precursors are dissolved in an organic solvent with a high boiling point and then heated to a high temperature while being exposed to a surfactant for a predetermined amount of time. Organic solvents with long chains and polar endings are frequently employed as surfactants and the precursors are used in the form of chlorides or acetates. The size and shape of the UCNPs will also be impacted by variations in reaction temperatures, reagent volumes, and reaction periods. Although the use of UCNPs in analytical or biological domains is hampered by the manufacturing process's inclusion of oily surfactants, surface modification, or functionalized UCNPs has become popular and important [25].

4.2 Hydro/Solvo Thermal Method

Heterogeneous chemical reactions that take place in a closed system at high temperatures and pressures when a solvent is present are referred to as the hydro/solvothermal technique. When water is utilized as the solvent, the term "hydrothermal" is used. The word "solvothermal" is used instead. When no organic additives or templates are utilized, straightforward hydrothermal procedures provide a lot of potential for producing one-dimensional rare-earth ion-doped nanowires, nanorods, or nanotubes. This is especially true for producing rare-earth hydroxides. For instance, at certain temperature and pH circumstances, hydroxide nanowires, nanotubes, and nanorods can be selectively produced in a water system containing rare-earth ions and NaOH (or KOH, $NH_3 \cdot H_2O$) [20, 54, 55]. In contrast to thermal breakdown and high-temperature coprecipitation procedures, Teflon-lined autoclaves, a specially-made vessel, are typically used for hydro/solvothermal synthesis [56]. This strategy is founded on the phase allocation and parting mechanism happening at the borders of the liquid, solid, and solution phases [6]. The additives have the potential to function as both complexing and structure-guiding agents during the nanoparticle production process [25]. It was stated that the pH of the solution, precursors ratio, concentration of doped ion, reaction duration, and temperature may all control the morphology, particle size and optical characteristics. To get the necessary phosphor, the sample must lastly be further cleaned, dried, calcined, and powdered. Because of its low synthesis temperature, mild circumstances, and steady reaction system, the hydrothermal/solvothermal method is an effective synthesis technique. Furthermore, this process has a high degree of reactivity in the atomic and molecular level reactions and the result is a very homogeneous product [25]. Nonetheless, the hydrothermal method's synthesis conditions are a little more intricate and sophisticated than those of other synthesis techniques.

4.3 Thermal Decomposition Method

Well-shaped, controllably sized particles can be produced via thermal breakdown. The precursors are dissolved and heated in solvents with high boiling points to initiate a chemical reaction. In thermal decomposition organic precursors are usually broken down with the help of one or more surfactants in a high-boiling organic solvent to produce matching metal fluorides at a high temperature [8]. It is important to remember that the experimental parameters, such as the temperature, heating rate, reaction time and concentration of organometallic precursors, can be accurately adjusted to control the crystal nucleation and growth processes. This method produces high-quality UCNPs with good optical characteristics, limited size distribution and crystallinity. The form and size of NPs may be altered by adding various organic solvents, such as oleic acid (OA) and octadecene (ODE), according to recent and previous findings. More significantly, ultrasmall UCNPs can be prepared via thermolysis synthesis. However, because they are made in an oily medium with hydrophobic molecules acting as capping agents, the synthesized UCNPs are often hydrophobic and must be stabilized in nonpolar solvents. The thermal decomposition approach has limitations even with the recent advancements in its ability to prepare UCNPs. The method's initial drawbacks include the costly and air-sensitive metal precursors, the toxic byproducts, the need for high reaction temperatures and lengthy reaction times, strict atmospheric conditions (oxygen-free and anhydrous) for synthesis, post-synthesis procedures and the requirement for hydrophilic and biocompatible coatings on materials in order to have hydrophobic capping ligands on the surface of UCNPs [57].

4.4 High-Temperature Coprecipitation Method

High-temperature coprecipitation, which does not require strict reaction conditions, is a facile and environmentally friendly way to synthesis UCNPs as compared to the thermolysis strategy. It is also among the most effective ways to make ultrasmall-sized (less than 10 nm) UCNPs. Once more, the size and form of NPs can be changed by using organic solvents like ODE and OA as a capping agent or solvent. The ligand's concentration can be changed to produce different sizes, shapes and forms of nanospheres, nanoplates, nano-ellipses, and three-dimensional nano-architectures. Furthermore, by extending this synthetic pathway, superior core–shell-structured UCNPs with amplified UC luminescence can be produced. To date, the most popular method for producing high-quality UNCPs is high-temperature coprecipitation synthesis. The resulting UCNPs, similar to those produced by the thermal breakdown process, are hydrophobic because OA and/or OM are typically used as the ligands in the high-temperature coprecipitation pathway. It's crucial to lessen the hydrophobic nature for biological applications [6, 7].

4.5 Sol–Gel Method

The sol–gel method, one of the wet chemical synthesis techniques, has greatly advanced research in the creation of luminous materials. The term "sol" describes the tiny solid particles, ranging in diameter from 1 to 100 nm, that are subjected to Brownian motion and form a suspension under the influence of the electric double layer and van der Waals force. According to Gibbs free energy, the atoms near the interface have more energy than the ones within the liquid. Thus, the phenomena of spontaneous condensation arise as the thermodynamic instability strives to reach equilibrium. Gel is a network structure made of liquid within a grid created by colloidal particles crosslinking. The sol becomes a gel if the intergranular forces outweigh the intergranular-solution interactions. To make sol–gel, one can either cool the sol or volatilize the medium to achieve supersaturation and gel formation. Then, add non-solvent to encourage gel formation, add electrolyte and use insoluble matter through chemical reaction. The following are some benefits of the sol–gel approach: improved product homogeneity; reduced calcination temperatures (in comparison to the high temperature solid phase method); resource conservation; increased product purity; increased luminescence intensity; and increased quantum yield [6].

5 Choice of Host and Dopants

We have experienced firsthand how crucial the UC process is to forensic applications. Currently, UCNPs consist of an inorganic host that has been doped with both a sensitizer and an activator. In theory, most lanthanide-doped host materials should produce UC emission; yet, careful consideration of the host material's type and the quantity of doped sensitizer and activator ions is necessary to produce effective UC luminescence. The electronic structure (of a part) of its $4f^N$ configuration is revealed by the peak position of the Ln spectral lines, as was covered in Sect. 2. The coordination polyhedron's shape and the rare-earth site's symmetry can both be inferred from the crystal-field splitting. The interaction between the Ln ion and its surroundings is also reflected in the intensity of spectrum transitions. When doped into distinct types of inorganic compounds, the effects of the Coulombic field, spin—orbital coupling and crystal field interaction are shown in Fig. 12 on the splitting of the RE configuration. Because the non-radiative relaxation from the excited state of doped rare earth ions has less quenching impact, it suggests that low phonon energy hosts, such as fluoride-based materials, are demonstrated to be more efficient than conventional oxide, molybdate, and phosphate-based materials. Maximized radiative emissions, great transparency in the excitation and emission wavelength and exceptional chemical stability to maintain the original crystal structure are the results of the reduced loss in the nonradiative transition. Less than 300 cm^{-1} is the typical low phonon energy of heavy halides, which include bromides, iodides, and chlorides. However, their limited chemical stability and hygroscopic characteristics restrict their uses. Fluoride-based systems typically have good luminescence efficiency because of their great chemical stability and comparatively low phonon energies of approximately 350 cm^{-1}. Oxides have a high degree of chemical stability, but because their phonon energies are often greater than 500 cm^{-1}, there will be a decrease in the UC quantum yield and a rise in non-radiative loss. Na^+, Ca^{2+}, and Y^{3+} fluoride-based materials are better host matrices for UC phosphors because they also have ionic radii of cations that are close to those of lanthanide dopant ions, which prevents the creation of crystal defects and lattice tensions. The $NaYF_4$ molecule is thought to be the most effective host to date because of its great chemical and thermal stability as well as low phonon energy. The extraordinary UC property in $NaGdF_4$, $NaLuF_4$ and $LiYF_4$ crystal systems is also suggested by some studies. Ln^{3+} doped KYF_4 nanocrystals, however, have been the subject of relatively few reports, which raises the possibility that they could serve as an extremely effective UC host matrix for the conversion of NIR photons to UV and visible photons. Because of their similar sizes, KYF_4 and the $NaYF_4$ crystal structure are both accommodating to all rare earth ions and, when doped with them, exhibit long fluorescence lifetimes [2, 6, 8, 18].

In addition to their optical characteristics, one of the main areas of research in this subject is the temperature-dependent UC emission behaviors, which may indicate that UC nanoparticles are suitable as thermal sensors. Some other most frequently used host crystals are $NaLaF_4$, LaF_3, GdF_3, CeO_2, LiNaF4, $Ca_3(PO_4)_2$, CaF_2, ZrO_2, GdOF, etc. [2, 6, 8, 18].

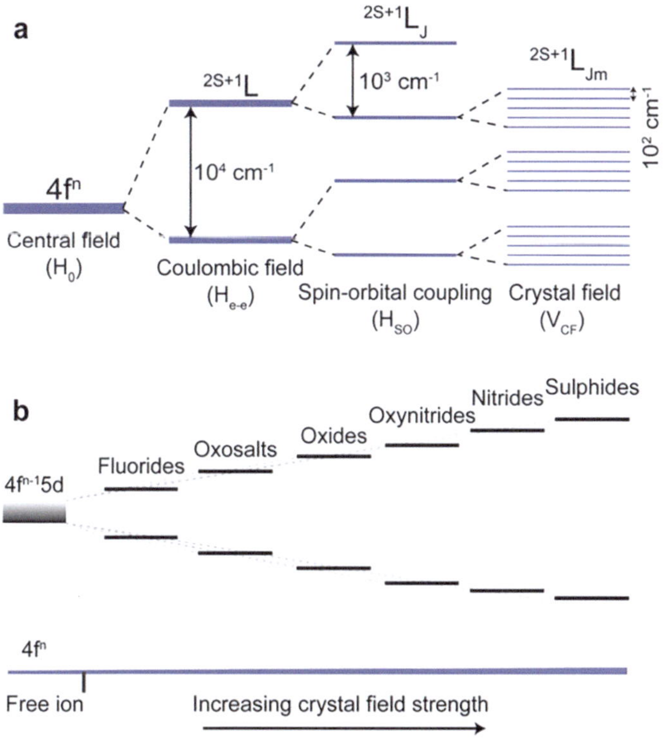

Fig. 12 **a** A simplified representation of how the crystal field interaction, spin—orbital coupling and Coulombic field affect the splitting of the RE configuration. **b** The diagram illustrates how much the energy splitting of the 5d and 4f levels of RE varies depending on which class of inorganic molecule is doped (Reprinted with permission [2]. Copyright 2017 American Chemical Society)

Usually, the host matrix is optically inert and hence does not participate in the UC process. Here, the sensitizer raises the UC luminescent efficiency while the activator acts as the luminescence center. In the host lattice, the sensitizers and activators are introduced at comparatively modest quantities (a few mole percent). Generally speaking, the UC efficiency will correlate with the amount of activator ions present in the host material. On the other hand, the ion-to-ion spacing of dopants in the host lattice would decrease with an increase in activator ions, leading to an increased non-radiation cross-relaxation. Furthermore, crystal flaws will surface when the nanocrystal is doped with lanthanide ions. Cation voids are created in order to preserve charge neutrality, which further reduces UC efficiency [2, 6, 8, 18]. Due to its large absorption cross section, a sensitizer's work typically consists of boosting NIR absorption and transferring the absorbed energy to the adjacent activator in the crystal lattice. The activator functions as the luminescence center in most UCNPs. With the exception of trivalent Yb, Ce, Lu, and La ions, the majority of RE^{3+} ions exhibit several excited states, making them appropriate candidates for UC activators. Due to their energy levels resembling a ladder, Tm, Ho, and Er ions are the most commonly utilized UC activators. With their energy levels arrayed in a ladder pattern, dopant ions Er, Tm, and Ho are often chosen as activators to produce UC emission during CW excitation. A comparatively low (< 2 mol %) amount of the activator is typically accepted to diminish the energy loss subsequent from cross relaxation. Yb^{3+} has a greater absorption cross-section than those of the lanthanide activators. Thus, Yb^{3+} is often co-doped with Er^{3+}, Tm^{3+}, or Ho^{3+} as a sensitizer to enhance UC emission. A CW 980 nm laser is applied as the excitation source to match the transition of Yb^{3+}. some often-reported combinations of RE are: Yb/Er, Yb/Ho, Yb/Tm, Er/Tm, Er/Yb, Yb/Nd co-doped UC materials [2, 6, 8, 18].

6 Tuning and Optimization of the UC Properties

In the present section, we are discussing the tuning and optimization of UCNP for forensic based application and recent literatures based on these techniques.

6.1 Changing the Choice of Excitation Spectra

A longer wavelength CW laser can be used as the excitation source for UC emission spectra measurement using a modified fluorescence spectrometer, much how xenon lamps are used in down conversion phosphors. A short-pass filter is positioned between the sample and the detector to block excitation photons in order to get a high-quality UC emission spectrum. Furthermore, for the TTA-based UC measurement, the sample cell in the optical path needs to be sufficiently thin [25]. In majority of the phosphors, the preferred excitation is ~ 980 nm which is near IR region. But currently new phosphors which can get excited using 1000–1200 nm are designed for the new research. Choosing the correct excitation wavelength is necessary for desired emission wavelength. Change in the excitation wavelength can cause broadening, shifting, or splitting in the emission peaks hence changing the CIE coordinates. It should be mentioned at this point that the multiphoton absorption and energy transfer mechanism is the reason why the UC emission intensities are directly correlated with the excitation power densities [39].

6.2 Tuning the Combination and Concentration of Doping

It is possible to adjust the UC emission color in both the TTA-based and lanthanide-based UC procedures. There are various methods for doing it. Adjusting the Doping Activator Ions and Their Concentration is the initial strategy. This is the very fundamental and basic approach which is usually followed in many or say every RE-doped materials. The concentration of ions significantly affects the emission intensity and their position. Using the compatibles combination of other dopant can do two things:

1. It can cause enhancement in the original emission intensity by increasing the absorption cross section in desired wavelength region.
2. It can rise its own emission peaks in its characteristic position which can tune the emission intensity significantly

The most often utilized lanthanide activator ions in UCNPs are Yb, Er, Tm, and Ho. These activators exhibit various UC emission bands in UV, blue, green, red, or NIR areas because of their varying energy levels. The emission wavelength can be tuned by adjusting the doping concentration. A single UCNP can produce tricolor green, red, and NIR UC emission by selecting the right mix. Here, are a few instances: As well as $NaYF_4$:Yb/Tm and $NaYF_4$:Er/Tm. The longer infrared second near-infrared (NIR II) spectral region has generated a lot of interest because to its enormous potential for high-resolution and deep-tissue imaging, especially at about 1530 nm. The spectral response of Er^{3+} is good between 1490 and 1550 nm. Due to harmful quenching interactions, it cannot perform well as the NIR II sensitizer in the widely used sensitizer–activator co-doping scheme. Thus, clever control of photon UC from a sequence of Ln ions following NIR II excitation has remained a difficulty. In addition to its regular function as a sensitizer in traditional UC materials, Yb^{3+} has recently shown good energy-migratory performance (e.g., in a Yb–Er linked system). By adjusting the rise and decay durations, the application of such energy migration can successfully improve Nd^{3+} sensitized UC, mediate UC, and downshift luminescence dynamics [39].

6.3 *Tuning NPs Size, Phase, and Surface Ligands*

The annihilation of quenching centers and materials selected to boost the absorption ability have been the main focus of efforts to improve UC efficiency since nonradiative decay and the short excitation cross sections of the lanthanide ions are the main factors influencing UC efficiency. As a result, altering the size, NP shape, crystal field symmetry, and host lattice structure can all increase UC efficiency. The ratio of multiple UC emissions can be influenced by the surface ligands of lanthanide UCNPs. Surface and small size effects are typically how morphology, size and surface ligands affect UC emissions. $NaYF_4$ is the most popular host for making effective UCNPs. $NaYF_4$ exists in two polymorphs at ambient pressure: a hexagonal phase (thermodynamically stable low-temperature phase) and a cubic phase (metastable high-temperature phase). UCNPs using the hexagonal-phase $NaYF_4$ as host often exhibit better UC efficiencies than those using the cubic phase because of its low phonon energy. Consequently, one effective method of increasing cubic $NaYF_4$'s luminescence effectiveness is to convert it into a pure hexagonal structure. Numerous techniques have been described to date for causing this cubic-to-hexagonal transition, including the use of ligands, controlled synthesis using hydrothermal technique, and heat breakdown. some other materials that exhibit phase transformation are: $NaLuF_4$, $LiYF_4$ etc. the size dependency-based research on the $NaYF_4$:Yb,Er NPs showed that the intensities of the green and red emissions decreased with decreasing NP size, implying that the emission intensity can be changed by modulating the NP size. The connected ligand refines the emission. Another work on the $NaYF_4$:Yb,Er NPs with the use of surface –OH groups displayed its contribution in quenching of emission properties of the lanthanide UCNPs [58].

6.4 Tuning the Emission Color of TTA-Based UC

Unlike the lanthanide-based UC method, photon UC based on the TTA process covers the singlet and triplet excited states of both the sensitizer and annihilator through absorption, energy transfer, fluorescence emission, and a variety of nonradiative processes. TTA-based UC systems can have tunable excitation and emission wavelengths by carefully selecting the sensitizer and acceptor separately. The bulk of known TTA-based UC systems require a high concentration of annihilator (chromophore-based dopant). As a result, there is a high risk of it aggregating and extinguishing the annihilator's fluorescence. To achieve efficient UC emission, the annihilator must meet certain requirements:

(i) The excitation light should not be absorbed by it.
(ii) The fluorescence quantum yield (QYFL) ought to be high.
(iii) In order to facilitate the TTA process, it must have a triplet state with an appropriate energy (typically less than the sensitizer's energy) but more than half of the energy of its own single excited state.
(iv) Its triplet state should have a long enough lifetime to guarantee that the TTA process takes place.

In compliance with the aforementioned specifications, a multitude of polycyclic aromatic hydrocarbons, encompassing benzene rings and heterocyclic compounds, as well as derivatives of anthracene, tetracene, pyrene, fluorene, and derivatives diimide-based dyes, have been effectively developed as annihilators for TTA-based uncoupling agents [7, 25].

6.5 Association of Energy Acceptors

As their name implies, the energy acceptors increase the excitation density, which results in increased emission. Ag or Au NPs with surface-plasmon-coupled emission characteristics can boost the rate of radiative decay or excitation efficiency. To increase the intensity of UC emission, Ag or Au NPs have been mixed with lanthanide UCNPs. We refer to them as energy acceptors. Other examples of energy acceptors besides Au and Ag NPs are: organic dyes, Ag_2S, quantum dots, dye molecules, metal–organic complexes, polymers, transition-metal oxides, noble metals, DNAs, RNAs, bioproducts, plant extracts, etc. These addition of energy acceptors make Ion detection (Hg^{+}, Pb^{2+}, Cr^{3+}), bacteria detection, glucose detection, formation of ATP sensors more sustainable, and precise [6, 7].

6.6 Other Methods

Using external magnetic fields to create matrix modulation, incorporating two sensitizers simultaneously to create a Sensitizer (I)-Annihilator-Sensitizer (II) system, incorporating two annihilators simultaneously to create a dual annihilator Single-Sensitizer system and adjusting the experimental conditions such as solvent, solvation properties, polarity, viscosity and boiling point, excitation power density, concentration of the sensitizer and annihilator, temperature and oxygen concentration are some other techniques to create tuning and optimization of the UC properties [7, 25].

7 Development of Core–Shell Structure

- A common method for raising the UC efficiency of lanthanide UCNPs is the use of core–shell nanostructures, one of several mechanistic attempts to control photon UC. The luminous centers of lanthanide UCNPs are efficiently isolated from the surface quenching species within their core–shell architectures. This section goes into great length about the core shell structure since it is a valuable phenomenon. To visualize the shell's fundamental composition, consider a little sphere that is divided into two parts:
- **Core**: This is the central particle, often made from a single material. It can be a metal, semiconductor or even a polymer. The core dictates some of the basic properties of the nanoparticle, like its size and magnetic, optical or electrical properties.
- **Shell**: This is the outermost layer that envelops the core and has the potential to provide more features. The most common method for increasing the UC emission intensity of lanthanide UCNPs is to utilize a homogeneous shell without any sensitizer or activator, as a comparable lattice constant minimizes any lattice tensions that can impede the creation of core–shell structures [25].

Numerous NPs with core–shell structures can be manufactured, including homogeneous core–shell nanocrystals, in which the doping concentration and dopant ions vary but the core and shell are both part of the same host crystal type. These artificial structures are intended to lessen the impact of surface ligands and imperfections for increased UC efficiency. A few instances of these structures include $NaYF_4$:Yb,Er/Tm/Ho@$NaYF_4$, $NaGdF_4$:Yb,Er/Tm@$NaGdF_4$, $KGdF_4$:Yb,Tm@$KGdF_4$, KYF_4:Yb,Er@KYF_4, α-$NaYF_4$:Yb,Er/

Tm@α-$NaYF_4$, YOF:Yb,Er@YOF, Y_2O_3:Yb,Ho/Tm@Y_2O_3, YF_3:Yb,Er@YF_3 and $NaGdF_4$:Er@$NaGdF_4$:Ho@$NaGdF_4$. Meanwhile, there is another approach, heterogeneous core–shell nanocrystals which is opposite of homogeneous core–shell structure as it includes different hosts in their discrete core and inert shell layers. These include examples such as $NaYF_4$:Yb,Er@$NaGdF_4$, $NaYF_4$:Yb,Tm@$NaLuF_4$, $NaYbF_4$:Tm@$NaGdF_4$, $NaYF_4$:Tm@$NaY_xGd_{1-x}F_4$, LaF_3:Yb,Er/Ho@LaF_3, $NaYF_4$:Yb,Er@CaF_2, BaF_2:Yb,Tm@SrF_2:Gd,Nd and $LaPO_4$:Er@Yb_2O_3 [6, 8, 25]. While the basic concept of a core–shell nanoparticle is a central core surrounded by a shell, there's complex and advanced applications of these structures [7]:

- **Multiple Shells**: Multiple shell layers are possible in core–shell nanoparticles. This makes it possible to control attributes even more. The epitaxial layer-by-layer growth on nanocrystals is utilized to implement the multilayered core shell structure. This technique allows you to adjust the thickness and composition by taking use of dynamics. It is discovered that thickening the shell might lower the luminescence quenching in water and significantly increase UC efficiency. For instance, a core of magnetic iron oxide surrounded by a silica shell and then a polyethylene glycol (PEG) outer layer might offer magnetic targeting, biocompatibility, and increased circulation time in the body for drug delivery applications.
- **Yolk–Shell Structures**: In this variation, the core is hollow, resembling a yolk in an egg. This design allows for increased drug loading capacity or the creation of unique optical properties.
- **Core–Shell–Shell Structures**: Here, two separate shells encircle a core, providing even more control over functionalities. Imagine a gold core with a polymer shell holding targeting molecules layered on top of a silica shell. This kind of structure might be utilized for targeted photothermal therapy, in which the gold core warms up in response to light exposure, killing cancer cells.

It is evident that properties can be tailored with the right core shell structure design. Since the shell's goal is to add functions to the core, it is possible to construct nanoparticles with specialized qualities and enhanced functionality by combining different materials for the core and shell. For instance, a magnetic core with a non-magnetic shell might preserve the other characteristics of the core while enabling particle manipulation with a magnetic field. The shell can also act as a protective barrier for the core, shielding it from degradation or unwanted interactions with the surrounding environment. Due to all these magnificent properties, core–shell nanoparticles are a becoming versatile tool in nanotechnology and there are some real-world advanced applications of core–shell nanoparticles are in LFP detection, drug delivery, biosensing, etc. [6, 25].

Despite the immense potential, challenges remain at a considerable place. Creating complex core–shell structures with exact properties requires further refinement of synthesis techniques. Another problem is scalability in production. Moving from lab-scale production to large-scale manufacturing needs to be addressed for widespread use. But with the solution of largescale production, comes a cost of polluting environment. Hence, the potential risks of nanoparticles entering the environment and their long-term effects need to be carefully evaluated. Along with it these studies are costly compared to other material science-based researches. To date, the best approach to characterize core shell structure-based NPs are high-resolution TEM (HRTEM), X-ray photoelectron spectroscopy (XPS), energy-dispersed X-ray spectrometry (EDS) and electron energy loss spectroscopy (EELS). But these studies are costlier to perform.

8 Surface Modification of NPs

The majority of UCNPs generated using the several techniques mentioned above are typically hydrophobic, which is acceptable because we are working with applications where solubility attributes are not required. However, the solubility of NPs is required for bioimaging-based forensic applications, pH sensing, blood sensing, saliva, and physiological liquid-based detection. In addition to solubility, the solvent impact should not reduce the emission intensity because it will prevent them from being used in vitro and in vivo in the future. In order to be considered for further usage as prospective candidates for bioassays and imaging, UCNPs must have acceptable surface wetting qualities in addition to high luminescence emission for other bio targeted detection. Surface modification of NPs is a rapidly evolving field with immense potential for diverse applications. By precisely controlling the surface properties, scientists can create NPs with tailored functionalities for advanced forensic-based application such as enhanced UC efficiency for LFP detection, biosensing, improved catalysis, and innovative sensing applications.

By introducing specific functional groups onto the NP surface, scientists can tailor their interaction with biological systems, environmental components, or other materials. For example, attaching targeting ligands to the surface of NPs allows for specific binding to diseased cells in drug delivery applications. It is important to remember that the NP size is also influenced by the type of surface coating materials. Hence, size-based modulation can be achieved. Modifying the surface with biocompatible materials like polyethylene glycol (PEG) can minimize immune system recognition and extend the circulation time of NPs in the body. By altering the surface chemistry or morphology, researchers can tune the light absorption, emission, and scattering properties of NPs. This has applications in the detection of biomolecules through fluorescence or localized surface plasmon resonance effects. The surface of UCNPs can be modified in two ways:

1. Surface modification using organic capping ligands ($-NH_2$, $-COOH$) or an inorganic shell layer (SiO_2).
2. The NP's surface coating with the proper polymer or surfactants.

To do so, Multiples techniques are employed. Some common methods include the Chemical vapor deposition (CVD) in which a thin film of desired material is deposited onto the NP surface in a controlled atmosphere. Because surface silanization employs proven chemical techniques for silica coating, it is the most widely utilized procedure. At the moment, the most common inorganic surface treatment is SiO_2 coating. It is easy to apply this silica coating to NPs that are hydrophilic or hydrophobic. Functional molecules may be able to be loaded into the open mesopores of mesoporous silica-coated nanoparticles. Thus, they can be used for both medication delivery and photodynamic treatment (PDT). The $-NH_2$, –COOH, or –SH functionalization required for additional chemical reactions with biological molecules and other functionalized substances is made possible by the SiO_2 shell. SiO_2 is therefore a great intermediate layer due to its adaptability, enabling the coupling of lanthanide UCNPs and other molecules that must be kept apart from one another [25]. The luminescence of UCNPs is somewhat influenced by variations in the silica shell's thickness (3–5 nm), which can be adjusted based on the needs of the application. While coated silica can increase the hydrophilicity of the UCNPs' surface, it cannot supply the functional groups needed for the application process under analysis. An integral component of the architecture of the UCNPs analysis platforms is the functionalization of the organic ligands of UCNPs, which are based on encapsulated silica. Apart from silica, some of the other coating materials are oleylamine, dextran, carboxydextran, poly (ethylene glycol) (PEG), polyethylenimine (PEI), poly (vinyl alcohol) (PVA), lauric acid, acrylic acid, oleic acid, precious metal, silane, carbon, etc. The partial sol–gel method and the solvothermal method are two other techniques for creating cire shell nanoparticles. Every preparation technique has proven to have advantages and disadvantages of its own and a comprehensive plan for synthesizing UCNPs to meet the needs of many applications has not yet been created [7, 8, 25].

Surface modification with hydrophilic (water-loving) molecules like polymers or surfactants creates a steric hindrance, preventing aggregation and facilitating their suspension in various media. Hydrothermal microemulsion synthesis and one-pot/multistep synthesis supported by hydrophilic ligands are two techniques for the direct production of water-dispersible lanthanide UCNPs. Hydrophilic ligands are directly capped on the surface of lanthanide UCNPs during the one-pot synthesis, making them water-dispersible. Until now, hydrophilic ligands have been employed in one-pot synthetic processes with polyols (diethylene glycol, glycol, glycerol and ethylene glycol), PEG, polyvinylpyrrolidone (PVP), 6-aminohexanoic acid, citrate, ethylenediamine tetra acetic acid (EDTA), poly(acrylic acid) (PAA), phosphino poly(acrylic acid), and polyethylenimine (PEI) [25]. Attaching passive dopant ions is chosen in many studies in order to reduce a variety of quenching mechanisms, including concentration quenching, solvent existence and quenching related surface defects.

A hydrophilic NPs are not always useful for some of the applications, sometimes a hydrophobic NPs are useful in order to determine selective solubility, toxicity-based tests, environmental assessments, etc. For hydrophobic lanthanide UCNPs, a two-step surface functionalization is necessary in order to transform them into hydrophilic NPs. ligand exchange approach is one of the most used ways as it has no substantial effect on parameters such as size, crystallization and shape of the lanthanide UCNPs. It is predicated on the idea that the ligand that is being added can coordinate with the lanthanide ion more strongly than the ligand that it is replacing. When a hydrophobic surface ligand is selected, another ligand with more strongly coordinating groups or with multiple coordination sites is adopted in order to improve its properties. A unique ligand exchange technique is the organic ligand-free production of hydrophilic lanthanide UCNPs. The idea behind this is that the ethanol and water's oxygen atoms can coordinate with the lanthanide ions on the UCNPs' surface to stabilize the NPs in water. Therefore, a hydrophobic nature can be produced by eliminating the surface ligand, utilizing additional organic components such nitrosonium tetrafluoroborate, BF_4-anions, folic acid, etc. and altering the pH of the solution [6–8, 25].

9 Solutions and Scope of Research

9.1 In the Field of Security Nano/Micro Barcodes

Although security barcodes are cutting edge technology, there are still certain unanswered or unsolved problems that need to be addressed. These include developing portable decoders, improving synthesis techniques, enhancing safety and establishing regulations for the use of nano-barcodes. There are several encoding methods that can limit the number of barcodes that can be obtained for a given use. For example, the most popular encoding approach with proven synthesis and decoding methods is fluorescence-based nano-barcoding. Fluorescent nano-barcodes, however, have a number of drawbacks, including spectral overlap restrictions, complicated processing requirements, trouble resolving signals, various emission bands, etc. Research is being done on activators with narrow spectra and single wavelength emissions to address the issue of multiple emission bands; these activators still require improvement in terms of intensity before they can be employed in multiplexing applications. When UC nanoparticles are combined with organic dyes, more excitation lasers are needed to increase the intensity, which raises the expense and complexity of detection. The limits of different discrete shapes and sizes also cause the morphological encoding technique to be unable to generate a huge number of codes. In order to overcome this difficulty, hybrid nano barcodes can be created by combining various encoding techniques, although doing so may raise the overall cost and complexity of the operation. On this front, simple decoders that are compact, transportable, affordable, and customizable must be created [59].

9.2 In the Field of Bioimaging

In forensic pathology, imaging may not always be able to identify the deceased's pathological conditions. Since forensic imaging cannot show every site that was damaged during an injury, direct evaluation of neuroimaging, forensic-pathological findings and forensic angiography are required. Thus, in these situations, attention is required. It is not practical for all forensic institutes to own the facility due to the high cost of forensic imaging. A low-cost forensic imaging resource has to be built in order to optimize resource utilization [43].

9.3 In the Field of Biomedicines

Because of their special photophysical qualities, UCNPs have a lot of potential for use in biological applications. However, further research will be needed to fully understand the difficulties in developing UCNPs for biomedical applications [45].

- Because of the low absorption cross section of sensitizers, UCNPs' poor UC efficiency typically limits their use in biomedical applications. High efficiency is still needed, but these luminescent UCNPs do improve the sensitivity for biosensing, bioimaging, and photo-initiated biological and chemical reactions, despite significant efforts being made to increase UC luminescence by investigating core–shell nanostructure (Sect. 7) and RE-doping concentration. Currently, one way to address the low conversion efficiency and narrow-band near-infrared absorption of Ln-doped UCNPs is to incorporate one or more dye sensitizers; however, there are not many known dyes that can sensitize organic–inorganic systems.
- Moreover, efficient UC can only be achieved through excitation using 808 nm and 980 nm lasers, which has significantly limited the practical uses of these lasers in biomedical applications.
- For further pre-clinical and clinical uses, systematic and thorough toxicity investigations of these NPs are still required.
- Hydrophobic ligands are typically used to stabilize the produced NPs' surfaces; however, these ligands are unsuited for use in biological settings. As a result, surface modification of Ln-UCNPs has been the subject of intense research in recent years, with numerous techniques having been proven effective.
- The majority of dye-sensitized ultraviolet curable (UC) is only achievable in organic solvents like DMF and chloroform, making aqueous solution applications extremely difficult. For dye-sensitized UCNPs in aqueous solutions, this challenge can be ascribed to several factors, such as (i) dye detachment from organic–inorganic nanocomposites; (ii) dye photo-bleaching that occurs quickly; and (iii) quenching of UC luminescence. Low energy transfers from dye to Ln^{3+} sensitizer is caused by these problems, which will need to be resolved in biological applications in the future.

TTA-based UCNPs exhibit a comparatively greater luminescence efficiency than Ln-doped UCNPs. However, the majority of current research is devoted to the design and synthesis of molecules that act as sensitizers and annihilators; hence, the development of TTA-UCNP systems that are both oxygen-tolerant and water-soluble is required for biological applications. Future studies utilizing TTA-based UCNPs should focus on photo-triggered drug delivery in aqueous solutions, biosensing, bioimaging, cancer detection, and phototherapy. Here, designing TTA-based UCNPs for biomedical uses requires careful consideration of a few important variables.

Since TTA UC is oxygen sensitive, a unique matrix is first required to develop TTA-based UCNPs that are oxygen-resistant. Second, although having a greater UC efficiency than previous TTA-based UCNPs, their shallow tissue penetration limits their use in biomedical applications because their excitation and emission wavelengths are mostly in the UV–visible region. Thus, it is desirable to construct annihilator molecules with red and NIR emission and sensitizers with strong NIR absorption. Third, the creation of photostable TTA-based UCNPs is required since the photostability of both sensitizer and annihilator molecules is relatively low under laser irradiation. Ultimately, it appears that TTA-based UCNPs are typically less hazardous than Ln-doped UCNPs; however, as some heavy metal ions may be liberated and taken up by living things following bodily breakdown, the toxicity of both sensitizer and annihilator needs to be carefully assessed [45].

9.4 In the Field of Security Ink

Because of their straightforward operations, low cost, minimal reagent usage, direct and high-throughput patterning capabilities, difficult to replicate and simple authentication features, luminous security inks work better than other conventional identification or printing processes. In addition, advanced anti-counterfeiting patterns can be created on a big scale by the inkjet printing of luminous security inks. But many synthesized phosphors suffer the problem of nonuniformity, problem of dispersing them in the solvents, low QY, background fluorescence interference, photobleaching, non-applicability on many surfaces, diminishment in intensity on pH, heat, water etc. Consequently, the background luminescence greatly obscures or even completely suppresses the detected encrypted information from fluorescent inks. This problem significantly postpones the widespread use of bright liquid ink for evidence encryption. In order to overcome this difficulty, alternative fluorescent security inks must be created in order to enhance whiteness, printability, and opacity while preventing background fluorescence and enabling practical information encryption. Furthermore, as was already covered in earlier parts, inkjet printing needs the fluorescent pigments in fluorescent security inks to be at the nanoscale in order to avoid any possible nozzle obstruction.

9.5 In the Field of Toxicology

The fact that typical UC micro phosphors are so small that they can pass through cells or veins causes concern. Because of their special characteristics, nanoparticles may unintentionally interact with biological systems. Additionally, UCNPs can be altered to a nanoscale size that allows them to possibly infiltrate cells and disrupt biological functions. The current situation involves the quick development of forensic science applications based on NMs for identifying lawbreakers, criminals and convicts as well as for detecting multiple signals. An increasing subset of the broader class of UCNMs, which include numerous materials with a size of less than 100 nm, are NPs for toxicity. UCNPs are categorized according to their physicochemical characteristics as soluble or insoluble. Because of their chemical makeup, soluble/biodegradable NPs have the potential to change into molecular byproducts during biological and photochemical destruction. Compared to their bulk equivalents, the resulting residues are less toxic since they are simpler to eliminate from the body. Not all UCNPs, however, are better suited for toxicology-based research because some of them have constraints that are still being looked into, such as the fact that they are poisonous and generally insoluble in most solvents. Furthermore, particles that are small (less than 10 nm) can readily infiltrate cells, harming the cellular environment through rapid disintegration and posing a long-term risk to the respiratory system. Recent studies on NPs with sizes range 10–50 nm, sphere shaped particles that can travel through the blood veins, pose low toxicity [13]. However, the solubility in biological fluid such as water can be modified using ligand attachments and thus can make a NP hydrophobic and hydrophilic. Existence of Hazardous materials such as drugs in human blood and respiratory system can be detected using UCNP interactions. However, professional forensic experts use other medical apparatus and procedure to detect toxics.

9.6 In LFP Detection

In general, the surface is crucial to the long-term visibility and stability of LFPs. Various biochemical, physical, and microbiological responses that rely on the immediate environment of the FP continue to alter the LFP identification. Certain LFP are discovered to be more delicate and immature and they could distort in experiments using chemicals. A large range of smooth, reflecting surfaces, including coins, ceramic tiles, leaves, glass, stainless steel pieces, coils, aluminum foil, mobile screens, marble, magazine covers, train tickets, etc., can be detected by UPNPs for LFP. Lifting LFP from smooth surfaces is simple; however, the main challenge is in lifting LFP from nonporous, semi-porous, and porous surfaces. The most common porous surfaces discovered in crime scenes are paper, wood, irregularly shaped blocks, curved pots, soil jars, certain kinds of rock, plasterboard, etc. Sometimes LFP are hard to identify because of insufficient details or traits. Therefore, there is a probability that more sophisticated UC doped NP or luminous materials will be developed. Changes in the synthesis settings might be made from a research perspective to obtain better points and patterns for detection [60, 61].

10 Conclusion

Because of their distinct optical characteristics, lanthanide-doped UCNPs have the potential to be effective chemosensors for fluorescence detection. The synthesis and surface modification of lanthanide-doped UCNPs have advanced recently and we discuss these advancements in this chapter. We also highlight the successes of UC-based fluorescent probes for the detection of pH, ions, molecules, bacteria, reactive species, temperature, and pressure. Even though lanthanide-doped UC nanophosphor-based fluorescent probes have advanced recently, there are still certain obstacles to be solved before UCNPs may be used in more real-world scenarios. First, the UCL efficiency of UCNPs is still low, typically less than 1%. Enhancing efficiency, quantum yield and luminescence decay time is therefore necessary. The creation of novel synthetic techniques to produce UCNPs with the appropriate size and brilliant emission for forensic-based applications presents the second hurdle. The most investigated hosts at the moment are RE-doped $NaYF_4$, $NaLuF_4$ and $NaGdF_4$. These hosts have surface alterations, cooping, core–shell structures, and other modifications. For in vivo detection, it is important to look into the stability and toxicity of the UC-based probes during extended exposure. Finally, but just as importantly, it would be ideal to create standard characterization methodologies or reference materials to facilitate the direct comparison of results from laboratories across the globe. Therefore, given that the field of UCNPs is now experiencing a high level of activity in research, many important breakthroughs in this area should still be expected in the near future. The aforementioned challenges are currently the focus of significant research and development and once they are solved, a wide range of further UCNP-based highly sensitive sensing methods will become accessible. It is quite probable that in the near future, forensics will use UCNPs for a wide range of sophisticated applications because to their extreme adaptability in sensing.

References

1. K.A. Gschneidner, L. Eyring, in *Handbook on the Physics and Chemistry of Rare Earths* (Elsevier, Amsterdam, 2000)
2. X. Qin, X. Liu, W. Huang, M. Bettinelli, X. Liu, Lanthanide-activated phosphors based on 4f–5d optical transitions: theoretical and experimental aspects. Chem. Rev. **117**, 4488–4527 (2017). https://doi.org/10.1021/acs.chemrev.6b00691
3. W.M. Yen, M.J. Weber, in *Inorganic Phosphors, Preparation and Compositions, Optical Properties* (CRC press, Taylor and Francis Group, 2000)
4. W.M. Yen, S. Shionoya, H. Yamamoto, in *Fundamentals of Phosphors* (CRC Press, 2007)
5. R.C. Ropp, in *Luminescence and the Solid State*, 2nd edn. (Elsevier B.V., 2004)
6. C. Sun, M. Gradzielski, Advances in fluorescence sensing enabled by lanthanide-doped upconversion nanophosphors. Adv. Colloid Interface Sci. **300**, 102579 (2022). https://doi.org/10.1016/j.cis.2021.102579
7. J. Zhou, Q. Liu, W. Feng, Y. Sun, F. Li, Upconversion luminescent materials : advances and applications, Chem. Rev. (2015). https://doi.org/10.1021/cr400478f

8. J. Yao, C. Huang, C. Liu, M. Yang, Upconversion luminescence nanomaterials: a versatile platform for imaging, sensing and therapy. Talanta **208**, 120157 (2020). https://doi.org/10.1016/j.talanta.2019.120157
9. D.R. Vij, in *Luminescence of Solids* (Springer Science + Business Media New York, Kurukshetra, n.d.). https://doi.org/10.1007/978-1-4615-5361-8
10. J. Prahlow, *Forensic Pathology* (Springer, New York, 2010)
11. B. Stuart, in *Forensic Analytical Techniques* (Wiley, USA, 2013)
12. C.M. Liu, L.Y. Zhang, L. Li, B.Y. Li, C.G. Wang, T.T. Wang, Specific detection of latent human blood fingerprints using antibody modified $NaYF_4$: Yb, Er, Gd fluorescent upconversion nanorods. Dye. Pigment. **149**, 822–829 (2018). https://doi.org/10.1016/j.dyepig.2017.11.050
13. A.A. Ansari, K.M. Aldajani, A.N. AlHazaa, H.A. Albrithen, Recent progress of fluorescent materials for fingermarks detection in forensic science and anti-counterfeiting. Coord. Chem. Rev. **462**, 214523 (2022). https://doi.org/10.1016/j.ccr.2022.214523
14. K. Tiwari, B.G. Sharma, N. Brahme, D.P. Bisen, T. Richhariya, A. Verma, S. Sahu, R. Tripathi, A. Sinha, Investigation of latent fingerprint detection and cheiloscopy development using Li_2SrSiO_4:Tb^{3+} phosphor for forensic-based applications. ACS Appl. Opt. Mater. **2**, 433–444 (2024). https://doi.org/10.1021/acsaom.3c00453
15. H.E. Barnes, History. Hist. Prospect. Soc. Sci. 1–54 (2011). https://doi.org/10.1037/13442-001
16. M.M. Houck, Forensic Fingerprints Advanced Forensic Science Series, Elsevier, n.d.
17. K. Abhishek, A. Yogi, A minutiae count based method for fake fingerprint detection. Proc. Comput. Sci. **58**, 447–452 (2015). https://doi.org/10.1016/j.procs.2015.08.061
18. K. Shwetabh, M.I. Sarkar, K. Kumar, Optical thermometry, latent fingerprint detection and anti-counterfeiting based on upconversion emission in Ho^{3+}/Yb^{3+} doped KYF_4 nanoparticles. Opt. Mater. (Amst). **145**, 114395 (2023). https://doi.org/10.1016/j.optmat.2023.114395
19. A. Baride, G. Sigdel, W.M. Cross, J.J. Kellar, P.S. May, Near infrared-to-near infrared upconversion nanocrystals for latent fingerprint development, ACS Appl. Nanomater. (2019). https://doi.org/10.1021/acsanm.9b00890
20. M. Wang, Y. Zhu, C. Mao, Synthesis of NIR-Responsive $NaYF_4$: Yb, Er upconversion fluorescent nanoparticles using an optimized solvothermal method and their applications in enhanced development of latent fingerprints on various smooth substrates. ACS Langmuir. (2015). https://doi.org/10.1021/acs.langmuir.5b01151
21. G.P. Darshan, H.B. Premkumar, H. Nagabhushana, S.C. Sharma, Blue light emitting ceramic nano-pigments of Tm^{3+} doped $YAlO_3$: applications in latent finger print, anti-counterfeiting and porcelain stoneware. Dye. Pigment. **131**, 268–281 (2016). https://doi.org/10.1016/j.dyepig.2016.02.015
22. A. Kumar, L. Kumar, K. Kumar, Comparative upconversion emission studies in AF_3 : Tm^{3+}/Yb^{3+} (A = La, Y and Gd) phosphor for latent fingermarks detection. Opt. Mater. (Amst). **145**, 114461 (2023). https://doi.org/10.1016/j.optmat.2023.114461
23. J. Li, X. Zhu, M. Xue, W. Feng, F. Li, Nd^{3+}-Sensitized upconversion nanostructure as a dual-channel emitting optical probe for near infrared-to-near infrared fingerprint imaging. ACS Inorg. Chem. (2016). https://doi.org/10.1021/acs.inorgchem.6b01536
24. G.P. Darshan, H.B. Premkumar, H. Nagabhushana, S.C. Sharma, B.D. Prasad, S.C. Prashantha, Neodymium doped yttrium aluminate synthesis and optical properties—a blue light emitting nanophosphor and its use in advanced forensic analysis. Dye. Pigment. **134**, 227–233 (2016). https://doi.org/10.1016/j.dyepig.2016.06.029
25. W. Jiang, J. Yi, X. Li, F. He, N. Niu, L. Chen, A comprehensive review on upconversion nanomaterials-based fluorescent sensor for environment, biology, food and medicine applications (2022)
26. H. Xie, Q. Wen, H. Huang, T. Sun, P. Li, Y. Li, Synthesis of bright upconversion submicrocrystals for high-contrast imaging of latent-fingerprints with cyanoacrylate fuming. RSC Adv. **5**, 79525–79531 (2015). https://doi.org/10.1039/C5RA15255K
27. K. Shwetabh, M.M. Upadhyay, K. Kumar, Synthesis and upconversion emission studies of $CaYF_5$:Ho^{3+}/Yb^{3+} phosphor and its applications in optical thermometry, fingerprint detection and security ink. RSC Adv. **13**, 9377–9386 (2023). https://doi.org/10.1039/d3ra00644a

28. X. Yang, R. Song, X. Gong, H. Deng, C. He, J. Liu, Multi-shell structured nanomaterials with strong red upconversion emission for trimodal biomedical imaging. Ceram. Int. **50**, 1601–1606 (2024). https://doi.org/10.1016/j.ceramint.2023.10.253
29. J. Dineshshankar, N. Ganapathi, T.R. Yoithapprabhunath, T. Maheswaran, M.S. Kumar, R. Aravindhan, Lip prints: role in forensic odontology. J. Pharm. Bioallied Sci. **5**, 95–98 (2013). https://doi.org/10.4103/0975-7406.113305
30. G.R. Mamatha, B.R. Radha Krushna, J. Malleshappa, B. Subramanian, B. Daruka Prasad, C. Srikanth, H. Nagabhushana, Designing orange-red emitting luminescent platform for data security and information encryption based Sm^{3+} doped BLAO phosphor. J. Photochem. Photobiol. A Chem. **439**, 114560 (2023). https://doi.org/10.1016/j.jphotochem.2023.114560
31. S. Li, J. Guo, W. Shi, X. Hu, S. Chen, J. Luo, Y. Li, J. Kong, J. Che, H. Wang, B. Deng, R. Yu, Synthesis and characterization of wide band excited yellow phosphor $LaNb_2VO_9$:Dy^{3+} and application in potential fingerprint and lip print detection. J. Lumin. **244**, 118681 (2022). https://doi.org/10.1016/j.jlumin.2021.118681
32. D.R. Lavanya, B.R.R. Krushna, K. Manjunatha, B.L. Lyu, H.H. Chiu, W.C. Lo, M.K. Ho, T.E. Hsu, S.Y. Wu, S.C. Sharma, B. Subramanian, J. Malleshappa, C. Srikanth, H. Nagabhushana, Enhanced down-conversion emission, high-level security and advanced latent fingerprint visualization in $La_2Zr_2O_7$:Er^{3+} nanophosphor through surface modification and deep learning analysis. ACS Appl. Opt. Mater. (2024). https://doi.org/10.1021/acsaom.4c00104
33. X. Zhou, L. Ning, J. Qiao, Y. Zhao, P. Xiong, Z. Xia, Interplay of defect levels and rare earth emission centers in multimode luminescent phosphors. Nat. Commun. **13**, 1–9 (2022). https://doi.org/10.1038/s41467-022-35366-3
34. J.J. Yanqing Hu, Q. Shao, X. Deng, D. Song, S. Han, Y. Dong, Thermally induced multicolor emissions of upconversion hybrids with large color shifts for anticounterfeiting applications. J. Mater. Chem. C (2019). https://doi.org/10.1039/C9TC03822A
35. L. He, J. Huang, Z. An, H. Huang, Y. Zhao, K. Zhong, B. Zhou, Achieving thermochromic upconversion of Tm^{3+} for high-sensitive nanoprobe and information encryption **266** (2024). https://doi.org/10.1016/j.jlumin.2023.120306
36. B.K.G. Pawan Kumar, S. Singha, Future prospects of luminescent nanomaterials based security ink: from synthesis to anti-counterfeiting applications. Nanoscale (2016). https://doi.org/10.1039/b000000x
37. T. Zhang, L. Liu, Multi-color luminescence and anticounterfeiting application of upconversion nanoparticle 9273–9280 (2023). https://doi.org/10.1039/d2ra07308k
38. N. Jaroch, J. Czajka, A. Szczeszak, Luminescent materials with dual-mode excitation and tunable emission color for anti-counterfeiting applications. Sci. Rep. 1–9 (2023). https://doi.org/10.1038/s41598-023-37608-w
39. B. Zhou, L. Yan, J. Huang, X. Liu, L. Tao, Q. Zhang, NIR II-responsive photon upconversion through energy migration in an ytterbium sublattice, Nat. Photonics (n.d.). https://doi.org/10.1038/s41566-020-00714-6
40. K. Tiwari, N. Brahme, Other emerging applications of mechanoluminescence and outlook, INC (2024). https://doi.org/10.1016/B978-0-323-95301-6.00008-1
41. X. Wang, H. Zhang, R. Yu, L. Dong, D. Peng, A. Zhang, Y. Zhang, H. Liu, C. Pan, Z.L. Wang, Dynamic pressure mapping of personalized handwriting by a flexible sensor matrix based on the mechanoluminescence process. Adv. Mater. 1–8 (2015). https://doi.org/10.1002/adma.201405826
42. R.S. Fontenot, W.A. Hollerman, S.M. Goedeke, Initial evidence of a triboluminescent wavelength shift for ZnS: Mn caused by ballistic impacts. Mater. Lett. **65**, 1108–1110 (2011). https://doi.org/10.1016/j.matlet.2011.01.043
43. M. Zhang, Forensic imaging: a powerful tool in modern forensic investigation. Forensic Sci. Res. **7**, 385–392 (2022). https://doi.org/10.1080/20961790.2021.2008705
44. M.T. Chaudhery Mustansar Hussain, D.R. Pandey, Atomic absorption and emission spectrometry in forensic analysis, in *Handbook on Analytical Techniques for Forensic Samples* (Elsevier, Amsterdam, 2021), pp. 75–90. https://doi.org/10.1016/B978-0-12-822300-0.00005-7

45. C. Duan, L. Liang, L. Li, Recent progress in upconversion luminescence nanomaterials for biomedical applications. J. Mater. Chem. B 192–209 (2018). https://doi.org/10.1039/c7tb02527k
46. Z. Yi, Z. Luo, X. Qin, Q. Chen, X. Liu, Lanthanide-activated nanoparticles: a toolbox for bioimaging. Ther. Neuromodulation (2020). https://doi.org/10.1021/acs.accounts.0c00513
47. Y. Zhang, X. Zhu, Y. Zhang, in *Exploring Heterostructured Upconversion Nanoparticles: From Rational Engineering to Diverse Applications* (2021). https://doi.org/10.1021/acsnano.0c09231
48. H. Yu, Y. Chai, A. Tiemuer, Y. Han, H. Wang, Y. Liu, Frequency upconversion luminescence: new frontiers in bioapplications. Coord. Chem. Rev. **496**, 215379 (2023). https://doi.org/10.1016/j.ccr.2023.215379
49. K. Krishan, T. Kanchan, A.K. Garg, Dental evidence in forensic identification—an overview, methodology and present status. Open Dent. J. 250–256 (2015)
50. J. Ata-ali, F. Ata-ali, Forensic dentistry in human identification: a review of the literature. Forensic Dent. Hum. Identif. **6**, 162–167 (2014)
51. S. Bhatia, in *Nanoparticles Types, Classification, Characterization, Fabrication Methods* (Springer, Berlin, 2016). https://doi.org/10.1007/978-3-319-41129-3
52. A.R. Azman, N.A. Mahat, R. Abdul Wahab, W.A. Ahmad, D. Ismail, Preliminary forensic assessment of the visualised fingerprints on nonporous substrates immersed in water using the green and optimised novel nanobio-based reagent. Sci. Rep. **12**, 1–10 (2022). https://doi.org/10.1038/s41598-022-18929-8
53. S.K. Sharma, J. James, UV-A,B,C emitting persistent luminescent materials. Materials (Basel) (2022). https://doi.org/10.3390/ma16010236
54. M. Han, L. Liu, D. Zhang, Y. Du, L. Zhao, Y. Wang, L. Lv, Tuning the morphology of Mg_2TiO_4: Mn^{4+} for luminescence performance and latent fingerprint visualization. J. Lumin. **252** (2022). https://doi.org/10.1016/j.jlumin.2022.119417
55. H. Xu, D. Ying, A. Lu, X. Wang, J. Hu, Surfactant-assistant solvothermal synthesis of $CaWO_4:Eu^{3+}$ phosphors and luminescence. Superlattices Microstruct. **83**, 668–675 (2015). https://doi.org/10.1016/j.spmi.2015.04.011
56. V.J. PAwde, S.J. Dhoble, in *Phosphors for Energy Saving and Conversion Technology*, 1st edn. (CRC press, Taylor and Francis Group, London, 2019)
57. P.P. Varghe, N.T. Kalyani, P.G. Shende, S.J. Dhoble, Synthesis and characterization of energy-efficient Mq_2 (M=Zn, Cd, Ca and Sr) organometallic complexes for OLED display applications. Energy Mater. Fundam. Appl. **2**, 505–521 (2021). https://doi.org/10.1016/B978-0-12-823710-6.00003-0
58. E. Andresen, F. Islam, C. Prinz, P. Gehrmann, K. Licha, J. Roik, S. Recknagel, U.R. Genger, Assessing the reproducibility and up—scaling of the synthesis of Er, Yb—doped $NaYF_4$-based upconverting nanoparticles and control of size, morphology and optical properties. Sci. Rep. 1–13 (2023). https://doi.org/10.1038/s41598-023-28875-8
59. S. Shikha, T. Salafi, J. Cheng, Y. Zhang, Versatile design and synthesis of nano-barcodes. R. Soc. Chem. 7054–7093 (2017). https://doi.org/10.1039/c7cs00271h
60. M.R. Hawthorne, in *Fingerprint Analysis and Understanding* (Taylor & Francis, 2009)
61. H. Moses Daluz, in *Fundamentals of Fingerprint Analysis* (Taylor & Francis, 2015)

A Comprehensive Study on the Thermoluminescence Dosimetry of Phosphors

Rodali Goswami, M. B. Latif, and Vikas Dubey

Abstract This review presents a detailed exploration of recent advancements in thermoluminescence dosimetry (TLD), with a particular focus on inorganic phosphor materials. Thermoluminescence, the light emitted by a material when heated after exposure to ionizing radiation, plays a vital role in measuring radiation doses. In this study, a wide array of phosphors—spanning chemical families such as aluminates, borates, fluorides, phosphates, silicates, sulfides, and titanates—are examined in terms of their synthesis techniques, structural characteristics, and suitability for radiation detection. Special attention is given to essential dosimetric attributes, including glow curve profiles, dose–response behavior, sensitivity, fading resistance, and biological tissue equivalence. Additionally, the article sheds light on the applications of these materials across various domains, such as medical, environmental, space, and personal radiation monitoring. By compiling insights from numerous studies, this review serves as a valuable resource for researchers and professionals seeking to optimize the performance of TLD materials for specific applications.

Keywords Thermoluminescence · Dosimetry · Phosphors · Tissue equivalent

1 Introduction

Exposure of ionizing radiation (x, β, γ rays and neutrons) to phosphors led to the emission of light during heating, the process, being called as Thermoluminescence (TL). When a material or phosphor is being irradiated by ionizing radiations, then

R. Goswami (✉) · V. Dubey
North Eastern Hill University, Shillong, Meghalaya, India
e-mail: rodaligoswamiing@gmail.com

V. Dubey
e-mail: jsvikasdubey@gmail.com

M. B. Latif
Division of Applied Nuclear Science and Technology, Centre for Energy Research and Development, Obafemi Awolowo University, Ile-Ife, Nigeria
e-mail: mlatif@oauife.edu.ng

V. Dubey et al. (eds.), *Upconversion Luminescence of Phosphors*, Progress in Optical Science and Photonics 39, https://doi.org/10.1007/978-981-95-1762-6_7

defects get created in the material's crystal structure. These defects act as traps for electrons. These trapped electrons, when get excited due to the radiation jumps from the conduction band to the valence band. These trapped electrons are confined to specific energy levels, which are known as trap centers or trapping centers [1–4].

Trapping centers are of two types:

(1) Shallow Trapping: When electrons are trapped by the energy levels near the conduction band, then a small amount of thermal energy is required for the release of trapped electrons by the energy levels, this process is known as shallow trapping. Trapped electrons are released faster in case of swallower trapping.
(2) Deep Trapping: When electrons are trapped by the energy levels far from the conduction band, then a much higher amount of thermal energy is required for the release of trapped electrons by the energy levels, this process is known as deep trapping. Trapped electrons are released slowly at high temperature in case of deep trapping.

When the material or persistent phosphor gets heated then the electrons get released from trap center, the released electrons then pair up or recombines with the positive charged defects of the crystal known as holes, resulting in the emission of light in the form of TL. The light emitted in the form of TL is detected and amplified by using a photomultiplier tube (PMT) or other light detection systems [5].

Thermoluminescence Dosimetry (TLD) is one of the most important methods in the field of science, which is used to measure the amount of dose proportional to the radiation. The use of TL in dosimetry was first proposed by an American scientist F. Daniels around 70 years back.

For a material to be a good thermoluminescent dosimeter (TLD), it should have a great trapping efficiency as well as a large number of electron–hole pairing should take place [5, 6]. The average time of the stay of electron in its trap is also an important factor for a material to be a good thermoluminescent dosimetry materials. All the above-mentioned factors can only be determined by using glow curve analysis. A glow curve is a plot of light intensity as a function of temperature. The glow curve analysis of a given material can be obtained by exposing the material to ionizing radiation. The glow curves intensity is directly proportional to the radiation dose absorbed by a material or a phosphor. Different distinct glow curves are produced by different types of radiation and its intensity [6–10].

The use of "Phosphors" as thermoluminescent dosimetry (TLD) materials starts emerging from the late 19th century and 20th century. The use of these phosphors as TLD materials was first done by Philip E. A. Lenard and his co-workers in Germany and they achieved impressive results. They performed the experiments on alkaline earth chalcogenides (sulfides and selenides) and investigated their properties as thermoluminescent dosimetry (TLD) materials [11].

Now-a-days, a lot TLD of commercial materials or Phosphors are available like LiF: Mg, Ti (TLD-100),CaF_2:Dy (TLD-200), CaF_2:Tm (TLD-300), CaF_2:Mn (TLD-400), Al_2O_3:C (TLD-500), LiF: Mg, Cu, P (TLD-700), and $CaSO_4$:Dy (TLD-900) [12].

These are highly useful as TLD materials because of its following properties:

1. Highly Sensitive: Phosphors are said to be highly selective TLD materials if they can detect small changes in dose of radiation.
2. Low Fading: A Phosphor is said to be a good TLD material, if the TL signal intensity is stable with relative to time and the decrease in TL signal intensity is very minimal.
3. Simple Glow Curve Structure: Phosphors are said to be good TLD material, if their Simple Glow Curve shows how accurately the radiation dose is measured.
4. Linear Dose Response: Phosphors are good TLD materials, if they have linear dose response, i.e. the TL signal emitted by a material is directly proportional to the radiation dose absorbed by the material, over a specific dose range.
5. Tissue Equivalent: A Phosphor having tissue equivalent composition is said to be a good TLD material. A Phosphor is said to be tissue equivalent if its composition is similar to that of a human tissue as well as the way it interacts with the ionizing radiation is similar to that of human tissue.
6. Easily Reproducible: A Phosphor is said to be an easily reproducible TLD material if its signal remains stable when it is exposed to same radiation dose as well as the variability in the signal is very low or less. Along with this, they can be used repeatedly by resetting the previous exposure.
7. Good Reusability: Good Reusability of a TLD material or Phosphors suggest that a TLD material can be reused several times without any significant loss of its thermoluminescence properties.

Due to the above-mentioned properties Phosphors are effectively used in Thermoluminescence Dosimetry (TLD).

2 Overview on Phosphors and Inorganic TLD Materials

2.1 General Overview on Phosphors

Phosphors that are generally used as TLD materials are actually a class of luminescent materials that emits light when they get excited by an external source of energy. A Phosphor generally consists of a host material and one or more activator ions.

(i) Host Material: It is a crystalline lattice that provides the main structure of the Phosphor. It is the place where activator ions of the Phosphor get embedded. Metal oxides, silicates, and aluminates are common host materials.
(ii) Activator Ions: Activator ions are the impurities that are introduced into the host materials and they create luminescence centers for absorption and emission of ions. Rare earth activated ions and transition metals are generally used as activator ions.

A number of organic as well as inorganic phosphors are used as TLD materials. A number of organic phosphors like Pyrene ($C_{16}H_{10}$), Acridine ($C_{13}H_9N$), Tetradentate

Pd(II) [13], Pyranone [14], and Pyrrolidone [15], are present in nature or are derived from other phosphors. All these organic phosphors are very useful and used in the field of medical as well as personnel dosimetry.

There are also a large number of inorganic phosphors that are used as TLD materials. Recently, researchers started using a large number of rare earth activated inorganic phosphors as TLD materials actively [16, 17].

Inorganic Phosphors can be further classified into different types. Some of the following types are mentioned below:

1. Aluminate Phosphors
2. Borate Phosphors
3. Fluoride Phosphors
4. Oxide Phosphors
5. Phosphate Phosphors
6. Rare Earth-Doped Phosphors
7. Silicate Phosphors
8. Sulfide Phosphors
9. Titanate Phosphors

2.2 Inorganic Phosphors as TLD Materials

Under this section, we will discuss about the various types of inorganic phosphors that are studied for Thermoluminescence Dosimetry purposes.

a. ***Aluminate Phosphors***: Aluminate Phosphors are the type of Phosphors that have aluminate compound as the host matrix. These Phosphors are often doped with rare earth elements (Eu, Dy) as well as transition metals (Mn, Cr, Zn, Ti, Y, Li, Ge, Sr), non-metals like C and lanthanides (Ce, Tm) to enhance their properties so that they behave as good TLD materials. Aluminate Phosphors are highly used as TLD materials because of their high stability, linear dose response and good thermoluminescence properties. A few examples of aluminates that are used as TLD materials are Al_2O_3 [18], $SrAl_2O_4$ [5], and $ZnAl_2O_4$ [19].
b. ***Borate Phosphors***: Phosphors that have BO_3 group as their host, are the Borate Phosphors. Borates are generally a combination of boron and oxygen. Transition metals like Ag, Cu, Mn, and nonmetals like P and post-transition metals like In as well as lanthanides like Tm are used as dopants. All these dopants are used with the host matrix so that they can be used as good TLD materials. Easily Reproducible, Wide Dose Range, Tissue Equivalence, Low Fading, and high sensitivity are the characteristics of Borate Phosphors that make them good TLD materials. $Li_2B_4O_7$ [20–31], MgB_4O_7 [32–42], $LiCaBO_3$ [43, 44], and $LiMgBO_3$ [43, 45] are some of the examples of borate phosphors.
c. ***Fluoride Phosphors***: Phosphors that have fluoride ions as their host which emit light when exposed to radiation are called as Fluoride Phosphors. Fluoride Phosphors are often doped with rare earth elements like Eu, Dy, transition metals like

Mn, Ti, Cu alkaline earth metals like Mg as well as lanthanides like Ce, Tm to enhance the properties of phosphors as TLD materials. Fluoride Phosphors are used as TLD materials because of their high sensitivity, linear dose range, wide dose response and tissue equivalent nature. A few examples of Fluoride TLD materials are $LiCaAlF_6$ [46], LiF [47], and CaF_2 [3].

d. ***Oxide Phosphors***: Oxide Phosphors are a class of luminescent materials that has metal oxide as its host. Oxide Phosphors generally have rare earth elements like Eu, Tb, Dy, transition metals like Cu, Mn, Ni, alkaline earth metals like Mg as well as lanthanides like Sm, Ce as their dopants. Oxide Phosphors are considered as good TLD materials because of their high sensitivity, linear dose response as well as wide dose range and reusable nature. Example of oxide phosphor as TLD material is BeO [48–51].
e. ***Phosphate Phosphors***: Phosphors having PO_4 group in their host are the Phosphate Phosphors. For a Phosphate Phosphor to be good TLD material they are doped with rare earth elements like Eu, Dy, Tb, Er, Gd then lanthanides like Sm, Ce, transition metals like Cu, Y. Being a material with wide dose response, low fading, reusability, and good thermoluminescence efficiency, Phosphate Phosphors are used in dosimetry purpose as TLD material. Some example of Phosphate Phosphors are $Sr_2P_2O_7$ [52], $LiCaPO_4$ [53], $NaCaPO_4$ [54], $KBaPO_4$ [55], $LaPO_4$ [56], and $NaLi_2PO_4$ [12, 57].
f. ***Rare Earth Phosphors***: Recently, the use of rare earth phosphors as TLD materials gains high significance. Rare Earth elements as host with dopants like Er, Eu, Dy, Yb, Tb, Gd (rare earth elements), Sm, Ce (lanthanides) and Mn, Cr (transition metals) combine to form the Rare Earth Phosphors. Rare Earth Phosphors are basically used in thermoluminescence dosimetry because of their high sensitivity, good tissue equivalence, linear dose response, and low fading. A few examples of rare earth phosphors are YAG [58–60], $Gd_2Zr_2O_7$ [61], Gd_2O_3 [62].
g. ***Silicate Phosphors***: Phosphors having silicon and oxygen in their chemical structure that can be used as host are called as Silicate Phosphors. They are usually doped with rare earth elements like Dy, Eu, Tb and lanthanides like Ce. In most of the literature, rare earth elements are used as dopants and co-dopants for Silicate Phosphors. Silicate Phosphors are used as TLD materials because of their good storage capacity. The general characteristics of Silicate Phosphors as TLD materials are high stability and versatile preparation method. Silicate Phosphors show persistence luminescence, so are highly preferable for thermoluminescence dosimetry. Some examples of Silicate Phosphors are Ba_2MgSiO_2 [63], Sr_2SiO_4 [64], $Sr_2MgSi_2O_7$ [65], Ba_2SiO_4 [66], and $Sr_3MgSi_2O_8$ [66, 67].
h. ***Sulfide Phosphors***: These Phosphors have sulfide (SO_4) phosphors in their chemical structure along with metals as their host. Sulfide Phosphors are mostly doped with rare earth element Dy. Some of the most used dopants for sulfide phosphors are rare earth elements like Dy, Tm, Eu, nonmetals like P, metals like Mo, Ag, Zn and lanthanides like Ce. Sulfide Phosphors are considered as good TLD materials because of their high sensitivity, linear dose response, low fading, and good tissue equivalent. Some examples of Sulfide Phosphors are $K_2Ca_2(SO_4)_3$ [68], $BaCa(SO_4)_2$ [69], $CaSO_4$ [70–82], $K_3Na(SO_4)_3$ [83].

i. ***Titanate Phosphors***: Titanium Phosphors have Ti ions in their chemical structure along with other metals as their host. The most common type of dopants for Titanium Phosphors are rare earth elements (Dy, Eu, Tb, Pr) and transition metal like Mn. Titanium Phosphors are used as TLD materials for their high sensitivity, good tissue equivalence, linear dose response, and low fading. Example of Titanate Phosphor is $CaTiO_3$ [84, 85].

3 Method of Synthesis of TLD Materials or Phosphors

There are various literatures that tells us about the various methods of synthesis of TLD materials or the Phosphors [43, 84, 86–111]. Here, we are discussing some of the method of synthesis.

1. **Wet Chemical Method**: In this method, materials in the form of nitrates are dissolved in water and when the solution gets formed the solution is first heated to form a solute and again the solute is heated to remove the remaining moisture from the solute. Wet Chemical Method is further classified into:

    1.1. **Main Wet Chemical Method**: In this method, the precursor or the starting material should be in stoichiometric ratio and we should dissolve them separately in different beakers and in minimum amount of water. Then the material should be constantly stirring on hotplate keeping it at 80 °C. After this, we have to add all this materials in hot beaker and had to done constant stirring and keep it on hotplate. And after that, a white precipitate gets formed and we have to keep it overnight to get dry at 90 °C in hot air oven and when the dry sample gets formed we have to anneal it for 2 h to remove the impurity present in the sample. Then we can collect the final product by cooling it [84, 89, 90].

    1.2. **Sol–Gel Method**: In this method, metal alkaloids are used for the production of main material or powder with a series of reaction that includes hydrolysis, gelation, drying, thermal treatment and annealing. In Sol–Gel Method, the precipitate formed during gelation is black in color and it has to be annealed at a temperature of 500 °C. After annealing the black precipitate turns into a white fine powder [84, 89, 90]. Dy^{3+} activated $Sr_5(PO_4)_3F$ a type of phosphor is prepared by sol–gel method [91].

    1.3. **Co-precipitation Method**: In this method, the chemical compounds or the salts are dissolved in double distilled water solution and the trace amount of activator or rare earth are dissolved in water solution with a slight amount of ethanol. Without ethanol in the solution the particle size cannot be controlled as well as there will be no consistency in the particle size [43, 84, 92]. The precipitated compound is then washed several times to remove impurity and then dried for 24 h at 100 °C, then again the sample is calcined for 2 h and we finally get the final product after cooling it at room temperature [93, 94].

1.4. **Combustion Method**: In this method, oxide-based phosphors of smaller size are prepared by a high exothermic reaction between an organic fuels and metal salts in an aqueous solution [43]. It can be further classified into two subcategories Simple Combustion Method and Solution Combustion Method. Different types of fuels like urea, citric acid, glycerine, ethanol, and DFH are used by both of these combustion methods for preparation of sample by both the methods.

1.4.1. **Simple Combustion Method**: In this method, first we have to weigh all the samples in stoichiometric ratio and then dissolve the sample and the fuel (like citric acid) in a mortar pestle using distilled water [95]. Then, the sample should be burnt in the furnace at 500–550 °C putting them in a china dish. After that the sample should be taken out from the furnace and let it to be cool at room temperature. After cooling the sample should be crushed for five minutes and then again annealed for 24 hours at the furnace for removing impurity and then again cooled down the sample at room temperature for further studies [84].

1.4.2. **Solution Combustion Method**: In this method, all the starting materials are dissolved in a beaker by keeping it on constant stirring at constant temperature on magnetic stirrer. Once the starting material and the fuel dissolved, the solution should be transferred into the china disc and have to repeat the same procedure as that in Simple Combustion Method [84].

2. **Solid State Diffusion Method**: In this method, the desired materials are mixed properly and sintered at high temperature for the formation of compound as well as uniform dispersal of the activator ions in the lattice takes place [43].
3. **Growing Single Crystals from Melt**: Through this method the single crystals of semiconductors, metals, oxides, halogenides, and chalcogenides are grown at large scale. This can be further classified into the following subcategories [43]:

3.1. **Czochralski Technique**: It is the process of growing a large, high quality single crystal by melt-growth techniques.

3.2. **Bridgman Technique**: In this technique, the directional solidification of a melt takes place. The slow moving of a crucible that contains the material through a temperature gradient is the basis for Bridgman Technique.

3.3. **Verneuil Technique**: It uses a high-temperature oxyhydrogen flame to melt finely powdered material. It was originally developed for the production of synthetic gemstones.

3.4. **Kyropoulos Technique**: This technique is used to produce large, high quality single crystals like sapphire. In this technique, a seed crystal is usually immersed in a molten material, and the temperature is lowered gradually allowing the crystal to grow downwards in the melt.

4. **Re-crystallization from Solution**: In this method, the sample consisting of the host atom and the activator ion is dissolved in a hot solvent to form a solution which is then evaporated to become dry to form the required TLD phosphor [43].

$CaSO_4$: Dy and $CaSO_4$: Tm are examples of phosphors prepared by this method [70].

Table 1 describes the various types of phosphors one from each family and their method of synthesis of as TLD materials.

Table 1 Some of the Phosphor materials prepared using various methods of synthesis

Materials	Precursors	Dopants	Method of synthesis	References
$ZnAl_2O_4$	$Zn(NO_3)_2.9H_2O$, $Al(NO_3)_3.6H_2O$,NH_4OH	Dy	Co-precipitation method	[19]
Al_2O_3	$Al(NO_3)_3$, $CO(NH_2)_2$	Sr, Li, Ge	Combustion method	[112]
SrB_4O_7	$SrCO_3$, H_3BO_3	Eu	Solid-state diffusion method	[112]
BaB_2O_4	$BaCO_3$, H_3BO_3	Tb	Solid-state method	[113]
$LiCaAlF_6$	LiCl, $CaCl_2$, $AlCl_3$, HF	Eu	Co-precipitation method	[46]
CaF_2	$Ca(NO_3)_2$, NH_4F	Tm	Hydrothermal method	[114]
YAG	$Ga(NO_3)_3$, $Al(NO_3)_3$, CH_6N_4O	Pr	Combustion method	[115]
ZnO	$Zn(NO_3)_2$, $C_2H_5NO_2$	–	Combustion method	[116]
ZrO_2	$C_{12}H_{28}O_4Zr$, C_2H_5OH, HCl	–	Sol–gel method	[117]
$NaLi_2PO_4$	LiOH, NaH_2PO_4	Eu	Solid state diffusion method	[12]
$KMgPO_4$	KNO_3, $Tb(NO_3)_3$, $NH_4H_2PO_4$, $Mg(NO_3)_2.6H_2O$, $Tb_4O_7 + HNO_3$	Tb	Solid state method	[118]
$BaSi_2O_5$	$BaCO_3$, SiO_2, HNO_3	Sm	Gel combustion method	[119]
$SrSiO_3$	$Sr(NO_3)_2$,Na_2SiO_3	Tb	Co-precipitation method	[120]
$K_2Ca_2(SO_4)_3$	$CaCl_2$, KCl, $CuCl_2$, $(NH_4)SO_4$	Cu	Chemical co-precipitation	[121]
$BaSO_4$	$Ba(NO_3)_2$, $(NH_4)SO_4$, CH_4N_2O	Eu	Solid state combustion method	[122]
$CaTiO_3$	$CaCl_2.2H_2O$, TiO_2, KOH	Ce	Hydrothermal method	[122]
$BaTiO_3$	$BaCO_3$, $Ti\{OCH(CH_3)_2\}_4$, CH_3COOH	Er, Yb	Co-precipitation method	[123]

4 Studies of Available Commercial TL Dosimeters

There are a number of commercially available TLD phosphors LiF: Mg, Ti (TLD-100), and CaF_2: Dy (TLD-200), CaF_2: Tm (TLD-300), CaF_2: Mn (TLD-400), Al_2O_3: C (TLD-500), LiF: Mg, Cu, P (TLD-700), and $CaSO_4$: Dy (TLD-900) and each phosphor has their own advantages and disadvantages. Under this section, we will discuss all the commercial dosimeters and their advantages and disadvantages.

LiF: Mg, Ti (TLD-100) was the first ever commercial TL dosimeter available. The study of this phosphor gained a lot of attention due to its tissue equivalence property, because of which it can be extensively used in personnel dosimetry. Due to its wide useful range, sensitivity to small doses, tissue equivalence, and limited fading it is highly considered an important TLD material [124, 125]. Due to high sensitivity, wide dose range, and low dose measurements of CaF_2: Dy (TLD-200), high sensitivity, linearity, and wide dose response of CaF_2: Tm (TLD-300) and high mechanical strength, single glow curve structure, and sensitivity to thermal stock of CaF_2: Mn (TLD-400), these phosphors are used as commercial dosimeters [124–126]. Al_2O_3: C (TLD-500) is another commercial dosimeter used because of its sensitivity to radiation, simplest TL curve, and linear dose response [125]. LiF: Mg, Cu, P (TLD-700) are used as commercial dosimeters because of their ability to measure accurate and low doses of radiation; as well as they also have linear dose response, low fading, and tissue equivalence because of which they are highly useful TL dosimeter. $CaSO_4$: Dy (TLD-900) is a commercial dosimeter known for its high sensitivity to low doses and wide dose response.

But each phosphor has its disadvantages like LiF-doped phosphors loses its sensitivity when heated above certain temperature and is used to detect only low doses of radiation; Al_2O_3: C and $CaSO_4$: Dy phosphors has rapid fading and are not tissue equivalent; CaF_2-doped phosphors cannot be used to detect high doses of radiation. Like this, every phosphor has its own drawback because of which there is always in need of some new TLD materials that has the ability to serve as the next generation commercial TL dosimeters.

5 Studies on Some of the Recent TLD Materials

Under this section, we study some of the TLD Phosphors that has the ability to be next generation commercial TL dosimeters.

a. **$NaLi_2PO_4$**

$NaLi_2PO_4$:Eu^{3+} synthesized by solid state method was earlier used for lighting purposes [127]. However, the same material when exposed to gamma radiation shows a sub-linear up to 10 Gy of the dose and later it becomes linear till it saturates beyond 1 kGy and it is more sensitive than most of the available commercial TLD phosphors (Fig. 1b) [12]. Another material $NaLi_2PO_4$: Ce^{3+} shows a sub-linear response at

0.1–1.0 Gy and saturates at 1.0 kGy [128]. Due to its high sensitivity and good reusability, Ce-doped $NaLi_2PO_4$ can be used for low dosimetric purposes while due to low fading and high sensitivity Eu-doped $NaLi_2PO_4$ can be used for high dosimetric purpose. And here it can be said that $NaLi_2PO_4$ is a novel phosphor which can be commercialized as advanced dosimetry devices because of its combination of high sensitivity, broad dose range, and thermal stability.

b. **LiCaAlF6**

$LiCaAlF_6$ is a versatile phosphor. $LiCaAlF_6$:Eu is found to be 5 times more sensitive than the commercially available $CaSO_4$:Dy (Fig. 2a) phosphor. And its low fading and high reusability make it a good candidate for high dosimetric purposes [46]. $LiCaAlF_6$: Ce prepared by Czochralski technique gives intensive thermostimulated luminescence in the UV region with peak at 350 °C (Fig. 2b) which is very

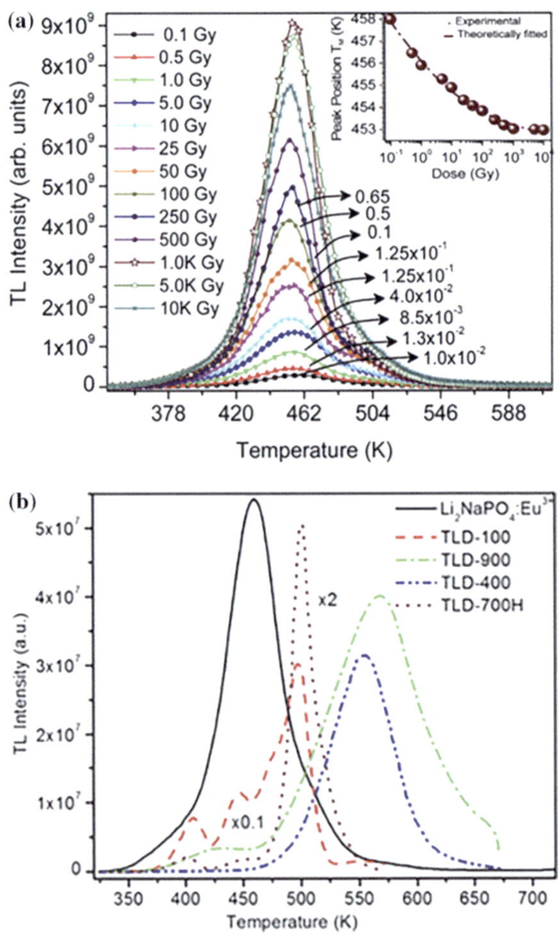

Fig. 1 **a** TL glow curves of $NaLi_2PO_4:Eu^{3+}$ exposed to various doses of γ-rays showing peak in the dose range of 0.1 Gy–10 kGy [12], **b** comparison of $NaLi_2PO_4:Eu^{3+}$ with some commercially available TLD phosphors [12]

useful for TL dosimetry [129]. When $LiCaAlF_6$ is doped with Eu it produces a prominent peak at 180 °C and a shoulder peak at 240 °C, while $LiCaAlF_6$:Eu is co-doped with Y it produces a more enhanced peak at 180 °C and remove the 240 °C peak whereas co-doping of La removes the 180 °C peak and produced a strong peak at 240 °C, TL intensities of both co-doped samples are nearly 3 times that of $CaSO_4$:Dy (Fig. 2c) and are highly used for personnel monitoring [130]. Due to its superior TL characteristics, $LiCaAlF_6$-based phosphors are highly suited for incorporation into next-generation dosimeter badges used in nuclear facilities, medical centers, and space missions.

c. **YAG**

YAG is highly used in dosimetry because of its ability of low fading. YAG: Ce^{3+} and YAG: Tb^{3+} synthesized by sol–gel method shows an increase in sensitivity and TL efficiency as well as linear dose response. The fading was high at first but latter get stabilized after 48 h [131–135]. These YAG-doped phosphors are used

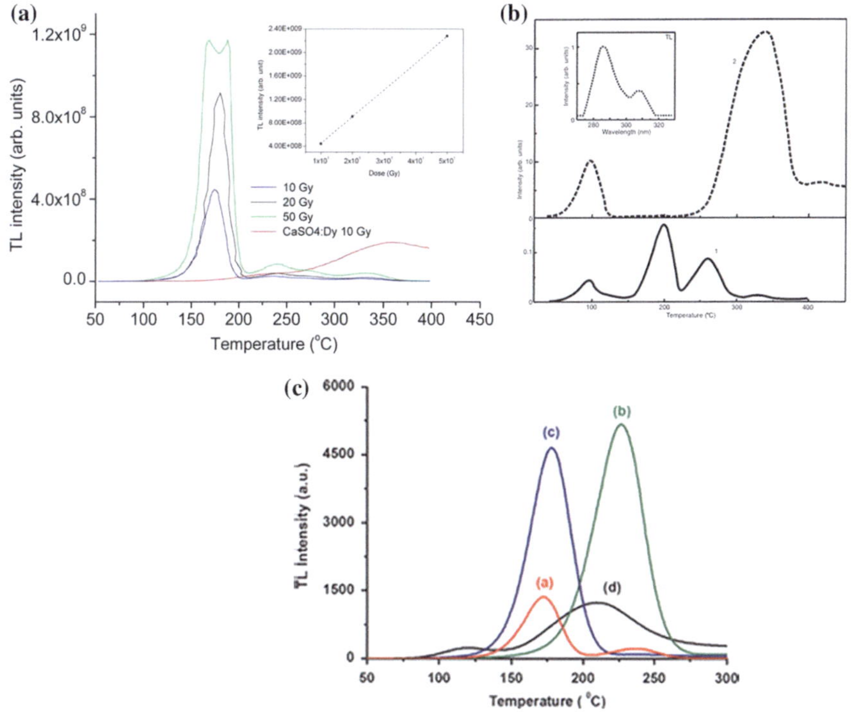

Fig. 2 **a** Comparison of dose response of $LiCaAlF_6$:Eu with commercially available standard phosphor $CaSO_4$:Dy(TLD 900) [46], **b** Thermostimulated Luminescence (TSL) of pure and Ce-doped $LiCaAlF_6$ crystal after X-ray irradiation. $LiCaAlF_6$ showing a peak at 350 °C [129], **c** Comparison of $LiCaAlF_6$ when (**a**) doped with Eu; (**b**) doped with Eu, La; (**c**) doped with Eu, Y and (**d**) $CasO_4$: Dy [130]

for high dosimetry purposes. Because of low sensitivity, a new C-doped YAG phosphor is developed by temperature gradient technique, where the dopant C increase the concentration of F^+ center and thus enhanced the TL sensitivity of the YAG phosphor [136]. And this result indicates that YAG phosphors can be used for both high as well as low dosimetric purposes by just tuning sensitivity through dopant choice (Figs. 3 and 4).

d. **BaTiO3**

Barium Titanate Phosphors are synthesized by solid state method. If $BaTiO_3$ is exposed with ^{60}Co between 1 and 9 kGy then the material shows good linearity to the applied dose as well as the material is reproducible. The fading was very fast initially but later it becomes stabilized which shows that $BaTiO_3$ as a good TLD material [137]. TL characteristics of undoped $BaTiO_3$ and $BaTiO_3$ doped with Gd^{3+}, Dy^{3+}, Tb^{3+}, and Lu^{3+} were characterized. All of them were prepared with the solid state method. Tb^{3+} and Gd^{3+} exhibit two glow peaks at 95 and 167 ° C and Dy^{3+} shows a strong peak at 95 °C. However, $BaTiO_3$ and $BaTiO_3:Lu^{3+}$ do not show any TL characteristics [138]. The linear dose response in the kilogray range positions $BaTiO_3$-based phosphors as suitable candidates for high-dose radiation monitoring, such as in industrial sterilization or radiation processing facilities (Fig. 5).

e. **CaSO4**

$CaSO_4$:Dy, Eu; $CaSO_4$:Dy, Tb; $CaSO_4$: Dy, Tm; $CaSO_4$:Tm are synthesized by the method of slow evaporation using sulfuric acid, calcium carbonate, and dopants Dy, Eu, Tb and Tm by Forner et al. [139], the glow curves of all the materials except $CaSO_4$:Eu shows maximum luminescent at 260 °C while among all the materials $CaSO_4$:Tm showed the highest thermoluminescent emission having the highest sensitivity with a fading of about 7.7% in two months making it a material for better dosimetric purposes [139]. $CaSO_4$: Dy, Cu samples are prepared by recrystallization, acid evaporation method [70, 140]. Cu suppresses effectively the growth of high temperature peak structure centered around 350 °C. However, Cu less than 0.1 mol % does not affect the TL sensitivity. Introduction of small traces of Cu to $CaSO_4$: Dy drastically enhances the intrinsic UV response of $CaSO_4$:Dy. Thus a small traces of Cu can make $CaSO_4$: Dy a good TL as well as UV dosimeter [140]. The TL characteristics and sensitivity of $CaSO_4$ strongly depend on the choice and concentration of dopants, which allows fine-tuning for specific dosimetric needs (Fig. 6).

f. **BaSi2O5**

$BaSi_2O_5:Sm^{3+}$(0.03 mol%) synthesized by gel combustion method by Alajlani et al. [119]shows two main peaks, one at 91 °C and another at 193 °C with a heating rate of 2 °C s^{-1} after 5 Gy β-irradiation. It shows a linear dose response from 1–10 Gy. The activation energy (E) and frequency factors (s) involved in the TL emission for CGCD method were found to be between 0.93 and 1.72 eV and 10^{10}–10^{13} s^{-1} respectively. All these result indicates that $BaSi_2O_5:Sm^{3+}$ can be used as a promising material for TL applications in environmental and retrospective dosimetry [119]. Another

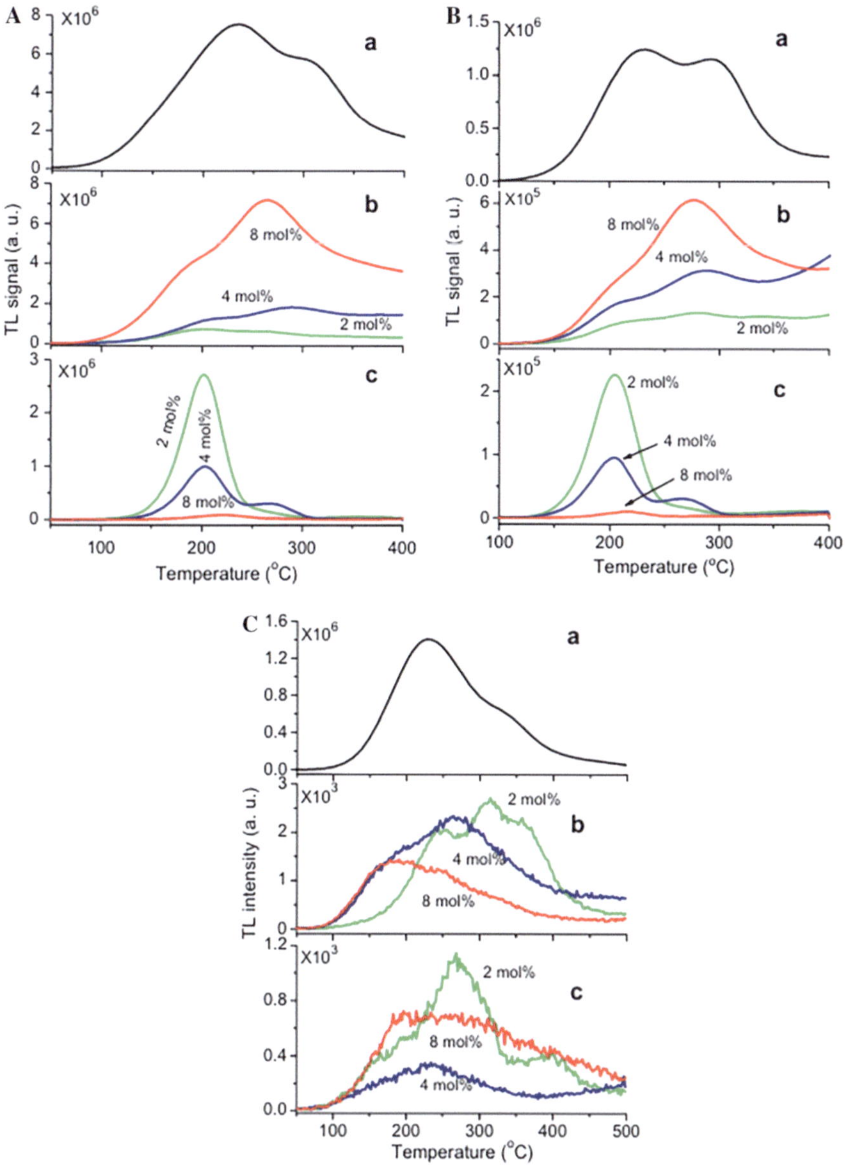

Fig. 3 **a** TL signal of YAG when irrdiated with UV (**a**) undoped; (**b**) doped with Tb^{3+}; (**c**) doped with Ce^{3+} [58], **b** TL signal of YAG when irradiated with X-ray (**a**) undoped; (**b**) doped with Tb^{3+}; (**c**) doped with Ce^{3+} [58], **c** TL signal of YAG irradiated with β-ray (**a**) undoped; (**b**) doped with Tb^{3+}; (**c**) doped with Ce^{3+} [58]

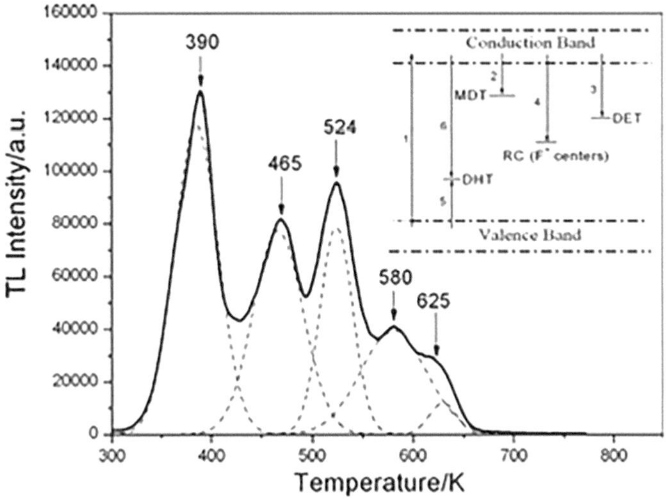

Fig. 4 β-irradiated TL glow curve of YAG:C for a dose of 50 Gy [136]

material Eu-doped $BaSi_2O_5$ glass ceramics (BS900 AND BS1050) synthesized by conventional melt quenching technique under a mixture of H_2, Ar, and heat treatment shows high sensitivity by showing a strong TL response in between 0.01 to 1000 mGy, while BS900 and BS1050 shows a strong peak around 180 °C and a shoulder peak around 100 °C while BS750 shows a relatively weak peak around 144 °C and due to all these response $BaSi_2O_5$ glass ceramics can be compared with the commercial dosimeters and are applicable for TL dosimetric applications such as imaging plates [141]. From these observations, we can conclude that $BaSi_2O_5$ serves as a very good TL material for both environmental and low-dose applications, with sensitivity and performance comparable to commercial dosimeters (Fig. 7).

g. **BeO**

BeO_R (used Optically Stimulated Luminescence technique to detect radiation) dosimeter was compared with the BeO_T dosimeter and BeO_T (used Thermoluminescence to detect radiation) dosimeter and it is observed that BeO_R shows a visible difference on the glow curve by showing a double peak in the region 100–300 °C at a heating rate of 2 °C s^{-1} when compared to BeO_T [142]. BeO_T has a low detection limit in OSL whereas BeO_R has a low detection limit in TL, so Askar et al. [142] suggested that the TL and OSL signals of the dosimeters should be examined before selecting them as a dosimeter for investigations. Altunal et al. [143] synthesized BeO: Na, Yb, Dy by the co-precipitation method and the luminescence characterization TL and OSL are performed. The OSL signal of peak is found to be high 1.3×10^{14} counts/mg for 0.2 Gy beta dose and a wide green emission band between 475 and 625 nm was detected for the first time in the RL emission spectra. BeO: Na, Dy, Yb showed a nearly linear wide range from 0.5 Gy to 100 Gy, reusability and negligible fading (89% in 30 days) it can be highly used as a personnel dosimeter [143]. These findings

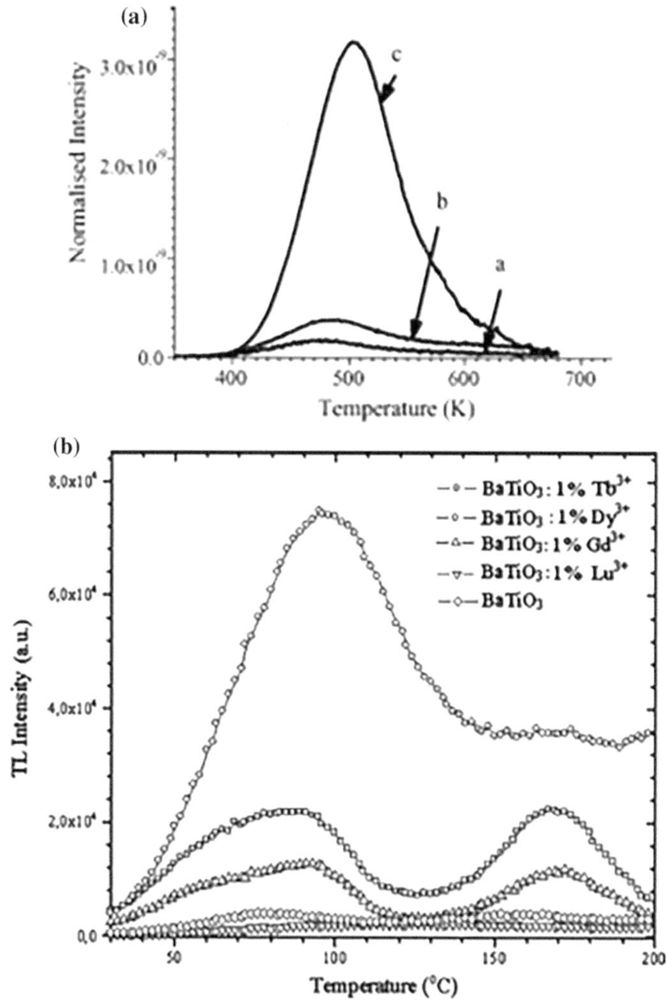

Fig. 5 **a** TL glow curve of undoped $BaTiO_3$ irradiated with (**a**) 2.27 kGy, (**b**) 4.54 kGy and (**c**) 9.08 kGy of gamma doses [137], **b** comparison between the TL glow curves of undoped $BaTiO_3$ and $BaTiO_3$ doped with Lu^{3+}, Dy^{3+}, Gd^{3+}, and Tb^{3+} when irradiated with a β-source [138]

collectively point to the growing potential of BeO-based materials in dosimetry, with tunable properties achievable through strategic doping and synthesis techniques, thus offering flexible options for both high- and low-dose radiation monitoring (Fig. 8).

h. **MgB4O7**

MgB_4O_7: Pr, Dy synthesized by Solution Combustion Synthesis showed four peaks at 85 °C, 180 °C, 265 °C, and 320 °C when recorded with a heating rate of 2 °C s^{-1}, linear dose range in between 0.1 to 100 Gy, good reusability by 2% deviation

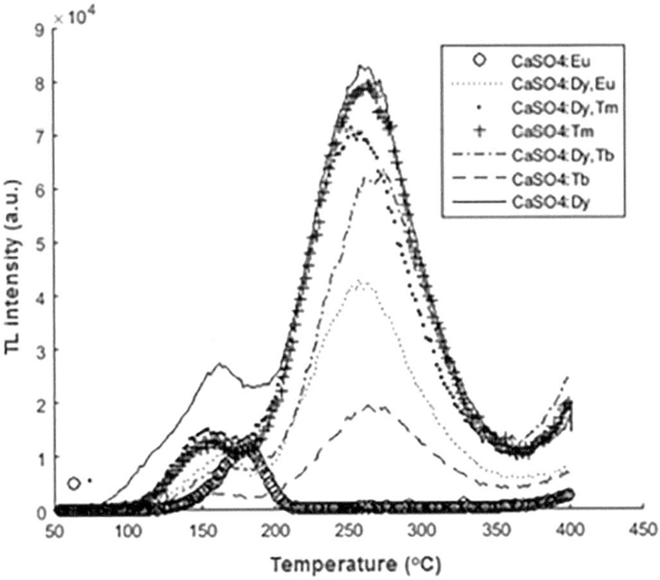

Fig. 6 Comparison between the glow curves of $CaSO_4$:Dy and $CaSO_4$ doped with Eu; Dy, Eu; Dy, Tm; Tm; Dy, Tb; Tb [139]

and a fading of 4–7% after 24 h of irradiation, all this property make it a suitable TLD material for personnel as well as medical dosimetry [144]. MgB_4O_7, MgB_4O_7:Mn, and MgB_4O_7:Cu are prepared by solid state reaction method and gamma irradiation to all of these materials shows a TL peak at 225 °C, 233 °C, 290 °C and the high intensity TL peak showed it can be used for dosimetric purposes [41]. From this part, we can conclude that MgB_4O_7:Pr, Dy emerging as a particularly strong candidate for advanced dosimetric applications (Fig. 9).

a. **CaAl2O4**

Madhukumar et al. [145] synthesized undoped $CaAl_2O_4$; $CaAl_2O_4$: Ce; $CaAl_2O_4$:Dy and $CaAl_2O_4$:Sm by combustion method and were irradiated with a ^{60}Co gamma dose of 1 Gy and it's been observed that undoped $CaAl_2O_4$ gave no emission, $CaAl_2O_4$:Dy shows a sharp intense peak at 110 °C with a minor peak at 282 °C, $CaAl_2O_4$:Sm shows a major peak at 182 °C and a minor peak at 276 °C while $CaAl_2O_4$: Ce shows an intense well defined peak at 295 °C. TL intensity of Ce-doped phosphor is very high compared to Dy- and Sm-doped phosphors, Dy-doped phosphor is not applicable for dosimetric purpose because of its high fading rate and Sm-doped phosphor is not good for dosimetry because of its low intensity while Ce-doped phosphor is useful for dosimetry because of its low fading (5% in two weeks) and linear dose response of 4 kGy and due to intense peak it can be used for high temperature dosimetry [145]. Tm-doped $CaAl_2O_4$ was synthesized by combustion method and the phosphor is irradiated by gamma ray and electron beam,

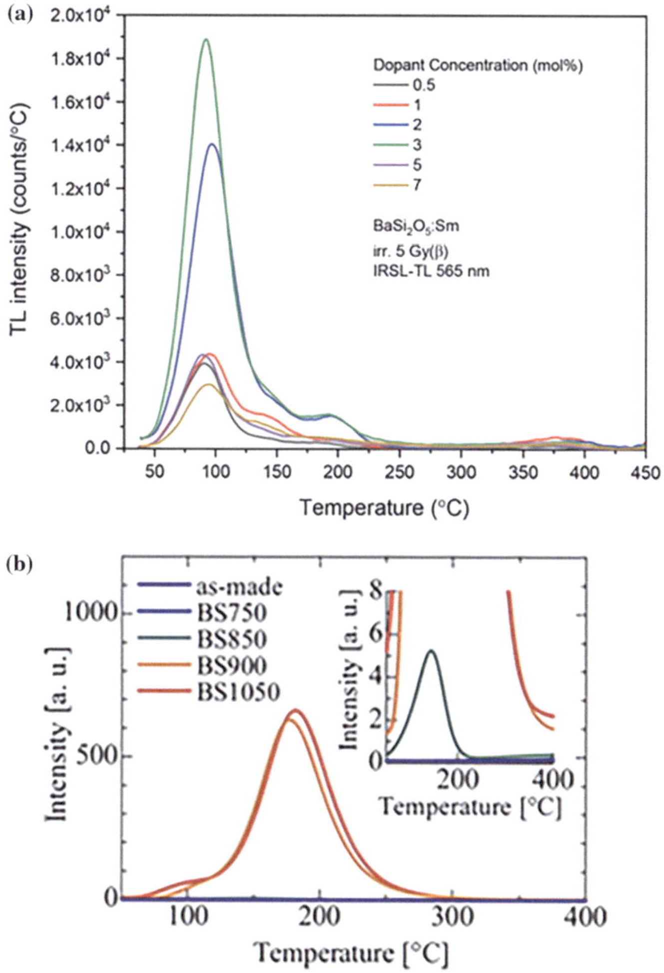

Fig. 7 **a** TL glow curve of Sm^{3+}-doped $BaSi_2O_5$ at various concentration of Sm^{3+} [119], **b** TL glow curve of X-ray irradiated glass and glass ceramics [141]

the gamma irradiated phosphor shows a linear dose response curves of three TL peaks at 190, 208 and 409 °C and electron beam irradiated phosphor shows two peaks at 188 and 432 °C showing their high usefulness in medicine and clinical dosimetry [146]. $CaAl_2O_4$: La was prepared by the sol–gel method and its TLD characteristics were studied, which shows its highest TL response at 0.1 mol % and when $CaAl_2O_4$: La was irradiated with β-rays it shows a linear dose response from 2.2 to 330 Gy and are highly reusable and due to all these characteristics $CaAl_2O_4$: La are highly commendable in environmental, personnel, and medical dosimetry [147]. Among the

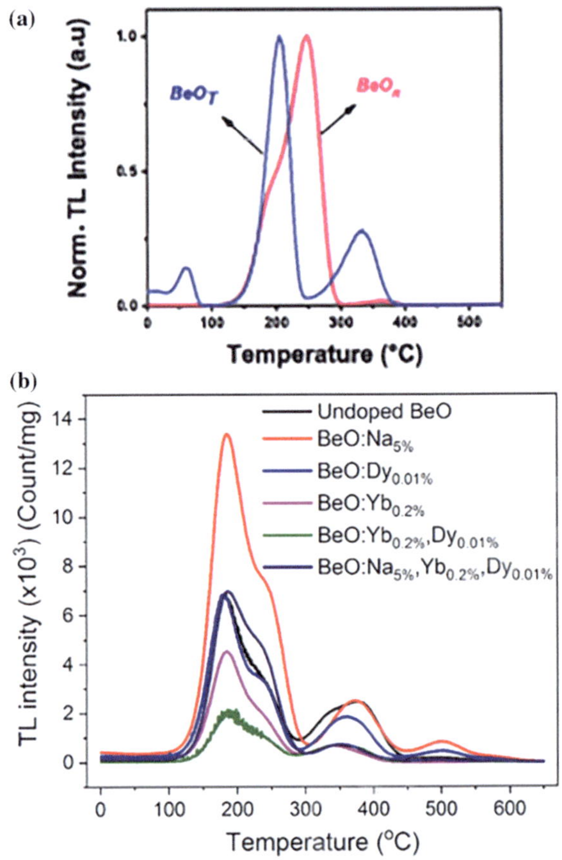

Fig. 8 **a** TL glow curves of BeO_R and BeO_T dosimeters [142], **b** TL glow curves for undoped BeO and BeO doped with Na; Dy; Yb; Yb, Dy; Na, Yb, Dy [143]

studied materials, $CaAl_2O_4$:Ce and $CaAl_2O_4$:La stand out for their high TL intensity, low fading, and broad linear dose ranges, making them particularly attractive for real-world radiation dosimetry (Figs. 10 and 11).

Table 2 represents a comparative study of the TL characteristics of some of the recently studied TLD phosphors.

In the above table we have written effective *Z* number or Z_{eff} of some of the TLD materials. Z_{eff} represents the tissue equivalence of a given TLD material and Z_{eff} for a good TLD material should be 7.42 and from the above table, among all the TLD materials or phosphors the tissue equivalence of BeO ($Z_{eff} = 7.1$) has nearly tissue equivalent property and can be used for dosimetric purpose.

6 Applications of TLD Materials or Phosphors

TLD materials are highly useful for the following areas:

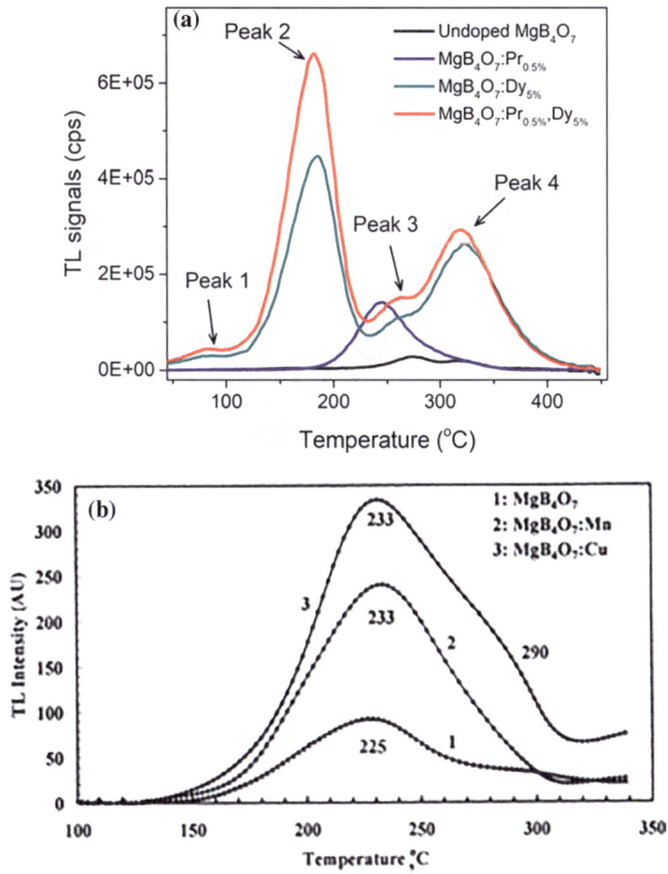

Fig. 9 **a** Comparing the TL signals of MgB_4O_7 (i) undoped, (ii) doped with Pr, (iii) doped with Dy, (iii) doped with both Pr and Dy [144], **b** Comparison of TL glows curves of MgB_4O_7 (1)undoped, (2) doped with Mn, (3) doped with Cu [41]

- Personnel Dosimetry
- Environmental Dosimetry
- Medical Dosimetry
- Retrospective Dosimetry
- Space Dosimetry
- **Personnel Dosimetry**

Personnel Dosimetry measures the radiation dose of individuals due to their occupational exposure [151]. The objectives of personnel monitoring includes the following points [43, 148, 151]:

1. Personnel Dosimetry plays an important role in providing information on the safety regulations of people working with radioactive devices and materials.

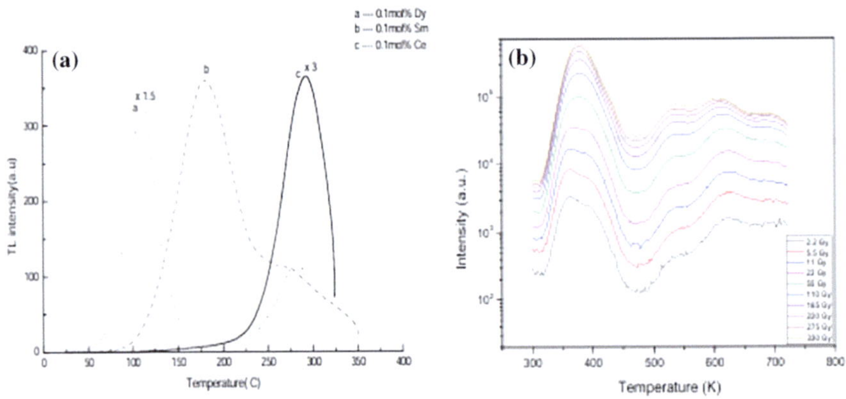

Fig. 10 **a** Comparison between the TL glow curves of $CaAl_2O_4$ for a dose of 1 kGy of γ irradiation doped with (**a**) Dy, (**b**) Sm, (**c**) Ce [145], **b** TL glow curve of β-irradiated $CaAl_2O_4$ in the dose range of 2.2 to 330 Gy [147]

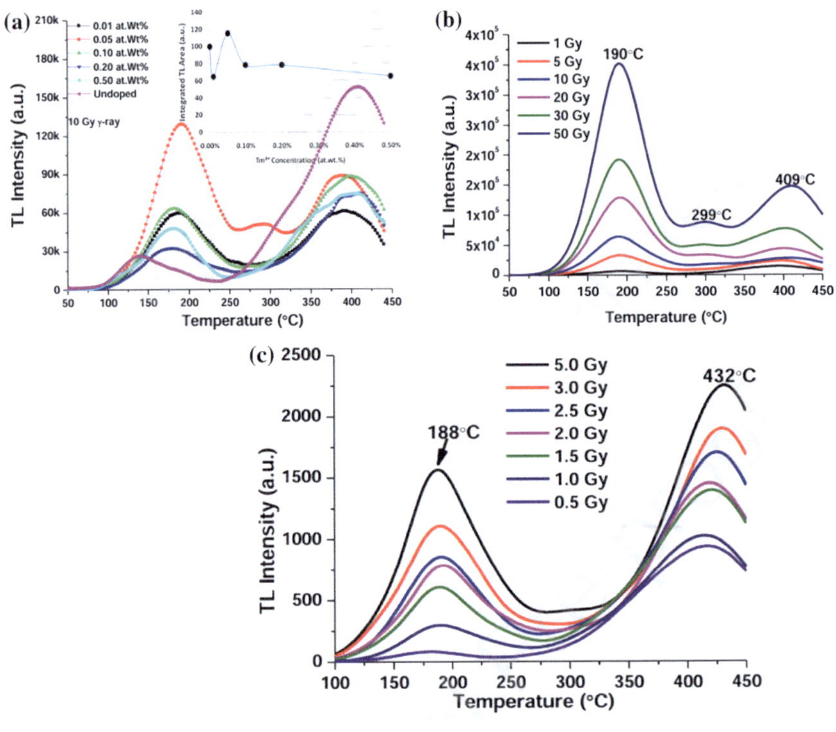

Fig. 11 TL glow curve of $CaAl_2O_4:Tm^{3+}$ **a** by varying dopant concentration, **b** by varying the amount of doses of γ-exposure, **c** exposed to 9 MeV electron beam at various doses [148]

Table 2 Comparison of TL characteristics of some TLD materials

Material name	Effective Z	Peak temperature	Dose response	Fading	Year of publication	References
$BaTiO_3$	–	480–530 K	Linear dose range	Stabilized after 16 days	2007	[137]
$NaLi_2PO_4$: Eu^{3+}	10.8	458 K	10 Gy-1.0 kGy	15% in 90 days	2013	[12]
YAG: Pr	–	370 K & 450 K	0.5–15 Gy	37% after 48 h	2017	[112]
$BaSi_2O_5$: Sm	–	364 K & 466 K	1–10 Gy	–	2020	[119]
BeO	7.1	373 K & 573 K	0.11–6.60 Gy	–	2021	[99, 100, 142]
$BaSO_4$:Eu	45–47	473 K	0.5–2 Gy	–	2021	[122, 149, 150]
$BaBO_4$:Tb	–	375 K, 485 K & 620 K	1–250 Gy	5–10% after one month	2021	[113]
$LiCaAlF_6$: Eu		453 K	10–50 Gy	16% in one month	2022	[19]
$ZnAl_2O_4$: Dy	–	446 K & 598 K	10 Gy-20 kGy	8% in 7 days	2024	[12]

2. It can also help the workers by measuring the amount of radiation exposure levels and assess potential health risks.
3. Personnel Dosimetry assists radiation exposure in As Low As Reasonably Achievable (ALARA).
4. Equivalent doses for assessment in effective dose range are measured.

Personnel Dosimetry is highly used for people working in nuclear fuel cycles, military services, medical sector and in industrial purposes. Apart from TLD materials, Optically Stimulated Luminescence Dosimeters (OSLDs) and Radiophotoluminescence Dosimeters (RPLDs) and film badges are used for personnel dosimetry purposes [152]. To use for personnel dosimetry, these dosimeters should follow the IEC (International Electrotechnical Commision) requirements [153, 154].

2. **Environmental Dosimetry**

Environmental Dosimetry is used to measure the amount of radiation exposure of the environment due to man-made hazards. Some of the man-made hazards for which environmental dosimetry needs to be used are: (i) Radiation exposure from the research reactors, (ii) Radiation exposure which gets produced when nuclear wastes are transported from place to place, (iii) Exposure of radiation produced due to nuclear fuel processing, (iv) Radiation Exposure from the place where nuclear wastes are disposed, (v) Radiation Exposure produced due to industrial radiation source such as radiography and sterilization. TLD materials are used in environmental dosimetry because of these advantages: (i) TLD materials can be used for passive measurement of the radiation of gamma exposure over time; (ii) TLD materials can provide us the accurate measure of the exposed radiation; (iii) TLD materials can be used for measuring the long-time radiation exposure of the environment. To measure the exposure of radiation the TLD materials should be placed in the nearer places of industrial as well as nuclear reactors. Important requirements of TLD materials so that they can be used for environmental dosimetry are:

- They should be tissue equivalent.
- They should be able to measure the long-term exposures.
- They should be able to detect the gamma radiation.

The minimum amount of dose that needs to be detected for environmental dosimetry is 10 μGy [153, 154]. Some of the Phosphors that are highly useful for environmental dosimetry are $CaSO_4$: Dy; LiF: Mg, Cu, P; CaF_2: Dy, $CaSO_4$: Tm [43].

3. **Medical Dosimetry**

The measurement of the amount of radiation dose delivered to a patient during the radiation therapies and diagnostic imaging is the medical dosimetry. TLD materials are highly recommended for medical dosimetry because of the following properties:

(i) TLD materials are tissue equivalent allowing accurate measurements of doses of radiation for human.
(ii) TLD materials are small and can be used to measure radiation doses in small areas.
(iii) TLDs are highly sensitive allowing precise dose measurements.
(iv) TLDs can be reused again and again if they are annealed properly.
(v) Even due to high stability of TLD materials they are a good choice for medical dosimetry.

Some of the TLD phosphors that are highly used in medical dosimetry are:

- LiF: These phosphors are highly used in medical dosimetry due to its tissue equivalence and high sensitivity. Examples includes LiF: Mg, Ti; LiF: Mg, Cu, P.
- CaF_2: These phosphors are widely used because of its high sensitivity and durability. Example include CaF_2: Dy.
- $CaSO_4$: Due to its high sensitivity and relatively low cost, they are used for dosimetric purposes. Example include $CaSO_4$: Dy.
- **Retrospective Dosimetry**

The method of estimating dose of an individual received in the past by that individual, due to the absence or insufficient availability of any dosimetry system at the time of exposure is called the Retrospective Dosimetry. TLDs are highly preferred for retrospective dosimetry purposes because (i) It can be used to estimate radiation exposure of already exposed materials; (ii) Can induce radiation signal over long periods; (iii) A correlated measureable signal of the radiation exposure is produced.

Use of TL techniques used in retrospective dosimetry includes the dose evaluation of the Hiroshima and Nagasaki Atomic bomb explosion in Japan and radioactive fallout from the Chernobyl nuclear reactor [43].

5. **Space Dosimetry**

The assessment and measurements of radiation doses of space travelers and astronauts is the basis of space dosimetry. Space radiation environment is a mixture of high energy ions, protons, electrons, alpha particles, and energy and fluxes [43, 155]. The radiation level of the astronauts or the space travelers is high due to the high cosmic radiation exposure levels in the space. Apart from this the Moon and Mars has higher radiation level as compared to the earth's atmosphere for which the astronauts and space travelers are at high risk of these radiation exposure. Due to all these factors there is a need of measurement of the radiation doses of the individuals travelling in space. TLDs are useful in space dosimetry purposes because of their small size, high sensitivity, passive detection and radiation hardness. However, it is very important to choose the correct TLD materials for space dosimetry because of the higher radiation levels of space.

6. **Future Advancements and Aspects of TLD Materials**

Emerging Phosphors used as TLD materials are not only used for detection of ionizing radiation but can also be used for LEDs and displays. TLD materials irradiated with UV, β, γ and electron are explored to far but the use of ion beam irradiated TLD materials are not yet explored, so they are also need to be explored so that can be used for dosimetry purposes. Researchers should focus on the development of new composite materials or hybrid systems like the polymer-phosphor mixes to improve biocompatibility and accuracy in medical dosimetry. Apart, from this with the rise of material informatics, AI and machine learning certain new TL dosimeters can be developed from the material databases. Furthermore, common TL materials saturate at high doses (~ 10^3 Gy-10^4 Gy), so need to work on residual TL and high temperature peaks to extend dose ranges for industrial and radiation core monitoring.

7 Conclusion

In summary, this review highlights the central role of inorganic phosphors in advancing thermoluminescence dosimetry. By analyzing a broad spectrum of materials and their fabrication methods, the study identifies key candidates with promising features—such as strong sensitivity to radiation, consistent reproducibility, minimal signal fading, and close similarity to human tissue in radiation interaction. These traits make them well-suited for use in critical areas like healthcare, environmental surveillance, nuclear safety, and space missions. Materials like $NaLi_2PO_4$, $LiCaAlF_6$, YAG, and $BaTiO_3$, in particular, emerge as strong contenders for next-generation TLD devices. Looking ahead, the integration of advanced synthesis techniques, ion beam activation, and AI-assisted material design could pave the way for even more effective and application-specific dosimeters. Continuous research in this direction is essential to meet the growing and diverse needs of radiation detection in both existing and emerging fields.

References

1. S. W. McKeever, M. Moscovitch, P. D Townsend, *Thermoluminescence Dosimetry Materials: Properties and Uses* (1995)
2. A.F. McKinlay, *Thermoluminescence Dosimetry-Medical Physics Handbooks 5* (Adam Hilger Ltd., Bristol, 1981)
3. S. W. McKeever, *Thermoluminescence of Solids*, vol. 3 (Cambridge University Press, 1985)
4. M.P. Chougaonkar, M. Kumar, B.C. Bhatt, Testing of phosphors for their use in radiation dosimetry: detailed procedure and protocol. Int. J. Lumin. Appl. **2**(3), 2277–6362 (2012)
5. X. Tang, E.D. Ehler, E. Brost, D.C. Mathew, Evaluation of $SrAl_2O_4$: Eu, Dy phosphor for potential applications in thermoluminescent dosimetry. J. Appl. Clin. Med. Phys. **22**(5), 191–197 (2021)
6. A.J. Bos, Theory of thermoluminescence. Radiat. Meas. **41**, S45–S56 (2006)
7. N. Dubey, V. Dubey, J. Saji, J. Kaur, Thermoluminescence glow curve analysis and trap parameters calculation of UV-induced $La_2Zr_2O_7$ phosphor doped with gadolinium. J. Mater. Sci. Mater. Electron. **31**(3), 1936–1944 (2020)
8. N. Juan Azorín, *Luminescence Dosimetry: Theory and Applications* (1990)
9. J. Azorín, A. Gutiérre, Determination of thermoluminescence parameters from glow curves—II. In $CaSO_4$: Dy. Int. J. Radiat. Appl. Instrum. Part D. Nucl. Tracks Radiat. Meas. **11**(3), 167–174 (1986)
10. R. Chen, S.W. McKeever, *Theory of thermoluminescence and related phenomena* (World Scientific, 1997)
11. S. Shionoya, Introduction to the handbook, in *Phosphor Handbook*, pp. 3–8 (2007)
12. M. Singh, P.D. Sahare, P. Kumar, Synthesis and dosimetry characteristics of a new high sensitivity TLD phosphor $NaLi_2PO_4$: Eu^{3+}. Radiat. Meas. **59**, 8–14 (2013)
13. L. Cao, K. Klimes, Y. Ji, T. Fleetham, J. Li, Efficient and stable organic light-emitting devices employing phosphorescent molecular aggregates. Nat. Photonics **15**(3), 230–237 (2021)
14. M. Wang, F. Li, Y. Lei, F. Xiao, M. Liu, S. Liu, H. Wu, Q. Zhao, Excitation-dependent organic phosphors exhibiting different luminescence colors for information anti-counterfeiting. Chem. Eng. J. **429**, 132288 (2022)
15. Yang, G., Hao, S., Deng, X., Song, X., Sun, B., Hyun, W. J., M. D. Li, L. Dang, Efficient intersystem crossing and tunable ultralong organic room-temperature phosphorescence via doping polyvinylpyrrolidone with polyaromatic hydrocarbons. Nat. Commun. **15**(1), 4674 (2024)
16. C. Shivakumara, R. Saraf, H. Nagabhushana, Red-emitting LaOF: Eu {sup 3+} phosphors: Synthesis, structure and their Judd–Ofelt analysis for LED applications. Mater. Res. Bull. **75** (2016)
17. L. Armelao, G. Bottaro, L. Bovo, C. Maccato, M. Pascolini, C. Sada, E. Soini, E. Tondello, Luminescent properties of Eu-doped lanthanum oxyfluoride sol– gel thin films. J. Phys. Chem. C **113**(32), 14429–14434 (2009)
18. M.S. Akselrod, V.S. Kortov, D.J. Kravetsky, V.I. Gotlib, Highly sensitive thermoluminescent anion-defect alpha-Al_2O_3: C single crystal detectors. Radiat. Prot. Dosimetry **33**(1–4), 119–122 (1990)
19. A. D. Vartha, P. R. Patankar, P. D. Sahare, L. Sharma, A. R. Kachere, P. M. Kakade, B. Bai, N. T. Mandlik, Synthesis and thermoluminescence study of Dy^{3+} doped ZnAl2O4 phosphor for high-dose dosimetry application. J. Radioanal. Nucl. Chem. **333**(9), 4503–522 (2024)

20. J. H. Schulman, R. D. Kirk, E. J. West, Use of lithium borate for thermoluminescence dosimetry, in *Proceedings of the International Conference on Luminescence Dosimetry, Stanford University, CONF-650637*, pp. 113–118 (1967)
21. M. Prokic, Progress in thermoluminescence dosimetry at the institute of nuclear sciences Vinca. Radiat. Prot. Dosim. **33**(1–4), 99–102 (1990)
22. O. Annalakshmi, M.T. Jose, G. Amarendra, Dosimetric characteristics of manganese doped lithium tetraborate–an improved TL phosphor. Radiat. Meas. **46**(8), 669–675 (2011)
23. M. Takenaga, O. Yamamoto, T. Yamashita, A new TLD phosphor based on $Li_2B_4O_7$, in *Proceedings of the Fifth International Conference on Luminescence Dosimetry* (1977)
24. A.S. Pradhan, R.C. Bhatt, K.G. Vohra, Some preparation parameters of $Li_2B_4O_7$: Cu TLD phosphor. Radiochem. Radioanal. Lett. **53**(2), 103–110 (1982)
25. M. Prokic, Lithium borate solid TL detectors. Radiat. Meas. **33**(4), 393–396 (2001)
26. C. Furetta, M. Prokic, R. Salamon, V. Prokic, G. Kitis, Dosimetric characteristics of tissue equivalent thermoluminescent solid TL detectors based on lithium borate. Nucl. Instrum. Methods Phys. Res. Sect. A **456**(3), 411–417 (2001)
27. M. Prokic, Dosimetric characteristics of $Li_2B_4O_7$: Cu, Ag, P solid TL detectors. Radiat. Prot. Dosimetry **100**(1–4), 265–268 (2002)
28. B. Tiwari, N. S. Rawat, D. G. Desai, S. G. Singh, M. Tyagi, P. Ratna, S. C. Gadkari, M. S. Kulkarni, Thermoluminescence studies on Cu-doped $Li_2B_4O_7$ single crystals. J. Lumin. **130**(11), 2076–2083 (2010)
29. N. S. Rawat, M. S. Kulkarni, M. Tyagi, P. Ratna, D. R. Mishra, S. G. Singh, B. Tiwari, A. Soni, S. C. Gadkari, S. K. Gupta, TL and OSL studies on lithium borate single crystals doped with Cu and Ag. J. Lumin. **132**(8), 1969–1975 (2012)
30. M. Martini, F. Meinardi, L. Kovacs, K. Polgar, Spectrally resolved thermoluminescence of $Li_2B_4O_7$: Cu single crystals. Radiat. Prot. Dosimetry **65**(1–4), 343–346 (1996)
31. B.M. Rzyski, S.P. Morato, Luminescence studies of rare earth doped lithium tetraborate. Nucl. Inst. Methods **175**(1), 62–64 (1980)
32. V. A. Kazanskaya, V. V. Kuzmin, E. E. Minaeva, A. D. Sokolov, Magnesium borate radio-thermoluminescent detectors, in *Proceedings of the fourth international conference on luminescence dosimetry, Krakow-Poland*, August 1974, pp. 27–31 (1974)
33. A.R. Lakshmanan, B. Chandra, A.S. Pradhan, R.C. Bhatt, Development of MgB_4O_7: Dy TLD phosphor. Radiochem. Radioanal. Lett. **37**(6), 377–382 (1979)
34. M.S. Prokic, New sintered thermoluminescent dosimeters for personnel and environmental dosimetry. Health Phys. **42**(6), 849–855 (1982)
35. G. Kitis, C. Furetta, M. Prokic, V. Prokic, Kinetic parameters of some tissue equivalent thermoluminescencematerials. J. Phys. D Appl. Phys. **33**(11), 1252 (2000)
36. M. Prokić, Individual monitoring based on magnesium borate. Radiat. Prot. Dosimetry **125**(1–4), 247–250 (2007)
37. M. Prokic, MgB_4O_7: Mn as a new TL dosemeter. Radiat. Prot. Dosimetry **47**(1–4), 191–193 (1993)
38. M. S. Prokic, Optimization of magnesium borate thermoluminescent material for radiation protection dosimetry, in *IRPA9: 1996 International Congress on Radiation Protection. Proceedings*, vol. 4 (1996)
39. C. Furetta, M. Prokic, R. Salamon, G. Kitis, Dosimetric characterisation of a new production of MgB4O7: Dy, Na thermoluminescent material. Appl. Radiat. Isot. **52**(2), 243–250 (2000)
40. L.L. Campos, O.O. Fernandes Filho, Thermoluminescent characterisation of MgB_4O_7: Dy sintered pellets. Radiat. Prot. Dosimetry **33**(1–4), 111–113 (1990)
41. M. R. Rao, B. S. Rao, N. P. Rao, K. Somaiah, K. V. R. Murthy, Thermoluminescence characteristics of MgB_4O_7, MgB_4O_7: Mn and MgB4O7: Cu phosphors (2009. Indian J. Pure Appl. Phys. **47**(6), 456
42. O. Annalakshmi, M.T. Jose, U. Madhusoodanan, B. Venkatraman, G. Amarendra, Synthesis and thermoluminescence characterization of MgB_4O_7: Gd, Li. Radiat. Meas. **59**, 15–22 (2013)
43. Bhatt, B. C., & Kulkarni, M. S. (2014, January). Thermoluminescent phosphors for radiation dosimetry, in *Defect and Diffusion Forum*, vol. 347. Trans Tech Publications Ltd., pp. 179–227

44. S.R. Anishia, M.T. Jose, O. Annalakshmi, V. Ponnusamy, V. Ramasamy, Dosimetric properties of rare earth doped $LiCaBO_3$ thermoluminescence phosphors. J. Lumin. **130**(10), 1834–1840 (2010)
45. S.R. Anishia, M.T. Jose, O. Annalakshmi, V. Ramasamy, Thermoluminescence properties of rare earth doped lithium magnesium borate phosphors. J. Lumin. **131**(12), 2492–2498 (2011)
46. A.S. Abhishek, S. Sahare, M. Singh, P.D. Sahare, Luminescence characteristics of $LiCaAlF_6$: Eu TLD Phosphor. J. At.Ic, Mol., Condens. Matter Nano Physic **9**(1), 85–92 (2022)
47. Harshaw Chemical Company, US patent, 1059518, published (1967)
48. G. Portal, Preparation and properties of principal TL products. Appl. Thermolumin. Dosim. 97–122 (1981)
49. G. Scarpa, The dosimetric use of beryllium oxide as a thermoluminescent material: a preliminary study. Phys. Med. Biol. **15**(4), 667 (1970)
50. K.W. Crase, R.B. Gammage, Improvements in the use of ceramic BeO for TLD. Health Phys. **29**(5), 739–746 (1975)
51. L. Lembo, G. Scarpa, Beryllium oxide, in *Thermoluminescent materials*, ed. by D. R. Vij (PTR Prentice Hall, Englewood Cliffs, New Jersey, 1993), pp. 264–280
52. I.B. Lozano, J. Roman-Lopez, M. Valverde, M. Garcia-Hipolito, J.A.I. Díaz-Góngora, J.G. Castañeda, G. Alarcon-Flores, Thermoluminescence properties of cerium and terbium doped alpha-strontium pyrophosphate. Radiat. Meas. **129**, 106192 (2019)
53. S.D. More, M.N. Meshram, S.P. Wankhede, P.L. Muthal, S.M. Dhopte, S.V. Moharil, Luminescence in $LiCaPO_4$. Physica B **406**(5), 1178–1181 (2011)
54. K.N. Shinde, S.J. Dhoble, Thermoluminescence and photoluminescence in the $NaCaPO_4$: Dy^{3+} phosphor. Radiat. Prot. Dosimetry **152**(4), 463–467 (2012)
55. S.K. Ramteke, A.N. Yerpude, S.J. Dhoble, N.S. Kokode, Luminescence characterization of $KBaPO_4$: RE (RE= Sm^{3+}, Eu^{3+}, Dy^{3+}) phosphors. Luminescence **35**(6), 969–973 (2020)
56. D.M. Pimpalshende, S.J. Dhoble, Evaluation of trapping parameters of γ-rays irradiated Dy^{3+}-doped $LaPO_4$ phosphors. Luminescence **29**(8), 1019–1026 (2014)
57. P.D. Sahare, M. Singh, P. Kumar, TL characteristics of Ce^{3+}-doped $NaLi_2PO_4$ TLD phosphor. J. Radioanal. Nucl. Chem. **302**, 517–525 (2014)
58. E. De la Rosa, R.A. Rodríguez, R. Melendrez, P. Salas, L.A. Diaz-Torres, M. Barboza-Flores, Thermoluminescence properties of undoped and Tb^{3+} and Ce^{3+} doped YAG nanophosphor under UV-, X-and β-ray irradiation. Nucl. Instrum. Methods Phys. Res. Sect. B **255**(2), 357–364 (2007)
59. S.A. Kumar, J. Senthilselvan, Photoluminescence and thermoluminescence investigations of Yb: YAG nanoparticles by dual-surfactant functionalization and microwave calcination method. Mater. Chem. Phys. **217**, 334–349 (2018)
60. G. Jovicic, L. Zigan, S. Will, A. Leipertz, Luminescence properties of the thermographic phosphors Dy^{3+}: YAG and Tm^{3+}: YAG for the application in high temperature systems. Z. Phys. Chem. **229**(6), 977–997 (2015)
61. V.K. Singh, S. Tripathi, M.K. Mishra, R. Tiwari, V. Dubey, N. Tiwari, Optical studies of erbium and ytterbium doped $Gd_2Zr_2O_7$ phosphor for display and optical communication applications. J. Disp. Technol. **12**(10), 1224–1228 (2016)
62. S. Delice, M. Isik, N.M. Gasanly, Low temperature thermoluminescence of Gd_2O_3 nanoparticles using various heating rate and Tmax—Texc. methods. Results in Physics **12**, 1809–1813 (2019)
63. S.K. Sao, N. Brahme, D.P. Bisen, G. Tiwari, Photoluminescence and thermoluminescence properties of Eu^{2+} doped and Eu^{2+}, Dy^{3+} co-doped $Ba_2MgSi_2O_7$ phosphors. Luminescence **31**(7), 1364–1371 (2016)
64. C. Manjunath, M.S. Rudresha, K.R. Nagabhushana, R.H. Krishna, B.M. Nagabhushana, B.M. Walsh, Thermoluminescence glow curve analysis of gamma irradiated Sr_2SiO_4: Dy^{3+} nanophosphor. Physica B **585**, 412113 (2020)
65. O. Hai, Q. Ren, X. Wu, Q. Zhang, Z. Zhang, Z. Zhang, Insights into the element gradient in the grain and luminescence mechanism of the long afterglow material $Sr_2MgSi_2O_7$: Eu^{2+}, Dy^{3+}. J. Alloy. Compd. **779**, 892–899 (2019)

66. V.E. Kafadar, T. Yeşilkaynak, R.E. Demirdogen, A.A. Othman, F.M. Emen, S. Erat, The effect of Dy^{3+} doping on the thermoluminescence properties of Ba_2SiO_4. Int. J. Appl. Ceram. Technol. **17**(3), 1453–1459 (2020)
67. P. Dewangan, D.P. Bisen, N. Brahme, S. Sharma, R.K. Tamrakar, I.P. Sahu, K. Upadhyay, Influence of Dy^{3+} concentration on spectroscopic behaviour of $Sr_3MgSi_2O_8$: Dy^{3+} phosphors. J. Alloy. Compd. **816**, 152590 (2020)
68. A. Pandey, S. Bahl, K. Sharma, R. Ranjan, P. Kumar, S. P. Lochab, V. E. Aleynikov, A. G.Molokanov, Thermoluminescence properties of nanocrystalline K_2Ca_2 $(SO_4)_3$: Eu irradiated with gamma rays and proton beam. Nucl. Instrum. Methods Phys. Res. Sect. B: Beam Interact. Mater. At. **269**(3), 216–222 (2011)
69. V.C. Kongre, S.C. Gedam, S.J. Dhoble, Photoluminescence and thermoluminescence characteristics of BaCa $(SO_4)_2$: Ce mixed alkaline earth sulfate. J. Lumin. **135**, 55–59 (2013)
70. T. Yamashita, N. Nada, H. Onishi, S. Kitamura, Calcium sulfate activated by thulium or dysprosium for thermoluminescence dosimetry. Health Phys. **21**(2), 295–300 (1971)
71. S. Bahl, S.P. Lochab, P. Kumar, $CaSO_4$: Dy, Mn: a new and highly sensitive thermoluminescence phosphor for versatile dosimetry. Radiat. Phys. Chem. **119**, 136–141 (2016)
72. YAMASHITA, T, in Calcium sulfate phosphor activated by rare earth, in *Proceedings of Second International Conference on Luminescence Dosimetry: Held at Mountain View Hotel, Gatlinburg, Tennessee, 23–26 September 1968*, vol. 8 ed. by T. Yamashita, N. Nada, H. Onishi, S. Kitamura (USAEC, Division of Technical Information Extension, 1968), p. 4
73. M. Prokić, Improvement of the thermoluminescence properties of the non-commercial dosimetry phosphors $CaSO_4$: Dy and $CaSO_4$: Tm. Nucl. Inst. Methods **151**(3), 603–608 (1978)
74. B. Chandra, R.C. Bhatt, Effect of dysprosium concentration on the TL characteristics of $CaSO_4$: Dy phosphor. Nucl. Inst. Methods **164**(3), 571–577 (1979)
75. J. Azorin, R. Salvi, A. Moreno, Improvement in the preparation of $CaSO_4$: Dy as a TL dosimeter. Nucl. Inst. Methods **175**(1), 81–82 (1980)
76. I. Kasa, Dependence of thermoluminescence response of $CaSO_4$: Dy and $CaSO_4$: Tm on grain size and activator concentration. Radiat. Prot. Dosimetry **33**(1–4), 299–302 (1990)
77. S.H. Li, P.C. Hsu, The role of annealing: effect on $CaSO_4$: Dy phosphor with manganese and sodium impurities. Radiat. Prot. Dosimetry **33**(1–4), 147–150 (1990)
78. A.R. Lakshmanan, Photoluminescence and thermostimulated luminescence processes in rare-earth-doped $CaSO_4$ phosphors. Prog. Mater. Sci. **44**(1–2), 1–187 (1999)
79. M.S. Atone, S.V. Moharil, T.K. Gundurao, Effective co-dopants for $CaSO_4$: Dy and $CaSO_4$: Tm phosphors. J. Phys. D Appl. Phys. **28**(6), 1263 (1995)
80. M. S. Atone, S. J. Dhoble, S. V. Moharil, S. M. Dhopte, P. L. Muthal, V. K. Kondawar, Sensitization of luminescence of $CaSO_4$: Dy. Phys. Status Solidi (A) **135**(1), 299–305 (1993)
81. U. Madhusoodanan, M.T. Jose, A. Tomita, A.R. Lakshmanan, New thermostimulated luminescence phosphors based on $CaSO_4$: Ag, RE. J. Lumin. **87**, 1300–1302 (2000)
82. M.T. Jose, U. Madhusoodanan, A.R. Lakshmanan, Influence of (group IIB) codopants on the thermostimulated luminescence sensitivity of $CaSO_4$: Dy/Tm. J. Phys. D Appl. Phys. **34**(5), 717 (2001)
83. S. J. Dhoble, S. V. Moharil, S. M. Dhopte, P. L. Muthal, V. K Kondawar, Preparation and characterization of the K_3Na $(SO_4)_2$: Eu phosphor. Phys. Status Solidi (A) **135**(1), 289–297 (1993)
84. R. C. Satkar, A. R. Kadam, D. A. Ovhal, S. J. Dhoble, Inorganic thermoluminescent phosphors in radiation dosimetry: an overview. J. Phys.: Conf. Ser. **1913**(1), 012023 (2021). IOP Publishing
85. A.R. Kadam, G.C. Mishra, S.J. Dhoble, Thermoluminescence study and evaluation of trapping parameters $CaTiO_3$: RE (RE= Eu^{3+}, Dy^{3+}) phosphor for TLD applications. J. Mol. Struct. **1225**, 129129 (2021)
86. T. Kalpana, B. Sanyal, Y. Gandhi, V. Ravi Kumar, G. S. Baskaran, P. Bragiel, M. Piasecki, N. Veeraiah, Thermoluminescence features of alumina-mixed borophosphate glasses with Tb^{3+} ions for dosimetric applications. Int. J. Appl. Glas. Sci. **8**(2), 188–195 (2017)

87. M.K. Biroon, M. Zahedifar, E. Sadeghi, F. Almasifard, Preparation, kinetic analysis and thermoluminescent dosimetry features of highly sensitive SrF_2: Dy phosphor. Radiat. Phys. Chem. **159**, 1–5 (2019)
88. C. Malik, R.K. Meena, P. Rathi, B. Singh, A. Pandey, Effect of dopant concentration on luminescence properties of a phosphor $KCaPO_4$: Dy. Radiat. Phys. Chem. **168**, 108561 (2020)
89. K.F. Oguz, E. Ekdal, M.A.A. Aslani, A. Canimoglu, J.G. Guinea, N. Can, T. Karali, Study of luminescence of Mn-doped CaB_4O_7 prepared by wet chemical method. J. Alloy. Compd. **683**, 76–85 (2016)
90. T.R. Shelke, S.J. Dhoble, V.M. Nanoti, Thermoluminescence in β-irradiated $SrZnF_4$: Dy TLD phosphor. Radiat. Eff. Defects Solids **169**(2), 157–164 (2014)
91. K.K. Gupta, R.M. Kadam, N.S. Dhoble, S.P. Lochab, V. Singh, S.J. Dhoble, Photoluminescence, thermoluminescence and evaluation of some parameters of Dy^{3+} activated $Sr_5(PO_4)_3F$ phosphor synthesized by sol-gel method. J. Alloy. Compd. **688**, 982–993 (2016)
92. M. Singh, P.D. Sahare, Redox reactions in Cu-activated nanocrystalline LiF TLD phosphor. Nucl. Instrum. Methods Phys. Res., Sect. B **289**, 59–67 (2012)
93. U. Madhusoodanan, M.T. Jose, A.R. Lakshmanan, Development of $BaSO_4$: Eu thermoluminescence phosphor. Radiat. Meas. **30**(1), 65–72 (1999)
94. M. E. Haghiri, E. Saion, W. S. wan Abdullah, N. Soltani, M. Hashim, M. Navasery, M. A. Shafaei, Thermoluminescence studies of manganese doped calcium tetraborate (CaB_4O_7: Mn) nanocrystal synthesized by co-precipitation method. Radiat. Phys. Chem. **90**, 1–5 (2013)
95. J. Malleshappa, H. Nagabhushana, B.D. Prasad, S.C. Sharma, Y.S. Vidya, K.S. Anantharaju, Structural, photoluminescence and thermoluminescence properties of CeO_2 nanoparticles. Optik **127**(2), 855–861 (2016)
96. V.R.O. Barrón, F.M.E. Ochoa, C.C. Vázquez, R. Bernal, Thermoluminescence of novel MgO–CeO_2 obtained by a glycine-based solution combustion method. Appl. Radiat. Isot. **117**, 86–90 (2016)
97. G. Sahu, A.S. Gour, R.K. Chandrakar, Thermoluminescence and optical properties of Dy^{3+} doped MgO nanoparticles, prepared by solution combustion synthesis method. J. Pure Appl. Ind. Phys. **7**(4), 115–127 (2017)
98. N.J. Shivaramu, K.R. Nagabhushana, B.N. Lakshminarasappa, F. Singh, Ion beam induced luminescence studies of sol gel derived Y_2O_3: Dy^{3+} nanophosphors. J. Lumin. **169**, 627–634 (2016)
99. S.W.S. McKeever, M. Moscovitch, P.D. Townsend, *Thermoluminescence dosimetry materials: properties and uses* (Nuclear Technology Publishing, Ashford, Kent, England, 1995)
100. K. Becker, *Solid State Dosimetry*, vol. 347 (CRC Press, Boca Raton, Fl., USA, 1973). Defect and Diffusion Forum, p. 213
101. M. Oberhofer, A. Scharmann, *Applied Thermoluninescence Dosimetry* (Adam Hilger, Bristol, 1981)
102. Y.S. Horowitz, C.R.C. Press (eds.), *Thermoluminescence and thermoluminescent dosimetry (Vol. 1)* (CRC Press, Boca Raton, 1984)
103. S.W. McKeever, Thermoluminescence materials. Radiat. Prot. Dosimetry **8**(1–2), 2–2 (1984)
104. S. Kumari, R. Verma, T.K. Gupta, Recent advances in the synthesis. Emerg. Trends Synth. Catal. Chem.: Proc. ETSC **2023**(51), 75 (2024)
105. G. I. Efenji, S. M. Iskandar, N. N. Yusof, J. A. Rabba, O. I. Mustapha, I. M. Fadhirul, S. A. Umar, F. A. Kamgba, P. O. Ushie, J. Munirah, H. K. Thair, S. E. Nabasu, S. N. Hayder, A. O. Oke, Structural properties of thermoluminescence dosimeter materials, preparation, application, and adaptability: a systematic review. J. Appl. Sci. Environ. Manage. **28**(4), 1129–1150 (2024)
106. K. Mahesh, P. S. Weng, C. Furetta, *Thermoluminescence in Solids and its Applications.* (No Title) (1989)
107. D.R. Vij (ed.), *Thermoluminescence materials* (Prentice Hall, Englewood Cliffs, 1993)
108. A.S. Pradhan, Thermoluminescence dosimetry and its applications. Radiat. Prot. Dosimetry **1**(3), 153–167 (1981)

109. J. Azorin, C. Furetta, A. Scacco, Preparation and properties of thermoluminescent materials. Phys. Status Solidi (A) **138**(1), 9–46 (1993)
110. G. Portal, Review of the principal materials available for thermoluminescent dosimetry. Radiat. Prot. Dosimetry **17**(1–4), 351–357 (1986)
111. A.J.J. Bos, High sensitivity thermoluminescence dosimetry. Nucl. Instrum. Methods Phys. Res., Sect. B **184**(1–2), 3–28 (2001)
112. P. R. Patankar, P. D. Sahare, L. Sharma, A. D. Vartha, A. R. Kachere, P. M. Kakade, S. D. Dhole, N. T. Mandlik, Optically stimulated luminescence, thermoluminescence and photoluminescence characteristics of SrB_4O_7: Eu phosphor. J. Radioanal. Nucl. Chem. **333**(8), 4397–4420 (2024)
113. S. Jakathamani, O. Annalakshmi, M.T. Jose, B. Venkatraman, Thermoluminescence dosimetric characteristics of terbium doped barium metaborate phosphors. Radiat. Phys. Chem. **187**, 109544 (2021)
114. M. Zahedifar, E. Sadeghi, Synthesis and dosimetric properties of the novel thermoluminescent CaF_2: Tm nanoparticles. Radiat. Phys. Chem. **81**(12), 1856–1861 (2012)
115. P. Sengar, H.A. Borbón-Nuñez, C.J. Salas-Juárez, E.M. Aguilar, C. Cruz-Vázquez, R. Bernal, G.A. Hirata, β-Irradiated thermoluminescence response of nanocrystalline YAGG: Pr^{3+} for radiation dosimetry. Mater. Res. Bull. **90**, 195–204 (2017)
116. V.R. Orante-Barrón, F.M. Escobar-Ochoa, C. Cruz-Vázquez, R. Bernal, Thermoluminescence of novel zinc oxide nanophosphors obtained by glycine-based solution combustion synthesis. J. Nanomater. **2015**(1), 273571 (2015)
117. T.R. Montalvo, L.O. Tenorio, J.A. Nieto, A.C. Celis, C.V. Ordoñez, R.S. Fonseca, Thermoluminescence and optical characteristics of ZrO_2 powder as a TL dosimeter. Radiat. Eff. Defects Solids **159**(11–12), 645–649 (2004)
118. C.B. Palan, N.S. Bajaj, A. Soni, S.K. Omanwar, A novel $KMgPO_4$: Tb^{3+} (KMPT) phosphor for radiation dosimetry. J. Lumin. **176**, 106–111 (2016)
119. Y. Alajlani, M. Oglakci, U.H. Kaynar, M. Ayvacikli, Z.G. Portakal-Uçar, M.U.S.T.A.F.A. Topaksu, N. Can, Thermoluminescence study and evaluation of trapping parameters of samarium doped barium silicate phosphor. J. Asian Ceram. Soc. **9**(1), 291–303 (2021)
120. R.A. Barve, N. Suriyamurthy, B.S. Panigrahi, B. Venkatraman, Dosimetric investigations of Tb^{3+}-doped strontium silicate phosphor. Radiat. Prot. Dosimetry **163**(4), 430–438 (2015)
121. N. Mandlik, P.D. Sahare, B.J. Patil, V.N. Bhoraskar, S.D. Dhole, Thermoluminescence study of $K_2Ca_2(SO_4)_3$: Cu nanophosphor for gamma ray dosimetry. Nucl. Instrum. Methods Phys. Res. Sect. B **315**, 273–277 (2013)
122. T. Hussain, V.K. Asfora, B.P. da Nobrega, V.S. de Barros, W.M. de Azevedo, H.J. Khoury, Thermoluminescence properties of nanocrystalline $BaSO_4$ doped with Eu^{2+} produced by solid state combustion synthesis. Radiat. Phys. Chem. **186**, 109531 (2021)
123. M.K. Mahata, K. Kumar, V.K. Rai, Structural and optical properties of Er^{3+}/Yb^{3+} doped barium titanate phosphor prepared by co-precipitation method. Spectrochim. Acta Part A Mol. Biomol. Spectrosc. **124**, 285–291 (2014)
124. M. N. Konopka, P. Bilski, B. Obryk, B. Marczewska, P. Olko, M. Kłosowski, W. Gieszczyk, Luminescence Dosimetry: Review of Methods, Detectors and Their Applications. Nonlinear Opt. Quantum Opt.: Concepts Mod. Opt. **48**(2) (2017)
125. V. Kortov, Materials for thermoluminescent dosimetry: current status and future trends. Radiat. Meas. **42**(4–5), 576–581 (2007)
126. R. Rodríguez, V. Correcher, J.M. Gómez-Ros, J.L. Plaza, Preliminary study on the thermoluminescence properties of CaF_2 and Tm_2O_3 pellets. Nucl. Instrum. Methods Phys. Res. Sect. A **959**, 163561 (2020)
127. K.N. Shinde, S.J. Dhoble, A novel reddish-orange phosphor, $NaLi_2PO_4$: Eu^{3+}. Luminescence **28**(1), 93–96 (2013)
128. W. Gieszczyk, P. Bilski, M. Kłosowski, T. Nowak, L. Malinowski, Thermoluminescent response of differently doped lithium magnesium phosphate ($LiMgPO_4$, LMP) crystals to protons, neutrons and alpha particles. Radiat. Meas. **113**, 14–19 (2018)

129. A. Getkin, S. Neicheva, N. Shiran, T. Fukuda, K. Shimamura, A new effective thermoluminescent material $LiCaAlF_6$: Ce. Radiat. Prot. Dosimetry **100**(1–4), 377–379 (2002)
130. S.R. Rahangdale, U.A. Palikundwar, S.P. Wankhede, B. Dhabekar, S. Kadam, S.V. Moharil, Effect of co-doping on luminescence of $LiCaAlF_6$: Eu phosphor. J. Lumin. **167**, 80–84 (2015)
131. A. Jain, P. Sengar, G.A. Hirata, Rare-earth-doped $Y_3Al_5O_{12}$ (YAG) nanophosphors: synthesis, surface functionalization, and applications in thermoluminescence dosimetry and nanomedicine. J. Phys. D Appl. Phys. **51**(30), 303002 (2018)
132. R.A. Rodriguez, E. De la Rosa, P. Salas, R. Meléndrez, M. Barboza-Flores, Thermoluminescence and optically stimulated luminescence properties of nanocrystalline Er^{3+} and Yb^{3+} doped $Y_3Al_5O_{12}$ exposed to β-rays. J. Phys. D Appl. Phys. **38**(21), 3854 (2005)
133. E. la Rosa, R. A. Rodríguez, R. Meléndrez, P. Salas, L. A. Diaz-Torres, M. Barboza-Flores, Thermoluminescence properties of undoped and Tb^{3+} and Ce^{3+} doped YAG nanophosphor under UV, X and β-ray irradiation. Nucl. Instrum. Methods Phys. Res. Sect. B: Beam Interact. Mater. At. **255**, 357–64
134. E. de la Rosa-Cruz, L.A. Diaz-Torres, W. Cordova-Martinez, R.A. Rodriguez-Rojas, M.A. Meneses Nava, O. Barbosa-Garcia, Photoluminescence and thermoluminescence of YAG:Ce^{3+}, Tb^{3+} nanocrystalline under UV. X B-Irradiat. Appl. Photonic Technol. **4833**, 948–955 (2003)
135. R.A. Rodrıguez, E. la Rosa, L.A. Dıaz-Torres, P. Salas, R. Meléndrez, M. Barboza-Flores, Thermoluminescence characterization of Tb^{3+} and Ce^{3+} doped nanocrystalline $Y_3Al_5O_{12}$ exposed to X and β-ray irradiation Opt. Mater. (Amst) **27**, 293–299 (2004)
136. X. B. Yang, J. Xu, H. J. Li, Q. Y. Bi, L. B. Su, Y. Cheng, Q. Tang, Thermoluminescence properties of carbon doped Y3Al5O12 (YAG) crystal. J. Appl. Phys. **106**(3) (2009)
137. A.Y. Fasasi, F.A. Balogun, M.K. Fasasi, P.O. Ogunleye, C.E. Mokobia, E.P. Inyang, Thermoluminescence properties of barium titanate prepared by solid-state reaction. Sens. Actuators A **135**(2), 598–604 (2007)
138. E. Korkmaz, N.I.L.G.U.N. Ozpozan Kalaycioglu, Synthesis and luminescence properties of $BaTiO_3$: RE (RE= Gd^{3+}, Dy^{3+}, Tb^{3+}, Lu^{3+}) phosphors. Bull. Mater. Sci. **35**, 1011–1017 (2012)
139. L.A. Forner, C. Viccari, P. Nicolucci, Dosimetric properties of thermoluminescent pellets of $CaSO_4$ doped with rare earths at low doses. Radiat. Phys. Chem. **171**, 108704 (2020)
140. J. K. Srivastava, B. C. Bhatt, J. S. Supe, *Radiation Protection Dosimetry*, vol. 40, no. 4 (Nuclear Technology Publishing, 1992), pp. 271–274
141. D. Nakauchi, K. Shinozaki, N. Kawaguchi, T. Yanagida, Photo-, radio-and thermoluminescence properties of Eu-doped $BaSi_2O_5$ glass-ceramics. Optik **185**, 812–818 (2019)
142. E. Aşlar, E. Şahiner, G.S. Polymeris, N. Meriç, Thermally and optically stimulated luminescence properties of BeO dosimeter with double TL peak in the main dosimetric region. Appl. Radiat. Isot. **170**, 109635 (2021)
143. V.O.L.K.A.N. Altunal, M. Jain, S. Hayat, V. Guckan, Z.E.H.R.A. Yegingil, Development of BeO: Na, Yb, Dy ceramics for optically stimulated luminescence dosimetry. Radiat. Meas. **158**, 106852 (2022)
144. A. Ozdemir, V. Guckan, V.O.L.K.A.N. Altunal, K. Kurt, Z.E.H.R.A. Yegingil, Thermoluminescence in MgB_4O_7: Pr, Dy dosimetry powder synthesized by solution combustion synthesis method. J. Lumin. **230**, 117761 (2021)
145. K. Madhukumar, K.R. Babu, K.A. Prasad, J. James, T.S. Elias, V. Padmanabhan, C.M.K. Nair, Thermoluminescence dosimetry of rare earth doped calcium aluminate phosphors. Bull. Mater. Sci. **29**, 119–122 (2006)
146. M.N. Singh, A.G. Barua, R.K. Gartia, Thermoluminescence studies of Tm doped nanocrystalline calcium aluminate ($CaAl_2O_4$: Tm^{3+}). Optik **228**, 166151 (2021)
147. A. Salah, S. Farouk, M. Ahmed, N. El-Faramawy, Synthesis and thermoluminescence characteristics of beta irradiated nanocrystalline calcium aluminate doped with lanthanum oxides. Radiat. Phys. Chem. **218**, 111571 (2024)
148. International Commission on Radiological Protection, in *ICRP Publication 75: General Principles for the Radiation Protection of Workers*, vol. 75. (Elsevier Health Sciences, 1997)

149. B.C. Bhatt, S.S. Sanaye, S.S. Shinde, J.K. Srivastava, A comparative study of the dosimetric characteristics of $BaSO_4$: Eu and $CaSO_4$: Dy teflon TLD discs. Radiat. Prot. Dosimetry **69**(2), 105–110 (1997)
150. P.R. Gonzalez, C. Furetta, B.E. Calvo, M.I. Gaso, E. Cruz-Zaragoza, Dosimetric characterization of a new preparation of $BaSO_4$ activated by Eu ions. Nucl. Instrum. Methods Phys. Res. Sect. B **260**(2), 685–692 (2007)
151. Smith, H, in *International Commission on Radiological Protection.* 1990 recommendations of the International Commission on Radiological Protection (ICRP Publication), p. 60
152. B. C. Bhatt, M. S. Kulkarni, Worldwide status of personnel monitoring using thermolu minescent (TL), optically stimulated luminescent (OSL) and radiophotoluminescent (RPL) dosimeters. J. ISSN No **2277**, 6362 (2013)
153. IEC, International Electrotechnical Commission. Radiation protection instrumentation. Passive integrating dosimetry systems for personal and environmental monitoring. *Part 1: General Characteristics and Performance Requirements* (IEC/CWI 62387-1, 2007)
154. International Electrotechnical Commission, *Radiation Protection Instrumentation—Passive Integrating Dosimetry Systems for Personal and Environmental Monitoring of Photon and Beta Radiation* (International Standard IEC, 2012), p. 62387
155. S. WS McKeever, New millennium frontiers of luminescence dosimetry. Radiat. Prot. Dosim. **100**(1–4), 27–32 (2002)

Thermoluminescence in Upconversion Materials: An Application to TL Dosimetry

Jyoti Singh and Vikas Dubey

Abstract Defect levels in upconversion phosphors play a significant role in its applicability as a suitable luminescent material. Thermoluminescence (TL) is a defect-related phenomenon and considered as an inevitable tool to understand the nature and distribution of traps/defects in the crystal lattice. This chapter reviews the physics of the thermoluminescence related to dosimetric applications in upconversion materials. Enormous research has been carried out to comprehend the fundamental mechanism and improve the materials characteristics as well as to evolve advance and novel thermoluminescence materials. Nowadays, thermoluminescence dosimetry (TLD) is a well-known dosimetric technique with a wide range of applications in personnel, environmental and clinical dosimetry. TLD is based on the materials which are previously exposed to ionizing radiations, emit visible radiations, when they are heated. The impurities in the upconversion TL materials encourage localized energy levels within the forbidden energy band gap which plays a vital role in TL process. By detecting the presence and nature of these defect levels, the sensitivity of TL is unrivalled. This chapter focuses on the investigation into the relation between the luminescence and defects involved in the process. Moreover, various models to understand the TL properties in the upconversion materials through energy band theory are discussed in detail. Different types of radiation exposure employed in TL studies display a remarkable impact on the trap distribution resulting in shallow and deep traps. Various techniques to determine different trapping parameters such as activation energy, order of kinetics, and frequency factor are discussed in detail which provides a comparative framework to interpret many observations. The knowledge of kinetic parameters namely order of kinetics equips one to understand recombination process and trapping profile comprehensively. The present chapter also discusses the behavior of TL glow peaks on the basis of order of kinetics which indicates the trapping/retrapping of charge carriers resulting to TL emissions in the phosphors.

J. Singh (✉)
Department of Applied Physics (SOVSAS), Gautam Buddha University, Greater Noida, Uttar Pradesh, India
e-mail: jyoti.singh.phy@gmail.com

V. Dubey
Department of Physics, North-Eastern Hill University (NEHU), Shillong, Meghalaya, India

V. Dubey et al. (eds.), *Upconversion Luminescence of Phosphors*, Progress in Optical Science and Photonics 39, https://doi.org/10.1007/978-981-95-1762-6_8

1 Introduction

The term TL stands for thermoluminescence: the first word thermo symbolizes heat and the later word luminescence symbolizes emission of light. These words may seem to indicate that the luminescence emission is originated by heat. However, heat plays only a marginal role in the present case [1–5]. Incandescence is the phenomenon primarily caused by heat and is responsible for the emission of light, light yield from hot filament lamp is the best example. The primary agents for the induction of TL in a material are the ionizing radiations, namely the X-rays, or radiations from radioactive elements to which the sample is pre-exposed [3, 6–11]. In some materials, ultraviolet light also may excite TL. When the sample is subsequently warmed up, luminescent light is emitted. The act of warming up stimulates the release of the stored energy in the sample due to pre-exposure to the ionizing radiations. The release of this stored energy produces the luminescence. Therefore, some authors have appropriately used an expanded expression for this emission calling it radiation induced thermally stimulated luminescence. Some others have abbreviated it to radiothermoluminescence to signify that the TL is induced by the radiations. Some authors named it thermally stimulated luminescence (TSL). Presently, however, most workers of this field have become accustomed to the abbreviated term TL [9, 12–15].

The phenomenon of thermoluminescence (TL) has been acknowledged for an extended period. The first documented application of TL for dosimetric purposes was by Daniel et al. [1]. Following this, significant research efforts have been made to deepen the understanding of this phenomenon and to refine material properties, alongside the development of novel TL materials. Currently, thermoluminescence dosimetry (TLD) is a recognized and established technique employed in personnel, environmental, and clinical dosimetry [3, 16–24].

Thermoluminescent dosimetry (TLD) operates on materials that release light upon heating following exposure to ionizing radiation. The presence of impurities in the thermoluminescent material is believed to generate localized energy levels within the forbidden energy band gap, which are essential for the thermoluminescence phenomenon [3, 25–29]. The ability of TLD to detect these defect levels is unmatched, with Townsend and Kelly [2] suggesting that the technique can identify as few as 10^9 defect levels in a specimen. To provide perspective, the detectable chemical purity in a sample is six orders of magnitude higher. This remarkable sensitivity facilitates the measurement of extremely low radiation doses. Conversely, this situation restricts our ability to investigate the relation between luminescence and the associated defects in this process. Additionally, it renders the properties of TLD highly sensitive to various parameters.

The objective of this chapter is to provide an understanding of the physics underlying the phenomenon of thermoluminescence (TL), with a focus on its dosimetric applications. Initially, the fundamental concepts of TL will be explained in detail. Subsequently, the general characteristics and methodology of TLD materials will be examined. Finally, the TL characteristics of the highly sensitive upconversion materials will be discussed in detailed way, which have garnered significant attention over the past decade due to their outstanding dosimetric properties.

2 Fundamental Principles of Thermoluminescence

Thermoluminescence (TL) is a luminescent effect observed in insulators or semiconductors when they are subjected to thermal stimulation. It is important to distinguish TL from the light that is emitted spontaneously by a material when it reaches incandescence due to heating. When temperatures exceed 200 °C, a solid emits infrared radiation, with the intensity of this emission rising in correlation with the temperature increase. This phenomenon is known as thermal or black body radiation. In contrast, thermoluminescence (TL) refers to the light emitted as a result of thermal stimulation, occurring after the solid has previously absorbed energy from radiation. This description allows for the identification of the three critical elements necessary for the occurrence of thermoluminescence [3, 10]. First and foremost, the material must be identified as either an insulator or a semiconductor, since metals do not demonstrate luminescent behavior. Moreover, it is crucial that the material has absorbed energy at some stage as a result of exposure to ionizing radiation. Finally, the luminescence emission is activated through the application of heat to the material.

A thermoluminescent material is characterized by its ability to absorb energy from ionizing radiation during exposure, which is subsequently stored within the material. Upon heating, this stored energy is released as visible light. It is crucial to note that thermoluminescence does not involve thermal excitation; rather, it pertains to the luminescence that occurs in a sample previously excited through a different mechanism [3, 30–34]. Consequently, a thermoluminescent material cannot simply emit light by cooling and reheating the sample, it requires re-exposure to ionizing radiation to emit light again. The energy storage capability of thermoluminescent materials renders them suitable for dosimetric applications.

2.1 *The One Trap-One Center Model*

The observed thermoluminescence (TL) properties can be elucidated through the energy band theory of solids. In a perfect crystalline semiconductor or insulator, the majority of electrons are found in the valence band. The conduction band, which is the next available energy level for electrons, is separated from the valence band by a region known as the forbidden band gap. The energy difference between these delocalized bands is referred to as E_g. However, the presence of structural defects or impurities within the crystal lattice can allow electrons to attain energy levels that would otherwise be prohibited in an ideal crystal [3, 10, 11]. A basic TL model assumes two energy levels: one level is positioned beneath the conduction band's

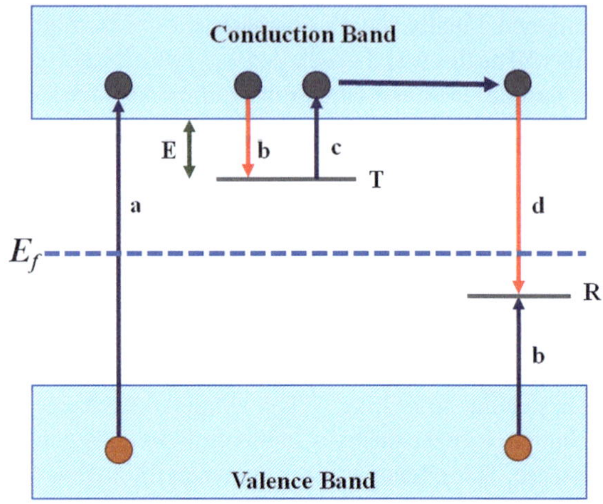

Fig. 1 Schematic representation of a simple two level model: **a** generation of electron and hole pairs; **b** trapping of electrons and holes; **c** Release of electron due to thermal stimulation; **d** recombination. Black circles represent electrons and orange circles are holes

lower edge, while the other is located above the valence band's upper edge (refer to Fig. 1). The uppermost level, denoted as T (electron trap level), lies above the equilibrium Fermi level (E_f) and remains unoccupied in the equilibrium state, prior to any radiation exposure that generates electrons and holes. Consequently, it serves as a potential trap for electrons. Conversely, the other level marked as R (recombination center), acts as a potential trap for holes and can also serve as a center for recombination.

When radiant energy with $h\vartheta > E_g$ is absorbed, it results in the ionization of valence electrons, generating energetic electrons and holes. After thermalization, these charge carriers transition into free electrons in the conduction band and free holes in the valence band (transition a). The free charge carriers may either recombine or become trapped. In instances of direct recombination, energy is released, which can excite a luminescent center, potentially coinciding with the recombination center. The luminescent center then relaxes back to its ground state, emitting light in the process.

This rapid recombination of free electrons and holes, occurring in less than 10^{-8} s and accompanied by light emission, is referred to as radioluminescence.

However, in semiconductors and insulators, a portion of the charge carriers becomes trapped: electrons at T and holes at R (transition b). The release probability of an electron from the trap, measured per unit time, is considered to be governed by the Arrhenius equation [3, 10]:

$$p = s\exp\left\{-\frac{E}{kT}\right\} \tag{1}$$

where p represents the probability per unit time, while the term s refers to the frequency factor or the attempt-to-escape factor. In this simplified model, s is regarded as a constant that does not vary with temperature, taking on a value in the vicinity of the lattice vibration frequency, approximately 10^{12} to 10^{14} s^{-1}. E is identified as the trap depth or activation energy, which signifies the energy necessary to liberate an electron from the trap into the conduction band (see Fig. 1). The remaining symbols are used in their conventional sense: k stands for Boltzmann's constant, which is 8.617 × 10–5 eV/K, and T represents the absolute temperature.

When the trap depth E exceeds kT_0, where T_0 is the irradiation temperature, any trapped electron will remain in that state for a prolonged time. This leads to a considerable number of trapped electrons remaining even after radiation exposure. Additionally, since free electrons and holes are generated and annihilated in pairs, there will be a corresponding population of trapped holes at level R. The normal equilibrium Fermi level E_f is positioned below level T and above level R, indicating that the populations of trapped electrons and holes are in a non-equilibrium state. Although the pathway for returning to equilibrium is always available, the perturbation from equilibrium caused by exposure to ionizing radiation occurs at a low temperature relative to E/k, resulting in a slow relaxation rate as described by Eq. (1). Therefore, this non-equilibrium state is metastable and can persist indefinitely, influenced by the rate parameters E and s.

The process of returning to equilibrium can be accelerated by increasing the temperature of the TL material beyond T_0. This elevation in temperature enhances the likelihood of detrapping, allowing electrons to escape from their traps and enter the conduction band. Once in the conduction band, the charge carriers move through the crystal until they reach a recombination center, designated as R. In this simplified model, the recombination center acts as a luminescent site where the electron and hole recombine, resulting in the center being left in an excited state. The transition back to the ground state is associated with the release of light quanta, known as TL. The intensity of TL, denoted as I(t), measured in photons per second at any given time t during the heating process, is directly proportional to the recombination rate of holes and electrons at the center R. If m (m^{-3}) represents the concentration of holes trapped at R, the TL intensity can be formulated as [3]:

$$I(t) = -\frac{dm}{dt} \tag{2}$$

In this context, we consider that each recombination leads to the generation of a photon, with all generated photons being detected. The rate at which recombination occurs is anticipated to be proportional to the concentration of free electrons in the conduction band, referred to as n_c, as well as the concentration of holes, represented by m [3],

$$I(t) = -\frac{dm}{dt} == n_c m A \tag{3a}$$

The recombination probability, denoted by the constant A, is expressed in terms of volume per unit time and is considered to be temperature-independent. The variation in the concentration of trapped electrons, n, is determined by the difference between the rate of thermal release and the rate of retrapping [3],

$$-\frac{dn}{dt} = np - n_c(N - n)A_\tau \tag{3b}$$

The concentration of electron traps is denoted as N, while Ar represents the probability of retrapping (m^3/s). Similarly, the rate of free electron concentration is calculated by taking the thermal release rate and deducting the rates of retrapping and recombination [3].

$$-\frac{dn_c}{dt} = np - n_c(N - n)A_\tau - n_c mA \tag{3c}$$

Equations (3a–3c) illustrate the movement of charge carriers during the release of a trapped electron from a single-electron trap and the subsequent recombination at a single center. The rate equations for thermoluminescence (TL) resulting from hole release are analogous to those in Eqs. (3a–3c). These equations serve as a foundation for numerous analyses of TL phenomena. It is important to note that a comprehensive analytical solution is not available. To formulate an analytical expression, specific simplifying assumptions must be adopted, with one significant assumption being that at any given time [3, 10, 11],

$$\left|\frac{dn_c}{dt}\right| \ll \left|\frac{dn}{dt}\right|, \left|\frac{dn_c}{dt}\right| \ll \left|\frac{dm}{dt}\right| \tag{4}$$

The assumption identified by Chen and McKeever is termed the quasiequilibrium assumption, which stipulates that the concentration of free electrons in the conduction band must be quasistationary. As irradiation occurs, pairs of trapped electrons and holes are created, thereby enforcing the principle of charge neutrality. Therefore,

$$n_c + n = m, \tag{5}$$

Which for $n_c \approx 0$ means that $n \approx m$ and

$$I(t) = -\frac{dm}{dt} \approx -\frac{dn}{dt} \tag{6}$$

Since, $\frac{dn_c}{dt} \approx 0$ one obtains from Eqs. 3.1 and 3.2 [3, 10, 11]:

$$I(t) = \frac{mAns \exp\left\{-\frac{E}{kT}\right\}}{(N - n)A_\tau + mA} \tag{7}$$

2.1.1 First-Order Kinetics

It is important to note that Eq. (7) cannot be solved analytically without further simplifying assumptions. Randall and Wilkins [8, 9] proposed that retrapping during the heating phase is negligible, specifically assuming that mA > > $(N–n)A_m$ [3, 10–12]. With this assumption, Eq. (7) can be reformulated as follows:

$$I(t) = -\frac{dn}{dt} = sn\exp\left\{-\frac{E}{kT}\right\} \tag{8}$$

This differential equation characterizes charge transport within the lattice as a first-order process, with the resulting glow peaks derived from this equation, referred as the first-order glow peaks. The solution to differential Eq. (8) is as follows.

$$I(t) = -\frac{dn}{dt} = n_0 s\ \exp\left\{-\frac{E}{kT}\right\} \times \exp\left\{-s\int_0^t \exp\left\{-\frac{E}{kT(t\prime)}\right\}dt\prime\right\} \tag{9}$$

Here, n_0 denotes the total count of trapped electrons at the time $t = 0$. Generally, the temperature is elevated as a linear function of time, as indicated by the subsequent expression [3],

$$T(t) = T_0 + \beta t \tag{10}$$

Considering β as the constant heating rate and T_0 as the initial temperature at $t = 0$, the intensity can be formulated in relation to temperature,

$$I(t) = -\frac{1}{\beta}\frac{dn}{dt} = n_0\frac{\mathrm{s}}{\beta}\ \exp\left\{-\frac{E}{kT}\right\} \times \exp\left\{-s\int_0^t \exp\left\{-\frac{E}{kT(t\prime)}\right\}dt\prime\right\} \tag{11}$$

The following expression attributed to Randall and Wilkins first-order equation for a single glow peak. This peak exhibits a distinct asymmetric profile, appearing broader on the lower temperature side compared to the higher temperature side. During the initial ascent of the glow peak, the intensity on the lower temperature side is primarily influenced by the first exponential term (exp (– E/kT)). Therefore, plotting I against $1/T$ is expected to yield a straight line in the initial temperature rise range, with a slope of – E/k, which facilitates the easy calculation of the activation energy E [3, 10–14]. The characteristics of the Randall–Wilkins equation are depicted in Fig. 2. In Fig. 2a, the variation of $I(T)$ is illustrated as n_0 changes from $n_0 = 0.25\ \mathrm{m}^{-3}$ to $n_0 = 2\ \mathrm{m}^{-3}$, while maintaining constant values for $E = 1$ eV, $s = 1.0 \times 10^{12}\ \mathrm{s}^{-1}$, and $\beta = 1$ K/s.

It is noteworthy that the temperature at the peak maximum, T_m, remains constant, which is a defining feature of all first-order TL curves. The condition for achieving the maximum can be derived by setting $dI/dt = 0$ (or, more conveniently, from d ln

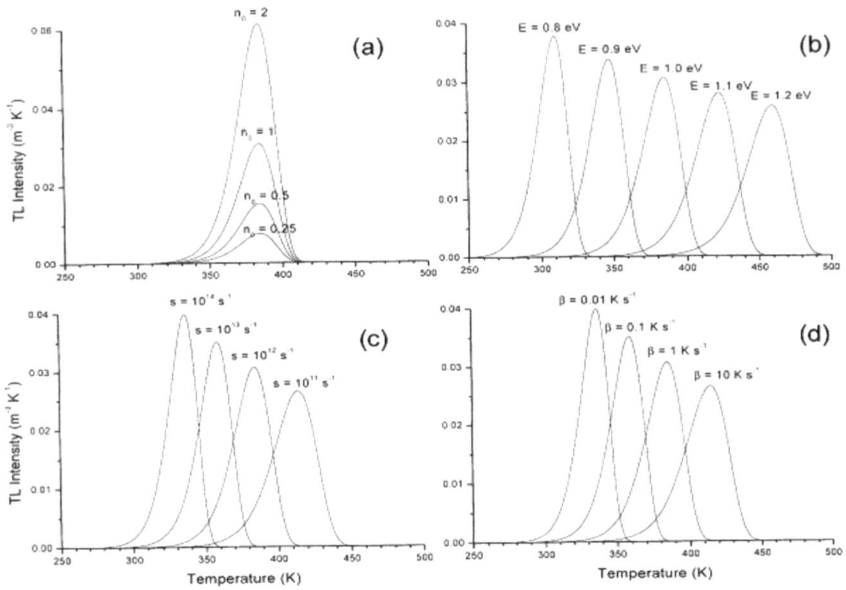

Fig. 2 Properties of the R-W first-order TL equation, showing: **a** variation with n0; the concentration of trapped charge carriers after irradiation; **b** the variation with E, the activation energy; **c** the variation with s, the escape frequency; **d** the variation with b, the heating rate. Parameter values: $n_0 = 1\ m^{-3}$; $E = 1$ eV; $s = 1 \times 10^{12}\ s^{-1}$; $\beta = 1$ K/s of which one parameter is varied while the others are kept constant. Figures reproduced (adapted) with permission from reference [11], Elsevier

$I(T)/dt = 0$). This leads to the following expression [3]:

$$\frac{\beta E}{kT_m^2} = s\ \exp\left\{-\frac{E}{kT_m}\right\} \tag{12}$$

The absence of n_0 in this equation indicates that T_m is independent of n_0. Additionally, as illustrated in Fig. 2a that every point on the curve, including the peak height at its maximum, is proportional to n_0. In dosimetry applications, n_0 is a critical parameter, as it correlates with the absorbed dose. It is evident that the area beneath the glow peak corresponds to n_0 [3].

$$\int_0^\infty I(t)dt = -\int_0^\infty -\frac{dn}{dt}dt = -\int_{n_0}^{n_\infty} dn = n_0 - n_\infty \tag{13}$$

where n_∞ is zero as $t \to \infty$.

In Fig. 2b, the activation energy E is varied from 0.8 to 1.2 eV. With an increase in E, the peak temperature shifts upward, while both the height decreases and the width increases, maintaining a constant area (n_0). A similar pattern is observed when adjusting s, as shown in Fig. 2c; however, in this scenario, increasing s causes the peak to move to lower temperatures, resulting in a height increase and a width decrease. Figure 2d illustrates the impact of changing the heating rate. As β rises, the peak shifts to higher temperatures, with a decrease in height and an increase in width, akin to the effects of decreasing s. This behavior is expected, given that s and β are expressed as a ratio s/β in Eq. (11). It is worth mentioning that, among the four parameters, the activation energy E and the frequency factor s are the key physical parameters, known as trapping parameters, which are dictated by the properties of the trapping center. The remaining two parameters are established by the experimenter through the selection of a particular dose (n_0) and the assessment of the signal at a specified heating rate β. Therefore, the study of a new TL material will initiate with an investigation into the behavior of the glow peak as it responds to variations in both the absorbed dose and the heating rate [3, 10–14].

The analysis of Eq. (11) faces challenges due to the integral on the right-hand side being non-elementary in the context of linear heating. Chen [3] has illustrated how this integral can be approximated through asymptotic series. For practical purposes, it is advantageous to express the glow peak in terms of experimentally accessible parameters, such as the maximum peak intensity, I_m, and the corresponding temperature, T_m. Kitis et al. [10] have shown that Eq. (11) can be effectively approximated by,

$$I(T) = I_m \exp\left[\left\{1 + \frac{E}{kT}\frac{T - T_m}{T_m} - \frac{T^2}{T_m^2}\right\} \times \exp\left\{\frac{E}{kT}\frac{T - T_m}{T_m}\right\}(1 - \Delta) - \Delta_m\right] \tag{14}$$

with $\Delta = 2kT/E$ and $\Delta_m = 2kT_m/E$. These expressions are advantageous for applications related to peak fitting.

2.1.2 Second-Order Kinetics

Garlick and Gibson [12] explored the possibility where retrapping is the predominant process, represented as mA < < (N–n) A_m. They also assumed that the trap remains well below saturation levels, indicated as $N > > n$ and $n = m$. Under these conditions, Eq. (7) simplifies to [3],

$$I(t) = -\frac{dn}{dt} = S\frac{A}{NA_\tau}n^2 \exp\left\{-\frac{E}{kT}\right\} \tag{15}$$

This leads to the conclusion that dn/dt is proportional to $n2$, signifying a second-order reaction. By further assuming that the probabilities of recombination and retrapping are equal, noted as $A = A_m$, the integration of Eq. (15) yields,

$$I(T) = \frac{n_0^2}{N}\frac{s}{\beta}\exp\left\{-\frac{E}{kT}\right\} \times \left\{1 + \frac{n_0}{N}\frac{s}{\beta}\int_0^T \exp\left\{-\frac{E}{kT'}\right\}dT'\right\} \tag{16}$$

The above Garlick–Gibson TL equation pertains to second-order kinetics. A prominent feature of this curve is its near-symmetrical appearance, with the high-temperature section being slightly wider than the low-temperature section. This can be understood by considering that in second-order reactions, a significant number of released electrons are re-captured prior to recombination. This process results in a delay in luminescence emission and a broader emission profile across various temperatures [3, 10–12]. In contrast to the first-order kinetics, the initial concentration n_0 is not merely a constant multiplier; its variation at different dose levels significantly alters the entire curve's shape. This is exemplified in Fig. 3a, which shows that T_m decreases as n_0 increases. The temperature shift can be approximated by

$$T_1 - T_2 \approx T_1 T_2 \frac{k}{E} \text{In} f \tag{17}$$

In this context, T_1 denotes the temperature at which maximum intensity is observed for a particular dose, and T_2 signifies the temperature at maximum intensity for a dose that is f times larger. Based on the parameter values presented in Fig. 3a, the observed shift is 25 K. When the energy E $=$ 1 eV, $T_1 = 400$ K, and an increase in the absorbed dose by a factor of 1000, an adjustment that can be easily achieved in experiments, results in an anticipated temperature shift of 77 K. Additionally, from

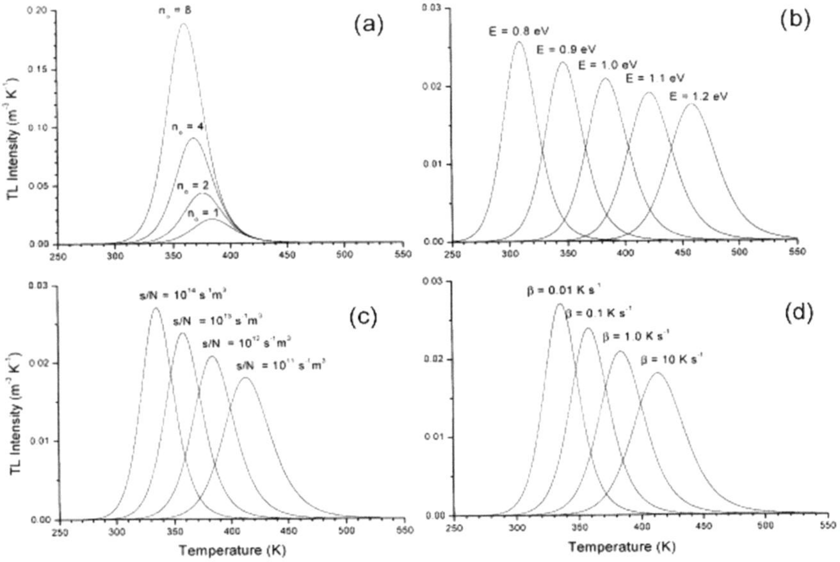

Fig. 3 Properties of the Garlick-Gibson second-order TL equation, showing: **a** variation with n_0, the concentration of trapped charge carriers after irradiation; **b** the variation with E, the activation energy; **c** the variation with s/N; **d** the variation with b, the heating rate. Parameter values: $n_0 = 1\ m^{-3}$; $E = 1$ eV; $s = 1 \times 10^{12}\ s^{-1}$; $\beta = 1$ K/s of which one parameter is varied while the others are kept constant. Figures reproduced (adapted) with permission from reference [11], Elsevier

Eq. (17), it can be inferred that for a specified increase in dose, a shallower trap (i.e., a smaller E) leads to a more significant peak shift. Figure 3b demonstrates how the size and position of a second-order peak vary with E, while Fig. 3c shows this variation in relation to s/N, and Fig. 3d illustrates it as a function of the heating rate.

Similar to first-order kinetics, the area under the curve correlates with the initial concentration n_0; however, the peak height does not maintain a direct proportionality to the peak area, although the difference is insignificant [2, 3, 5, 10–14].

It is worthwhile to note that, similar to the first-order case, the term that dominates the temperature dependence in the initial rise is exp (– E/kT). As a result, the "initial rise method" for determining trap depth is also applicable in this situation. Moreover, for second-order kinetics, the glow peak shape, as indicated in Eq. (16), can be approximated by a function that relates to the maximum peak intensity I_m and the maximum peak temperature T_m [10].

$$I(T) = 4I_m \exp\left(\frac{E}{kT}\frac{T - T_m}{T_m}\right) \times \left[\frac{T^2}{T_m^2}(1 - \Delta)\exp\left\{\frac{E}{kT}\frac{T - T_m}{T_m}\right\} + 1 + \Delta_m\right]^{-2} \tag{18}$$

2.1.3 General-Order Kinetics

The derivation of the first- and second-order forms of the TL equation relies on certain simplifying assumptions. However, when these assumptions are not applicable, the TL peak will not conform to either first- or second-order kinetics. May and Partridge [14] proposed an empirical expression to address this issue, which represents general-order TL kinetics.

$$I(t) = -\frac{dn}{dt} = n^b s' \exp\left\{-\frac{E}{kT}\right\} \tag{19}$$

In this formulation, s' is expressed in the dimension of $m^{3(b-1)}$ s^{-1}, with b representing the general-order parameter, which is not restricted to the values of 1 or 2. The integration of Eq. (19) for $b \neq 1$ yield the following result:

$$I(T) = \frac{s''}{\beta} n_0 \exp\left\{-\frac{E}{kT}\right\} \times \left\{1 + (b-1)\frac{s''}{\beta}\int_{T_0}^{T} \exp\left\{-\frac{E}{kT}\right\} dT'\right\}^{-\frac{b}{b-1}} \tag{20}$$

In this formulation, $s'' = s' n^{b-1}$, measured in units of s^{-1}. In this context, Eq. (20) addresses the second-order case ($b = 2$) and reverts to Eq. (11) when b $\rightarrow$ 1.

It is important to highlight that, as per Eq. (19), the dimension of s' is $m^{3(b-1)}$ s^{-1}, indicating that the dimension varies with the order b, which complicates physical interpretation. Nonetheless, the general-order case is advantageous, as it facilitates

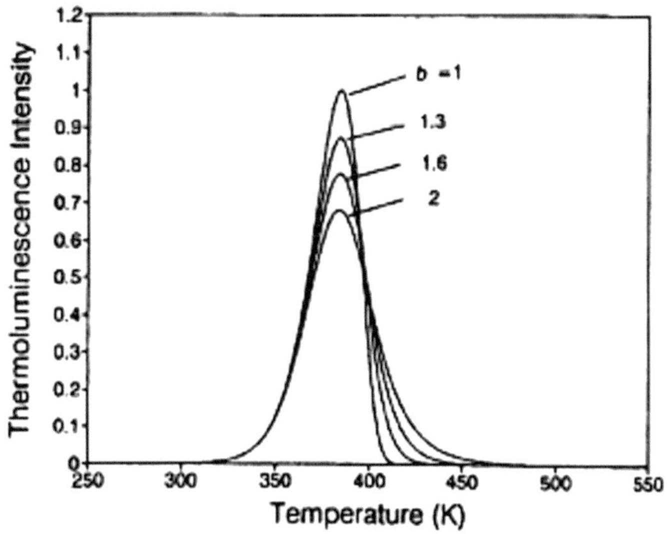

Fig. 4 Comparison of first-order ($b = 1$), second-order ($b = 2$), and intermediate- order ($b = 1.3$ and 1.6) TL peaks, with $E = 1$ eV, $s = 1 \times 10^{12}$ s^{-1}; $n_0 = N = 1$ m^{-3} and $b = 1$ K/s. Figures reproduced (adapted) with permission from reference [11], Elsevier

the examination of intermediate cases and transitions seamlessly to first- and second-order cases as b approaches 1 and 2, respectively (see Fig. 4).

2.2 *Advanced Model*

The one trap–one center model encapsulates the essential characteristics of thermoluminescence (TL) and elucidates the behavior of glow peak shapes in response to variations in dose and heating rate. However, no TL material has been identified that can be precisely described by this simplistic model. This does not imply that the model lacks significance; rather, it serves as a valuable tool for interpreting various features that can be viewed as extensions of the one trap–one center framework. A detailed examination of more advanced and realistic models is beyond the scope of this discussion. This section will briefly introduce a few models to illustrate the complexity inherent in real TL materials [3, 10–14].

Typically, a genuine thermoluminescent (TL) material will exhibit multiple electron traps. It is important to note that not all traps will be active within the specific temperature range during which the specimen is subjected to heating. A thermally disconnected trap is characterized by its ability to be filled with electrons during irradiation, yet it has a trap depth that is considerably greater than that of the active trap. As a result, when the specimen is heated, only the electrons trapped in the active trap (AT) and the shallow traps (ST) are released (refer to Fig. 5a). Electrons that are

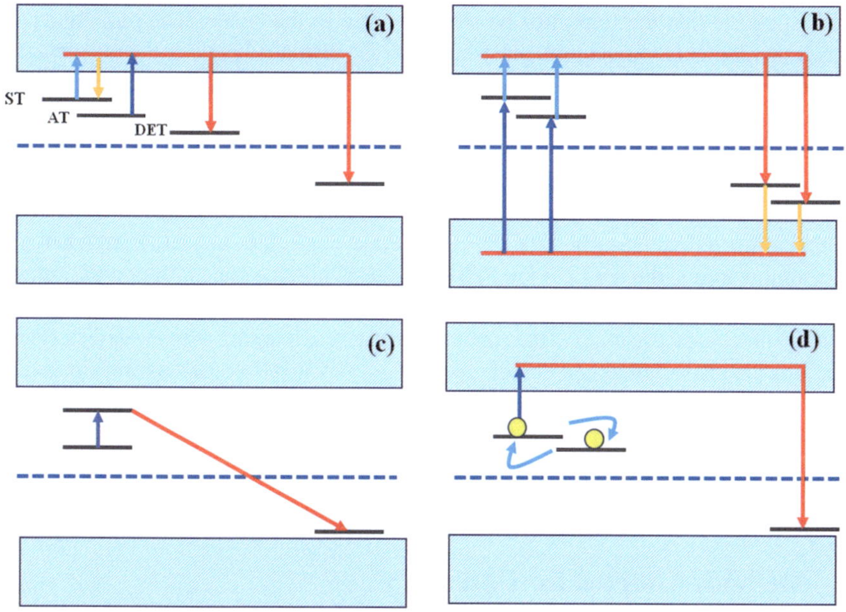

Fig. 5 Schematic representation of advanced model: **a** a shallow trap (ST), a deep electron trap (DET), and an active trap (AT); **b** two active centers and corresponding recombination centers; **c** localized transitions; **d** defect interaction

trapped in deeper levels remain unaffected, which is why this deeper electron trap (marked as DET in Fig. 5a) is considered thermally disconnected. Nevertheless, its presence influences the filling of traps and ultimately affects the shape of the glow peak [15].

In the present section, it was suggested that trapped electrons are liberated during the heating process, while the trapped holes remain stable within the recombination center. An alternative case, where holes are released and recombine at a center where electrons remain stable during heating, is mathematically equivalent. However, the dynamics shift if both electrons and holes are simultaneously released from their traps within the same temperature range, with holes being thermally released from the same centers that serve as recombination sites for the released electrons, and vice versa (refer to Fig. 5b). Under these conditions, Eq. (2) becomes invalid, necessitating the formulation of new differential equations. The analysis of this intricate kinetic model indicates a thermoluminescence (TL) glow curve that maintains the basic shape of either the Randall–Wilkins (Eq. (11)) or Garlick–Gibson (Eq. (16)) equations, contingent upon the selected parameter values. However, the energy (E) and frequency factor (s) values used in Eqs. (11) and (16) to fit this complex kinetic model require additional interpretation [3, 10–14].

Another potential process is recombination occurring without the electron moving into the conduction band (refer to Fig. 5c). In this case, the electron is thermally excited to an elevated state, from which it can transition to the recombination center.

This suggests that the trap must be situated close to the center. The probability of this transition may vary significantly based on the distance between the two centers. Under specific conditions, a relationship for the thermoluminescence (TL) intensity can be derived [16], which takes a form similar to Eq. (11), but with "s" replaced by a term that reflects the recombination probability. Thus, these localized transitions are governed by first-order kinetics. Finally, we consider the possibility that the defect capturing the electron may not be stable and could be participating in a reaction with another defect (see Fig. 5d). This interaction might lead to a situation where, at low temperatures, the depth of the trap fluctuates while the concentration of trapped electrons remains constant. As the temperature increases, electrons engage in two distinct processes: escaping to the conduction band and reacting with the defect. Piters and Bos [17] have integrated defect reactions into their rate equations and simulated glow curves. The results indicate that the simulated glow curves align closely with the predictions of Eq. (11). It is important to note that the fitting parameters do not simply represent trap depth and escape frequency.

3 Methods Adopted for Dosimetry

3.1 *Ionization Method*

The most common technique used for dosimetry is ionization method. The population of produced ion pairs is [3].

$$I = \sum_{i} \int_{B}^{\infty} \frac{\varepsilon_i n_i(E) d\varepsilon_i}{w_i(\varepsilon_i)} \tag{21}$$

where B denotes the lowest limit of energy loss and $w_i(\varepsilon_i)$ is the energy needed for ith particle with energy ε to procreate ion pairs.

When dose measurement is required to examine at a specific position, the detector size must be small in comparison with the attenuation dimension of primary radiation. If improbability of the former case occurs, the first collision dose should be identical in the detector as remains in the medium, or, minimal, the first collision doses ratio for two different materials must not depend on energy. It is an inevitable necessity to always maintain the ratios between stopping powers of two materials free from the respective energies.

3.2 Chemical Method

In certain systems, the absorbed radiation energy is competent enough for alteration of chemical composition of the concerned systems. If (Y) represents the observed change in chemical composition, then [3]

$$Y = \sum_i \int_0^\infty \varepsilon_i n_i(\varepsilon_i, E) G_i(\varepsilon_i) d\varepsilon_i \tag{22}$$

where $G_i(\varepsilon_i)$ denotes the yield per unit energy absorbed. If G is free from the particle type and (ε_i), then: $Y = G\,\varepsilon\,T$.

3.3 Calorimetric Methods

The energy from radiation that is captured by the dosimeter is converted into thermal energy, resulting in an increase in the temperature of the dosimeter. The variation in temperature can be expressed as follows [3]:

$$\Delta T = \frac{1}{c} \sum_i \int_0^\infty \varepsilon_i n_i(\varepsilon_i, E) F_i(\varepsilon_i) d\varepsilon_i \tag{23}$$

where $\varepsilon_i n_i(\varepsilon_i,\ E) d\varepsilon_i$ represents the amount of energy absorbed per unit mass of a substance. $F_i(\varepsilon_i)$ denotes the proportion of energy from charged particles that is converted into heat. (c) indicates the thermal capacity of the material. $F_i(\varepsilon_i)$ is considered to be nearly constant and close to unity so that,

$$\Delta T = \frac{\varepsilon T}{c} \tag{24}$$

3.4 Thermoluminescence Methods

The absorption of radiation by an impure crystal results in the trapping of certain electrons within the energy levels established in the forbidden gap. The application of heat can induce these electrons to return to the valence band, resulting in the emission of energy as light. The intensity of the emitted light correlates directly with the amount of radiation absorbed by the crystal [3, 10, 11]:

$$L = \sum_i \int_B^{\infty} \varepsilon_i n_i(E) d\varepsilon_i \quad (25)$$

where L refers to the total amount of light, ε_i is the energy associated with light photon, and n_i represents the total number of light photons present. The parameter B establishes the lower limit for the detection of light.

4 Examples of Upconversion TL Materials

4.1 Sr_2CeO_4:Er^{3+}/Yb^{3+}

Crystal Structure: Sr_2CeO_4:Er^{3+}/Yb^{3+} phosphor exhibits orthorhombic structure with space group Pbam (55). In the SCO:Er, Yb, the substitution of Sr atoms by Er atoms is more feasible due to the smaller ionic radius of Er^{3+} in comparison to Sr^{2+}. For the Yb^{3+} dopant, it can be accommodated in the Ce^{4+} sites, given their similar ionic radii [30]. Additionally, Yb^{3+} ions have the potential to occupy the Sr^{2+} sites, which are inter-chain, while the Ce^{4+} sites are positioned at the center of the octahedra. This substitution of Ce and Sr with Er and Yb will create a charge imbalance, resulting in defects that are beneficial for its excellent thermoluminescent characteristics [30–32].

TL Characteristics

The performance of SCO and SCO: Er, Yb phosphors as thermoluminescent emitters were conducted with respect to different Yb concentrations. This evaluation involved the use of two irradiation sources to induce traps in their structure: a UV lamp emitting at 360 nm and solar simulators.

The thermoluminescence (TL) glow curves for the SCO (undoped), SCO1, SCO3, SCO7, SCO9, and SCO12 samples subjected to solar irradiation are illustrated in Fig. 6a. An increase in Yb concentration from 0 (SCO) to 12 mol% (SCO12) results in a shift of the TL signal from 91 °C to 122 °C. The relatively low temperature necessary for generating the TL signals suggests that solar irradiation facilitates the filling of shallow traps, specifically the oxygen vacancies created by the introduction of the Yb ion, within the SCO matrix [30]. These traps lead to the formation of TL emission bands that are centered between 91 °C and 122 °C following the thermal excitation-induced release of electrons. Conversely, when the SCO:Yb, Er samples with lower Yb concentrations (ranging from 0% to 3 mol%) are exposed to higher energy radiation, such as UV light, both shallow and deeper traps (oxygen vacancies) become filled with electrons. The TL signals generated by the shallow and deep traps, following their thermal excitation, were observed at temperatures of 64 °C and between 150 °C and 350 °C, respectively (refer to Fig. 6b). Notably, samples with elevated Yb concentrations (7–12 mol%) exhibited TL signals ranging from 60 °C

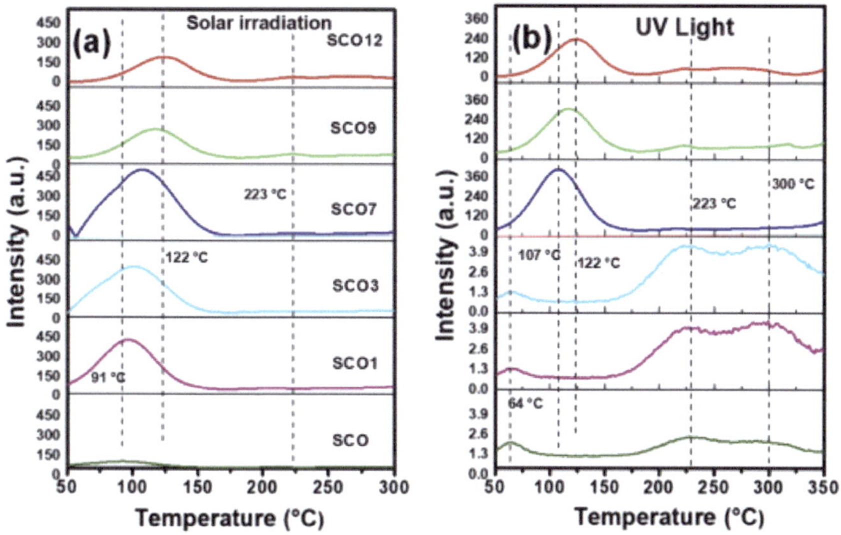

Fig. 6 TL curves for SCO and SCO: Er, Yb samples obtained after exposing them for 10 min with: **a** solar irradiation and **b** UV light (360 nm). The heating rate was 10 °C/min. Figures reproduced (adapted) with permission from reference [30], Elsevier

to 150 °C, indicating that these signals originate from shallow defects. Generally, an observable shift in the TL peaks occurs with increasing Yb concentration, as illustrated by the shift of the band from 91 °C to 122 °C for samples SCO–SCO12 (excited with solar light) in Fig. 6a, and from 107 °C to 122 °C for samples SCO7–SCO12 (also excited with solar light) in Fig. 6b. This shift in bands is attributed to the continuous increase in Yb concentration, which leads to the formation of additional defects, thereby requiring more energy for their complete release. As a result, the TL bands are shifted to higher temperatures [30–32]. The study revealed that SCO: Er, Yb phosphors are capable of emitting a thermoluminescent (TL) signal following exposure to UV or solar light, indicating their suitability for use as detectors for both radiation types.

5 Conclusion

Upconversion materials stand out as promising alternatives for thermoluminescence (TL) dosimetry, providing unique benefits that render them particularly effective for accurate radiation dose evaluation. The present chapter has delved into the essential principles of upconversion and thermoluminescence, the features of upconversion materials, and their possible applications in TL dosimetry, underscoring their significant ability to enhance the accuracy, sensitivity, and flexibility of radiation detection.

The fundamental principles of thermoluminescence (TL) have been illustrated through the widely recognized one trap–one recombination center model. This simple model can qualitatively account for all the essential characteristics of luminescence production. It emphasizes that the existence of traps is crucial for TL generation; therefore, ideal materials devoid of traps do not exhibit TL. Radiant energy absorption is required to populate the traps, while heating is necessary to liberate the charge carriers from these traps. This model provides valuable insights into the storage of radiant energy. Additionally, it clarifies that glow peaks at lower temperatures, which correspond to shallower traps, are more prone to fading. It also elucidates the typical asymmetric glow peak that is frequently observed in measurements.

The simple model is fundamentally phenomenological. It does not provide insights into the characteristics of trapping and recombination centers. In practical materials, identifying these centers can be quite challenging. For certain rare earth elements, the emission wavelength is distinctly characteristic, allowing for straightforward identification of luminescent centers. However, in many instances, significant effort is necessary to uncover the structure and nature of these centers. Most centers do not appear to be highly localized point defects, and truly isolated trapping and recombination centers are more of an exception than the norm. This suggests that real thermoluminescent materials exhibit complex defect structures. Additionally, the stability of defects may be compromised during heating, further complicating the process. Consequently, it is crucial to meticulously control each step in the dose determination procedure, including annealing, irradiation, and read-out. The thermoluminescence characteristics of $Sr_2CeO_4:Er^{3+}/Yb^{3+}$ upconversion materials for various radiations have been discussed in detail.

In conclusion, upconversion materials signify a notable progress in thermoluminescent (TL) dosimetry, providing improved sensitivity, tunability, stability, and versatility.

References

1. F. Daniels, C.A. Boyd, D.F. Saunders, Thermoluminescence as a research tool. Science **117**(3040), 343–349 (1953)
2. P.D. Townsend, J.C. Kelly, *Colour centres and imperfections in insulators and semiconductors* (Sussex University Press, London, 1993), p. 70.
3. S. Horowitz, Yigal, C.R.C. Press, eds, *Thermoluminescence and thermoluminescent dosimetry*, vol 1 (Boca Raton, CRC Press, 1984)
4. S.W.S. McKeever, *Thermoluminescence of Solids* (Cambridge University Press, 1985)
5. D.R. Vij, *Thermoluminescence Materials,* vol 7632 (PTR Prentice-Hall, Inc. A Simon a Schuster Company, Englewood Cliffs, New Jersey, 1993)
6. S.W.S. McKeever, M. Moscovitch, P.D. Townsend, *Thermoluminescence dosimetry materials: properties and uses* (Nuclear Technology Publishing, Kent, 1995)
7. R. Chen, S.W.S. McKeever, *Theory of thermoluminescence and related phenomena* (World Scientific, 1997)
8. J.T. Randall, M.H.F. Wilkins, Phosphorescence and electron traps-I. The study of trap distributions. Proc Royal Soc London. Series A. Mathemat Phys Sci **184.999**, 365–389 (1945)

9. J.T. Randall, M.H.F. Wilkins, Phosphorescence and electron traps II. The interpretation of long-period phosphorescence. Proc. Royal Soc. London. Series A. Mathemat. Phys. Sci. **184.999**, 390–407 (1945)
10. K. Georgios, J.M. Gomez-Ros, J.W.N. Tuyn, Thermoluminescence glow-curve deconvolution functions for first, second and general orders of kinetics. J. Phys. D: Appl. Phys. **31.19**, 2636 (1998)
11. A.J.J. Bos, High sensitivity thermoluminescence dosimetry. Nucl. Instrum. Methods Phys. Res., Sect. B **184**(1–2), 3–28 (2001)
12. G.F.J. Garlick, A.F. Gibson, The electron trap mechanism of luminescence in sulphide and silicate phosphors. Proc. Phys. Soc. **60**(6), 574 (1948)
13. A.J.J. Bos, J.B. Dielhof, The analysis of thermoluminescent glow peaks in CaF_2: Tm (TLD-300). Radiat. Prot. Dosimetry **37**(4), 231–239 (1991)
14. C.E. May, J.A. Partridge, Thermoluminescent kinetics of alpha-irradiated alkali halides. J. Chem. Phys. **40**(5), 1401–1409 (1964)
15. M. Bhattacharya, A critical appraisal of the determination of activation energy of TL peaks by Kirsh method. Mod. Phys. Lett. B **19**(27), 1387–1396 (2005)
16. C. Reuven, S.A.A. Winer, Effects of various heating rates on glow curves. J. Appl. Phys. **41.13**, 5227–5232 (1970)
17. T.M. Piters, A.J.J. Bos, A model for the influence of defect interactions during heating on thermoluminescence in LiF: Mg, Ti (TLD-100). J. Phys. D Appl. Phys. **26**(12), 2255 (1993)
18. A.J.J. Bos, T.M. Piters, Study of ageing effects in LiF: Mg, Ti by analysis of thermoluminescence glow curves. Nucl. Tracks Radiat. Meas. **21**(1), 163–167 (1993)
19. Y.S. Horowitz, M. Moscovitch, Computerized glow curve deconvolution applied to high dose (102–105 Gy) TL dosimetry. Nucl. Instrum. Methods Phys. Res., Sect. A **243**(1), 207–214 (1986)
20. A.J.J. Bos et al., Effects of cooling and heating rate on trapping parameters in LiF: Mg, Ti crystals. J. Phys. D Appl. Phys. **25**(8), 1249 (1992)
21. P. Dorenbos et al., Afterglow and thermoluminescence properties of Lu_2SiO_5: Ce scintillation crystals. J. Phys. Condens. Matter **6**(22), 4167 (1994)
22. D.W. Cooke et al., X-ray-induced thermally stimulated luminescence of cerium- doped gadolinium oxyorthosilicate. Radiat. Meas. **33**(4), 403–408 (2001)
23. Y.S. Horowitz, Theory of thermoluminescence gamma dose response: the unified interaction model. Nucl. Instrum. Methods Phys. Res., Sect. B **184**(1–2), 68–84 (2001)
24. Y.S. Horowitz, O. Avila, M. Rodriguez-Villafuerte, Theory of heavy charged particle response (efficiency and supralinearity) in TL materials. Nuclear Instrum. Methods Phys. Res. Sect. B: Beam Interact. Mater. Atoms **184.1–2**, 85–112 (2001)
25. R.L. Kathren, External beta-photon dosimetry for radiation protection. The Dosimetry of Ionizing Radiat. **2**, 321–370 (1987)
26. C.R. Hirning, Detection and determination limits for thermoluminescence dosimetry. Health Phys. **62.3**, 223–227 (1992)
27. P. Guy, in *Ionizing RadiationProtection and Dosimetry* (CRC Press, 1988)
28. T. Kron et al., A checklist for reporting of thermoluminescence dosimetry (TLD) measurements. Phys. Med. Biol. **44** (1999)
29. N. Toshiyuki et al., Development of a new highly sensitive LiF thermoluminescence dosimeter and its applications. Nuclear Instrum. Methods **157.1**, 155–162 (1978)
30. E. Viesca-Villanueva et al., Effect of Yb^{3+} codopant on the upconversion and thermoluminescent emission of Sr_2CeO_4: Er^{3+}, Yb^{3+} phosphors. J. Phys. Chem. Solids **145**, 109547 (2020)
31. H.A. Höppe et al., Luminescence in Eu^{2+}-doped $Ba_2Si_5N_8$: fluorescence, thermoluminescence, and upconversion. J. Phys. Chem. Solids **61.12**, 2001–2006 (2000)
32. V. Singh et al., NIR to visible up-conversion, infrared luminescence, thermoluminescence and defect centres in Y_2O_3: Er phosphor. J. Phys. D: Appl. Phys. **42.6**, 065104 (2009)
33. N.O. Etafo et al., The effect of a Yb Co-dopant on the blue upconversion and thermoluminescent emission of $SrLaAlO_4$: Yb^{3+}, Tm^{3+} phosphors. Coatings **13.6**, 1003 (2023)

34. A.M. Voiculescu et al., Optical thermometry through infrared excited green upconversion emissions of Er^{3+}-Yb^{3+} co-doped $LaAlO_3$ phosphors. J. Lumin. **242**, 118602 (2022)

Upconversion Phosphors for Targeted Drug Delivery

Shireen Aman and Vikas Dubey

Abstract This article focuses on lanthanide-based upconversion phosphors that have drawn significant interest in the field of material science and biomedicine for targeted drug delivery. The unique optical and chemical properties of lanthanide ions for having at least one electron in the 4f orbital make it suitable for upconversion. Upconversion nanoparticles (UCNPs) possess distinctive attributes, including autofluorescence, a high signal-to-noise ratio, minimal photodamage to living organisms, and deep penetration in biological samples, which render them highly suitable for drug delivery applications. UCNPs appear to be a suitable drug-making technique that can be used to treat cancer with nearly minimal side effects, among other applications. Upconversion materials play a critical part in sensing applications. Utilizing these nanoparticles allows for the conversion of longer wavelength photons into shorter wavelength fluorescence, enabling the detection of temperature, pH, ions, and trace biomolecules. Delivering the drug to the tumor site stands out as a primary issue, as the traditional approach of drug distribution in the biological context leads to various side effects in individuals. Researchers from many fields are interested in photodynamic treatment because of its capacity to activate the manufactured medicine during the upconversion process using light. Hence, there is a growing prominence of Ln^{3+}-based upconversion materials, indicating their potential as groundbreaking options for new biomedical applications. This article review specifically examines the chemical makeup and development of upconversion nanoparticles, as well as the methods for creating targeted drug delivery systems. Additionally, the article delves into the application of these particles in different cancer treatment approaches, including therapies such as photodynamic therapy (PDT).

Keywords Upconversion phosphors/nanoparticles (UCNPs) · PDT (photodynamic therapy) · Targeted drug delivery

S. Aman (✉) · V. Dubey
Department of Physics, North-Eastern Hill University, Shillong, India
e-mail: amanshireen123@gmail.com

V. Dubey et al. (eds.), *Upconversion Luminescence of Phosphors*, Progress in Optical Science and Photonics 39, https://doi.org/10.1007/978-981-95-1762-6_9

1 Introduction

Nanoparticles with upconversion luminescence (UCNPs) represent an exciting advancement in luminescent nanomaterials. They can be activated by low-energy radiation, such as NIR light, to produce higher energy emission at higher electronic levels, which aligns with an anti-Stokes mechanism. The processes through which we can synthesize UCNPs are thermal decomposition, hydrothermal synthesis, ionic liquid-based synthesis process, etc. [1, 2]. UCNPs comprise an inorganic host matrix (such as $NaYF_4$, $LiYF_4$, and Gd_2O_3) doped with lanthanide ions like Eu, Sm, Nd, Yt, etc., which are called luminous centers. Fluorescent materials have recently found widespread use in the biomedical sciences as a high-tech, highly effective tool for treating and diagnosing malignant disorders like cancer. NIR light has the fewest negative impacts on physiological cells as well as tissues while treating cancer.

Therefore, it has recently gained significant attention for enhancing penetration depth and achieving successful applications in cancer diagnosis and treatment [3, 4]. Host matrices doped using Lanthanide ions showed efficient upconversion (UC) emission characteristics. When exposed to ultraviolet (UV) light, materials doped with different Ln^{3+} ions, like Ce^{3+}, Nd^{3+}, Sm^{3+}, Eu^{3+}, Tb^{3+}, and Dy^{3+}, display a range of colors in the visible spectrum based on the specific doping types. These UCNPs exhibit favorable characteristics due to their lower energy NIR excitation, including the capacity to penetrate deeply into tissues upon NIR light stimulation [1, 2]. Under continuous single-particle stimulation, UCNPs provide remarkable photostability and unusual chemical stability with no detectable photobleaching as well as off-on blinking [5–13]. Nd^{3+} ions at 808 nm with large absorption can act as a sensitizer for the process of upconversion [14–19]. A promising approach that can considerably minimize the overheating coupled to standard 980 nm stimulation is to employ Nd^{3+} ions. This could potentially serve as a promising approach to minimize the likelihood of living cell wrecks brought about by near-infrared (NIR) excitation lasers. Moreover, this strategy proves particularly suitable for low-energy (NIR) photoactivation of biomolecules or the photo-triggered delivery of targeted drugs. Besides, compelling evidence from multiple investigations confirms that a solid matrix encasing gadolinium (Gd), ytterbium (Yb), and barium (Ba) emerges as potential contrast agent for computed X-ray tomography (CT) [20–28]. Thus, Ln^{3+}-based upconversion materials have emerged as the new stars of biomedical applications, exhibiting the ability to revolutionize drug/gene delivery, photodynamic therapy, and multimodal bioimaging. The creation and application of drug delivery systems that can react to external factors such as the potential of hydrogen (pH), temperature, magnetic zone, light, and redox gradients, along with internal stimuli, has brought out the significant benefits of nanocomposites based on upconverting nanoparticles (UCNPs). Figure 1 displays various UCNP-based designs for different cancer therapy applications. This review will include the UCNPs' manufacturing processes and potential benefits in delivering drugs to cancer cells and combating the disease.

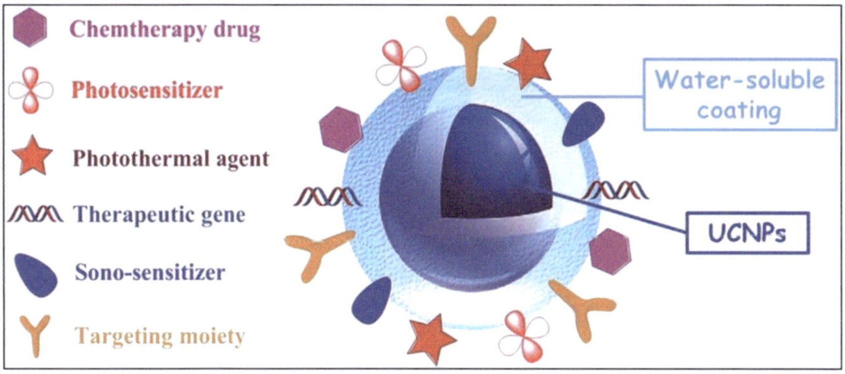

Fig. 1 Diverse UCNP-based topologies for diverse cancer treatments (taken permission from Elsevier, Jafari and Rezvanpour [29], license no:5798851286382)

2 The Chemical Composition and Fabrication Methods for UCNPs

2.1 UCNPs Fundamental Structure

The fifteen lanthanides (Ln) from ^{57}La to ^{71}Lu along with ^{39}Y and ^{21}Sc, are included in the category of rare earth elements [3, 4]. A notable characteristic of rare earth-activated light conversion in the material science area that has gained significant traction in recent years is Ln^{3+}-based activated light conversion. UC is a non-standard light interaction that creates a stronger signal at a specific frequency. The mechanisms take advantage of the unique properties of lanthanide ions, which are the extended excited state lifetimes and energy levels arranged like a ladder. The mechanism involves upconversion via a non-linear interaction, where multiple continuous wave NIR photons are combined to generate a single, high-energy photon, like ultraviolet, visible, or NIR. To increase the power of capturing near-infrared light by the upconverting lanthanide ions (Ln^{3+}), the surrounding crystal structure (host lattice) is co-doped with Er^{3+}, Ho^{3+}, and Tm^{3+} ions, etc. These co-dopants act as sensitizers for the UC mechanism [1, 2]. Except for La^{3+} and Lu^{3+}, they have remarkable chemical and optical properties that make them known as UC materials. These characteristics include high tissue penetration, strong resistance to photobleaching, low photodamage, and compatibility in biological contexts. To create a UCNP, two distinct ions must be co-doped. Out of the two, one ion behaves as a light absorber, and the other behaves as a light emitter that affects the ion. Three processes can cause upconversion, as shown in Fig. 2. The simplest one is excited-state absorption (ESA). The next process is called "energy transfer upconversion (ETU)", and the third one is known as "photon avalanche (PA)." Out of these three processes, the "energy transfer upconversion process" is the most frequently utilized.

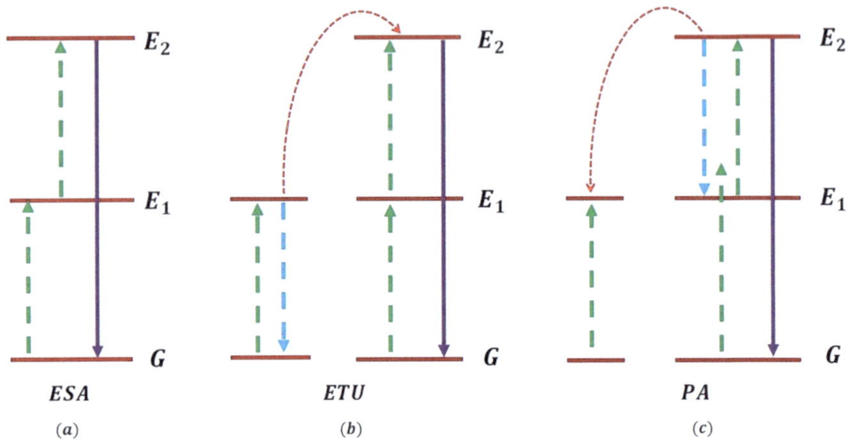

Fig. 2 The three processes of upconversion: (i) ESA, (ii) ETU, (iii) PA (taken permission from Elsevier, Jafari and Rezvanpour [29], license no:5798850080111)

Energy transfer upconversion consists of two ion systems. The sensitizer absorbs light, while the activator emits light. It means it has two lanthanide ions. One sensitizer absorbs light, and it can transfer that energy to the activator. Once that happens, the activator stores the energy until the following photon comes in. Then the sensitizer can sense it and transfer it again to the activator so that it can have enough energy to emit light.

There are two distinct types of luminous upconversion nano phosphors based on neodymium(III) and gadolinium(III), that can be used extensively in several investigations. Nd^{3+}-doped upconversion nanoparticles (UCNPs) are activated by 808-nm wavelength energy radiation. Due to their ability to be triggered at a lower wavelength, Nd^{3+} doped UCNPs result in decreased photodamage from the light source. Furthermore, the properties of Gd^{3+}-based UCNPs that include mainly magnetic make them appropriate for use in the targeted drug delivery system for magnetic resonance imaging (MRI). Studies in recent times have provided evidence for the use of upconversion nanoparticles (UCNPs) as carriers for well-known anti-cancer drugs like doxorubicin (DOX) to effectively treat in vivo malignant diseases such as lung cancer [30]. Along with the aforementioned benefits, the upconversion approach brings additional advantages in the biological setting, including minimal photodamage to normal tissue and the ability to penetrate deeply into tumor tissue. It is feasible to combine UC with specific substances to ensure the controlled release of the medication in response to various stimuli, encompassing a range of in vivo and in vitro triggers such as pH, temperature, magnetic fields, and more [31].

2.2 *Synthesis Approaches for Targeted Drug Delivery Based on UCNPs*

Since UCNPs are hydrophobic by nature, coating their surface is necessary to convert them to hydrophilic qualities and allow for proper blood circulation. UCNP-based drug delivery relies on leveraging the distinct characteristics of UCNPs by combining them with other inorganic and organic substances to form a unified nano platform for integrated disease diagnosis, treatment, and monitoring progress in therapy. To achieve this purpose, three considerations must be made. In order to effectively integrate additional functional nanostructures, it is crucial to attain a high-quality, pure-phase, uniform, well-formed synthesis of UCNP. Furthermore, the combined elements should serve as an appropriate porous structure or connecting point to accommodate the anticancer drug molecules through covalent bonding or physical absorption. Finally, efficient modification processes need to be done to develop stimuli-responsive medication delivery systems that enhance therapeutic efficacy in cancer therapy while lowering cytotoxic medication side effects. Leveraging the UltraViolet and visible light emitted by upconversion luminescence nanoparticles to activate the photo-responsive species affixed to the UCNP surface creates a tailored drug delivery vehicle that is activated by NIR light. The upper layer of the UCNPs has to be coated. The layering of upconversion particles can be achieved through four methods, namely coating with silica or mesoporous silica (UCNPs@SiO_2), polymer grafting or self-assembly (UCNPs@polymer), creating hollow UCNPs with a mesoporous surface (UCNPs@hollow), and incorporating UCNPs into electrospun fibers (UCNPs @fibers). These methods are depicted in Fig. 3 [30–43].

The production of upconversion luminescent nanoparticles can be achieved through different methods, including the sol-gel approach, the hydrothermally assisted template technique, the polymer grafting or self-assembly approach, and the electrospinning technique.

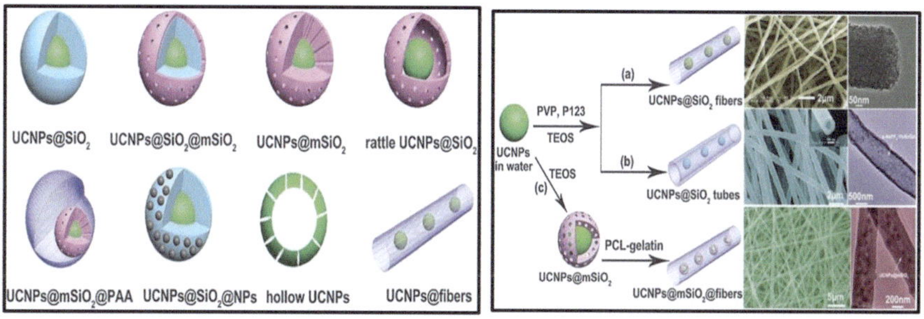

Fig. 3 The four ways to layer UCNPs and the synthesis process [31]

2.2.1 Sol-Gel Method for Synthesizing UNCPs

A method for UCNP-based systems to deliver drugs that don't require complex tools or equipment is the sol-gel method. The conventional sol-gel approach, also referred to as the "Stober method," uses inorganic precursors (metal salts or metal-alkoxides) [44] mixed with water to generate 3D oxide networks [45] via condensation and hydrolysis. Furthermore, OsseoAsare and Arriagada's reverse micro-emulsion, which produces more homogeneous-sized silica nanoparticles, is included in the category of sol-gel methods [44–48]. The process involves the polycondensation of metals and the hydrolysis of metal alkoxide precursors, which are commonly known as wet chemicals [33]. In order to achieve high uniformity and single dispersion of UCNPs within mesoporous silica (SiO_2), a thin, dense silica layer was applied to the UCNPs using the microemulsion method. This method involved using the surfactant Trix100 or Igepal CO-520 in a mixture of cyclohexane and an aqueous solution [49, 50]. Shi et al. [51–56] have shown that the resulting UCNPs@SiO_2 can be combined with other functional building blocks, like ultra-small CuS for photothermal therapy and Au nanoparticles for CT imaging. The result is multifunctional upconversion nanoparticles coated using silica and gold (SiO_2, Au) or upconversion nanoparticles coated using silica and copper(III)sulfide (SiO_2, CuS), which exhibit the special qualities of each constituent to enable radiotherapy and photothermal destruction for therapeutic purposes as well as multimodal bioimaging. Additionally, a mesoporous silica ($mSiO_2$) coating is applied to the functional nanoparticles using a modified Stober approach. The processes of Sol-gel can be categorized into two groups: (i) one-step methods and (ii) two-step techniques [1, 5]. Two types of structure-directing agents are typically used to produce a mesoporous silica layer: (1) organosilanes like octadecyltrimethoxysilane (C18TMS) and (2) surfactants such as cetyltrimethylammonium bromide (CTAB) [11, 57, 58]. Park and his colleagues were the first to report the synthesis of $NaYF_4$ using the wet chemical process (sol-gel) [59]. The sol-gel method was utilized by Patra and his team to produce Er^{3+}-doped ZrO_2 nanocrystals. In order to create a water-in-oil (w/o) micro-emulsion, they employed cyclohexane as an oil phase and a non-ionic surfactant in a sol-emulsion-gel procedure. Their final product had respective wavelengths of 550 and 675 nm [60], which allowed it to absorb light and produce dazzling green and red lights. Additionally, the sol-gel method, as illustrated in Fig. 4, can coat nanoparticles with metal [61, 62].

Using CTAB (cetyltrimethylammonium bromide) (as a core, tetraethyl orthosilicate (TEOS) can be degraded catalytically to help form the mesoporous silica shell. After that, when CTAB is eliminated by the mechanism of ion exchange, the hollow core-shell particle may form. Furthermore, the "protein corona" issue can be solved by poly(ethylene glycol) (PEG). Subsequently, the upper layer of the nanoparticle would be coated with a certain amino acid, which would notify the immune system and ultimately destroy the particle. PEG has the ability to mitigate this impact and resist the particle from being removed. The flow of blood flow in the living environment could potentially be enhanced through the motion of nanoparticles. Conjugated folic acid (FA) stimulates cellular absorption through the endocytosis process using

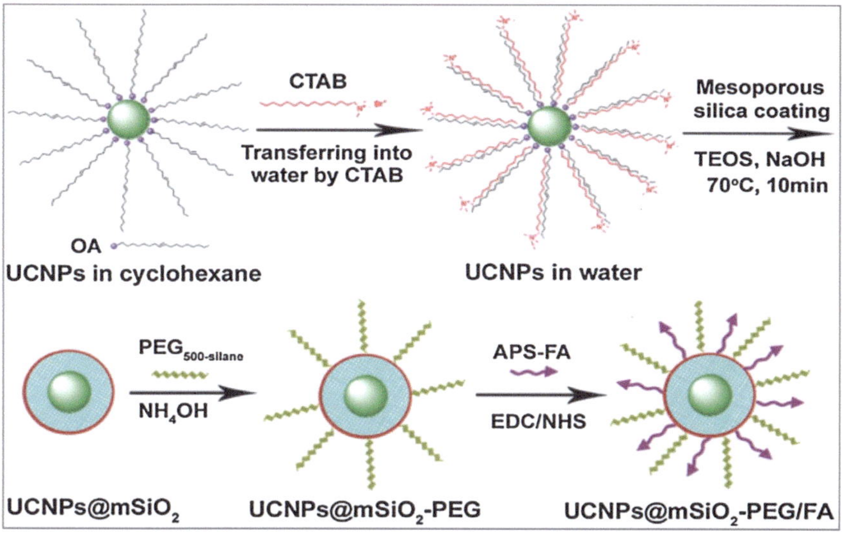

Fig. 4 The process of making nano-particles by the sol-gel technique (taken permission from Elsevier, Jafari and Rezvanpour [29], license no:5798850463444)

the folate receptor on the upper layer of the cell membrane. It can enhance the delivery of DOX as an anti-cancer drug [62]. In this condition, the complex emits green and yellow light with a 980-nm wavelength. Gai and his team performed how to coat the upper layer of Fe_3O_4 with silica using the two-step sol-gel method, resulting in a core-shell nanoparticle that may be utilized as an Ibuprofen (IBU) drug carrier. A demonstration of this method is presented in Fig. 5.

This process comprised two steps: first, utilizing CTAB to coat the surface of Fe_3O_4 with 8 nm thick silica, and second, using a modified Stober method to fabricate the UCNP. They concluded that the magnet is an appropriate target and that the UCNP's radiation wavelength can be used to alter the amount of medication released [63]. Furthermore, polyacrylic acid (PAA) has the potential to be used in the production of a biodegradable superabsorbent, enabling the creation of pH-responsive drug delivery nanomolecules. It has the capability to serve as a template and nanoreactor for CTAB-stabilized UCNPs. Using a concentrated double shell and core, Wang et al. have devised a distinctive and unconventional multifunctional nanostructure, denoted as (UCNPs@mSiO_2)@PAA [64]. These materials exhibit two noteworthy qualities: an exceptionally high drug carrier capacity and pH-responsive properties for sensitive medication release [65].

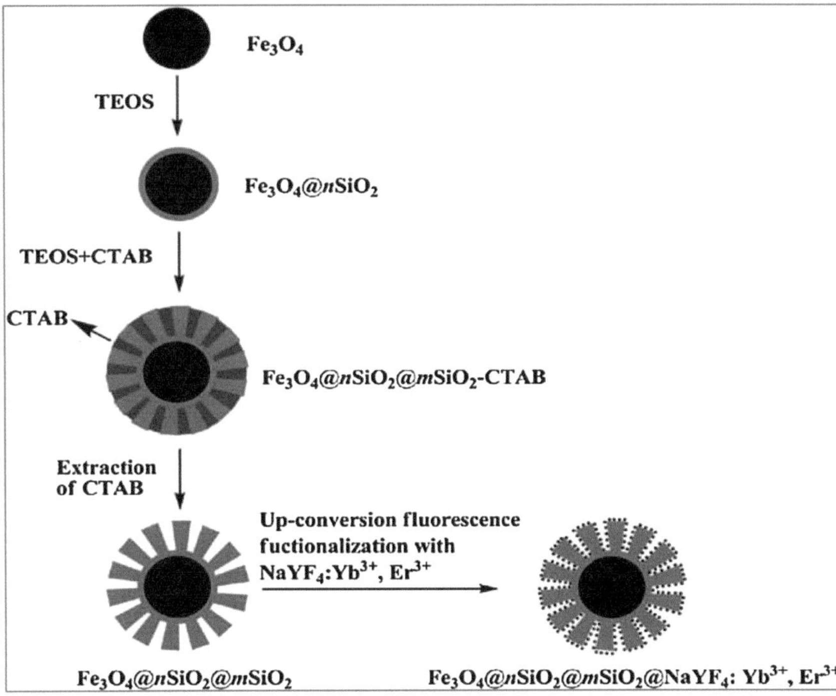

Fig. 5 The coating of Fe_3O_4 with silica using the two-step sol-gel method (taken permission from Elsevier, Jafari and Rezvanpour [29], license no:5772570022387)

2.2.2 The Hydrothermally Assisted Template Method for Synthesizing UNCPs

One of the most often techniques for creating hollow materials is the template method, which is mostly accomplished with the use of soft-directing agents or hard templates [31, 66]. The hydrothermal assisted template approach allows for the employment of two different types of agents: (i) soft directing agents like sodium poly (4-styrene sulfonate) [67], (ii) hard template agents like poly (acrylic acid sodium salt) [68]. According to recent studies, sodium yttrium fluoride doped with ytterbium and erbium, yttrium(III) oxide doped with ytterbium and erbium, gadolinium(III) oxide doped with ytterbium and erbium, etc., samples are created using melamine formaldehyde, carbonaceous nanospheres, acrylic acid sodium salt, 4-styrene sulfonate as templates.

Kang et al. showed how to use the hydrothermal approach to create a hollow nanoparticle. In order to create these particles in the process of ion exchange, the particles must be in an acidic condition. They used $Gd(OH)CO_3$ as a template for gadolinium orthovanadate. After that, within the medication, they gave the capillary force of acrylic acid (AA), including DOX, as seen in Fig. 6 [69].

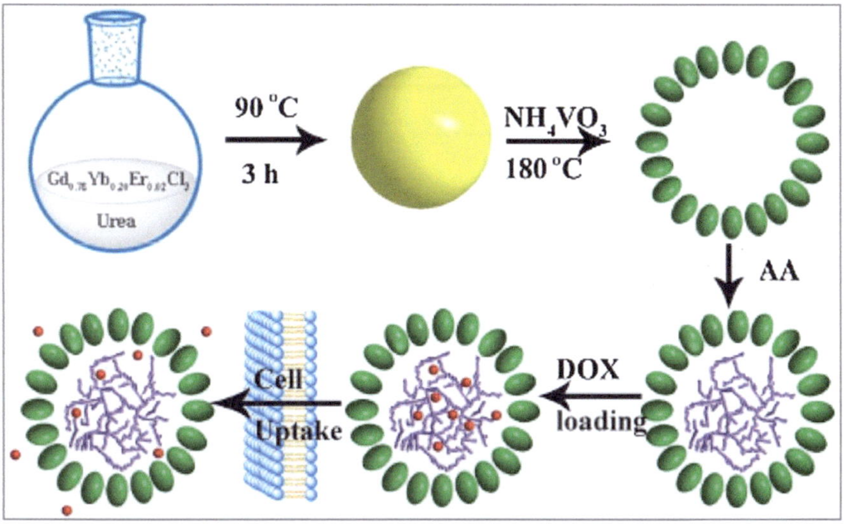

Fig. 6 The use of the hydrothermal approach by Kang et al. (taken permission from Elsevier Jafari and Rezvanpour [29], license no:5772990768523)

Another efficient technique for creating hollow upconversion materials is lithography using a beam of electrons. An organic substance that may form a micelle like OA (oleic acid) can be utilized to demonstrate this process. The organic substance keeps breaking down into smaller pieces, such as carbon, through this process. The created voids now stabilize at a certain size [70]. Using an electron beam for irradiation, Yan et al. synthesized beta-$NaYF_4$:Yb/Er hollow-sphere nanocrystals $NaREF_4$ (RE = ND-LU, Y) with the help of a heat-induced etching acid mechanism [70, 71]. However, the procedure was restricted to a small field of view with TEM, which does not allow for the potential of materials like hollow structures. Additionally, it was discovered that during electron-beam irradiation, a significant number of hollow arrangements of $NaREF_4$ (RE = Y, Yb, and Lu) nanocrystals were formed unchanged. Also, hollow structures are not formed by all rare earth ions. The only species that the electron beam affected were Y, Yb, and Lu. It is unclear exactly why this occurrence occurs at the moment. The hollow-structured nanoparticles in their manufactured state can be used for upconversion cell imaging and medication storage/release as anti-cancer medication carriers.

2.2.3 The Polymer Grafting or Self-Assembly Method for Synthesizing UNCPs

As seen in Fig. 7, the UCNPs may be encased inside a polymer shell using polymer grafting, themselves with a surfactant such as OA (oleic acid), or self-assembly implementing a covalent link [1, 31]. The superior UCNPs generated

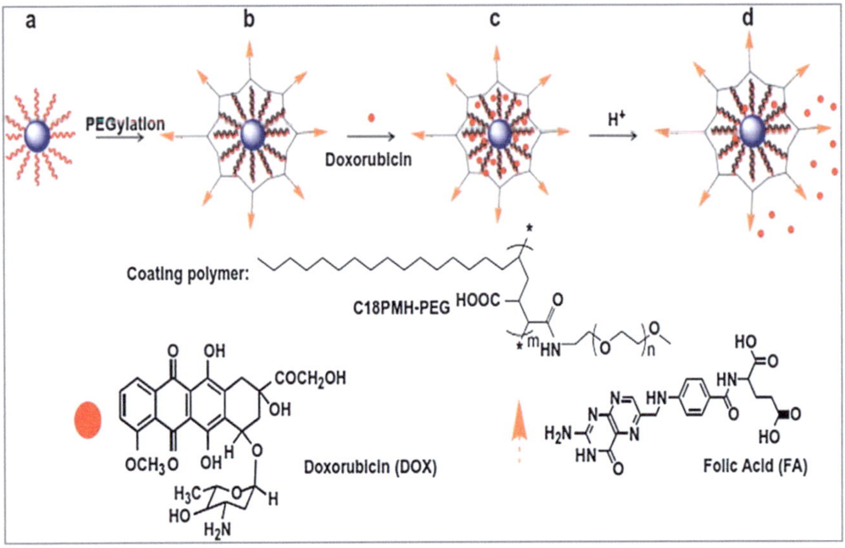

Fig. 7 The UCNPs encased inside a polymer shell using polymer grafting with a surfactant such as OA (oleic acid) (taken permission from Elsevier, Jafari and Rezvanpour [29] license no:5778841233444)

by current synthesis methods are hydrophobic, rendering them unsuitable for biological applications because they employ capping surfactants, such as oleic acid. UCNPs are connected or enclosed by suitable inorganic or organic materials via covalent association or self-assembly. An example is an amphiphilic tri-block copolymer named mPEG-b-PCL-b-PLL{methoxy poly ethylene glycol-block-poly(ε-caprolactone)block-poly(L-lysine)} [72, 73], coupled with a prodrug cisplatin(IV), may be combined with hydrophobic UCNPs to form core-shell nanocomposites. This might be utilized to trace the transport route using in vivo and in vitro imaging and administer cisplatin to cancer cells.

As per Zhao and Gu, it is viable to modify hydrophobic UCNPs into insoluble in water and biocompatible UCNPs via the use of PEG phospholipids and TWEEN [31, 74, 75]. It may be viable to co-load DOX and the cancer-fighting agent camptothecin (CPT) into the “hydrophobic pockets” since they have aquaphobic interlinkage [31].

Additionally, Liu and his colleagues on a unique multifunctional nanoparticle that was created using the layer-by-layer assembly technique. Furthermore, Liu’s group reported on the creation of a novel multifunctional nanoparticle via layer-by-layer assembly [76, 77]. The nanoparticle had a thin coating of gold as the shell, a densely packed Fe_3O_4 interlayer, and an inner core made of UCNP. These multipurpose nanomaterials are helpful for magnetically targeted medication drug delivery, multi-modal diagnostic imaging in situ and in vitro conditions, as well as cancer treatment methods, including chemotherapy and photothermal therapy. This kind of construction enables the particle to be sensitive to the magnetic and pH environments. It is possible to regulate the particle’s entry into the tumor location with the aid of an outside force. As a result, the effects of the medication can be avoided on normal tissue [78].

2.2.4 The Electrospinning Method

The electrospinning method offers the capability to effortlessly produce a range of one-dimensional luminescent materials, including fibers, wires, belts, and tubes, by dispersing luminescent nanoparticles of inorganic rare earth elements into polymer hybrid precursors. This holds potential for applications in fluorescent lamps and color displays [79–81]. This technique entails employing a syringe pump to inject a polymer via a needle while simultaneously offering the needle an adequate voltage for generating an electrostatic force. The chemical solution rate of flow, applied voltage, needle-to-collector distance, and type of polymer in the solution are vital parameters for synthesizing acceptable fibers [74]. Dong and colleagues demonstrated the manufacture of an elemental-shell luminous nanoparticle, known as polyvinyl pyrrolidone (PVP), applying a voltage of 13 kV at the needle spike during electrospinning. Another study developed a method for delivering DOX by using UCNPs@mSiO_2 nanospheres. This method involved a coating process that utilized phase transfer-assisted surfactant-templating. The details of this study can be found in reference [79]. The thereby formed DOX using mesoporous silica upconversion nanoparticles were subsequently merged with an electrospinning suspension that contained both the antiinflammatory medication indomethacin (IMC) and poly(e-caprolactone)–gelatin (PG). This led to the development of composite fibers loaded with both drugs, designated as UCNPs@mSiO_2@fibers [82]. The in vitro release patterns of the two drugs exhibited notable distinctions, highlighting their unique release characteristics. The optimal drug release was characterized by a sustained and extended process, offering an effective solution to partially alleviate the issue of sudden drug release. Furthermore, Park and his research team discovered that nanofiber patches containing curcumin (Cur) and constructed from polylactic acid (PLA) demonstrate significant wound healing capabilities in a live mouse model. A comparison between Cur-PLA nanofibers and PLA nanofibers revealed an 87% increase in the rate of wound closure within seven days of the experiment [83]. Additionally, a separate study by Schneider et al. found that an in vivo test using silk nanofiber electrospun with epidermal growth factor (EGF) resulted in a 90% improvement in scar closure in mice [84].

3 Targeted Drug Delivery System

In recent times, there has been a noticeable increase in the interest in creating versatile nanomedical platforms. The primary aim is to establish a link between biomaterials and clinical theranostics, facilitating the concurrent execution of a single nanocarrier, illness detection, and treatment. Sickness diagnosis and therapy within a mono-nanocarrier. To satisfy this requirement, diverse minimal composites utilizing upconversion nanoparticles (UCNP) have been investigated as drug delivery systems (DDS). By enhancing the pharmaceutical index of pharmaceutical compounds and reducing their adverse effects on sound tissues and cells, targeted drug delivery systems serve as a valuable tool. These systems aim to deliver medications precisely to specific tissues or cells, maximizing therapeutic efficacy while minimizing the unintended impact on surrounding healthy tissues. The existing targeted drug delivery systems are mainly divided into two groups: (i) systems that target specific molecules by linking upconversion nanoparticles with specific biomolecules through covalent bonds, and (ii) targeting guided by magnetic fields for drug release.

3.1 Specific Molecule-Targeted Drug Delivery System

The common approach for designing targeted drug delivery systems (DDSs) involves integrating specific ligands or biomolecules onto the drug delivery vehicle, which can include folic acid, antibodies, RGD (Arginine, Glycine, and Aspartic Acid peptide), TAT (a short chain of amino acids, specifically eight amino acids long) peptide, and aptamers. These nanocarriers are reclassified with acceptors, and they can distinguish and uniquely the receptors on the upper layer of target cells. This recognition triggers a receptor-mediated endocytotic pathway, facilitating site-specific drug release. Folic acid has gained prominence as a key factor for targeted anticancer drug release due to the widespread overexpression of folate receptors (FR) in numerous cancer cells in humans while being missed in normal cells. This includes cancers affecting the ovary, lung, breast, kidney, brain, endometrium, and colon [44, 85, 86]. High stability, cost of scale, immunity-inhibition, heterogeneity in conjugation, and long-term high binding affinity for FR are the characteristics that define FA. FA-coupled UCNPs have been thoroughly studied for therapeutic and diagnostic purposes, capitalizing on the benefits of upconversion of tiny particles (UCNPs) [62, 75, 87–90]. Utilizing FR-mediated active targeting, the FA-modified nanocarriers may be specifically accumulated in tumor cells that overexpress folate receptors. This significantly improves treatment effectiveness over conventional photodynamic therapy (PDT) and increases near-infrared (NIR) fluorescence at the tumor site as seen in Fig. 8.

For cancer therapy, photodynamic therapy (PDT) has attracted a lot of interest. Unlike other forms of treatment, such as surgery or chemotherapy, PDT enables targeted application to certain areas where the tumor is located. A further benefit of PDT over other therapy methods is its quick recovery time; in contrast, surgery, for example, requires a longer time for wound healing [33, 91]. The construction and development of targeting dual upconversion nanomaterials platform for bicolor luminescence were demonstrated by Yan et al. and this is known as photodynamic therapy (PDT). The nano platform combined hyaluronate fullerene (HAC60) with 3-aminophenyl boronic acid (APBA) functionalized upconversion nanocrystals (APBA-UCNPs) through a specific diol-borate condensation. The nano platform was specifically created to harness the combined targeting effects and potent targeting

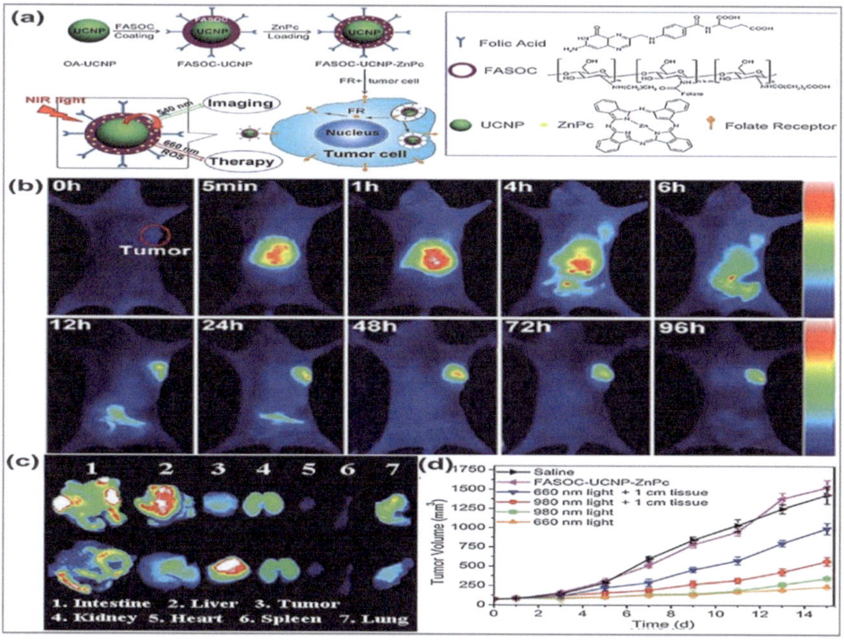

Fig. 8 **a** The synthesis of the FA-chitosan-coated UCNPs as carriers. **b, c** Targeting tumor in vivo. **d** Differences in the therapeutic efficacy of deep tissue PDT provoked by 980 and 660 nm light demonstrate the tumor development in mice from various treatment groups within fifteen days [31]

capabilities of two particular ligands, hyaluronic acid (HA) and APBA, resulting in enhanced uptake by cancer cells. The abundant production of singlet oxygen (1O2) through multiplexed resonance energy transfer between APBA-UCNPs serving as the donor and HAC60 (hyaluronate fullerene) serving as the acceptor allowed for an efficient treatment approach [92]. This field of nanotechnology shows great potential for highly specific, image-guided, tumor-targeted PDT. However, there are also some drawbacks associated with utilizing PDT. The first problem with PDT is that it doesn't penetrate the PS sufficiently, especially when using the older approach. The other restriction is the risk of skin problems caused by exposure to the light source. Also, the fluorescent protein known as "KillerRed," a novel biological photosensitizer, was employed by Liang et al. It is acknowledged as an effective substitute for the current photosensitizer used in PDT. The UCNPs were used by Liang et al. to get around this constraint. In Fig. 9, the UCNPs shown have the ability to activate KillerRed in a localized fashion by transforming NIR light, which penetrates deeply, into green light. This transformation leads to the production of ROS (Reactive Oxygen Species), ultimately leading to the destruction of malignant tissue in deeper parts of the tissue [93].

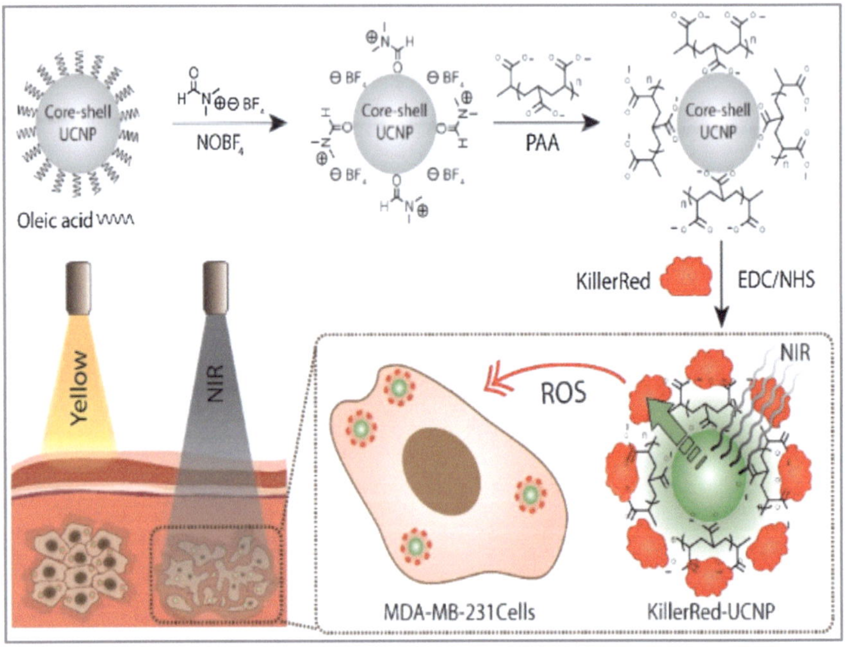

Fig. 9 An illustration of KillerRed UCNP for deep penetration in photo dynamic therapy (taken permission from Elsevier, Jafari and Rezvanpour [29], license no:5778850792248)

3.2 Magnetic Field-Guided Targeting for Drug Release

Proposing magnetic-field-guided drug delivery as a convenient and effective method to accurately target specific areas without impacting surrounding areas. The important advantages of superparamagnetic iron oxide (Fe_3O_4), such as its biocompatibility, biodegradability, and responsiveness, make it one of the most promising targeting agents [77, 94, 95]. Because of these characteristics, there has been an abundance of interest in creating magneto-optical multifunctional nanoplatforms that combine UCNPs and Fe_3O_4 in a single nanocarrier for simultaneous therapeutic and diagnostic applications. Liu and colleagues developed a nanocomposite encapsulated in polymer, consisting of upconversion nanoparticles (UCNP), superparamagnetic iron oxide (Fe_3O_4), and the anticancer drug DOX (doxorubicin). This nanocomposite was designed for targeted medicine delivery and multi-modal imaging. The DOX-loaded UCNPs/Fe_3O_4 [77] nanocomposites effectively eliminated a significant portion of cancer cells near the magnet in the presence of the magnetic field, while those located farther away from the magnet predominantly survived. In addition, a core with a multishell system was created by Ai and collaborators to improve the efficiency of a Pt-based drug when used for treating ovarian cancer with chemotherapy. The upconversion nanoparticle (UCNP) served as the central platform of the system, facilitating high-efficiency loading of both the drug and the RB (as a photosensitizer)

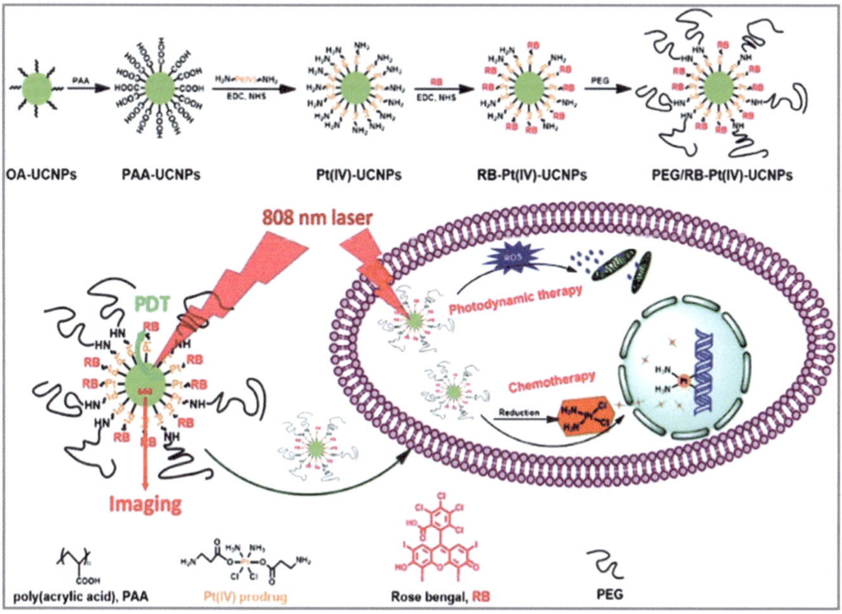

Fig. 10 Formation of the various phases of UCNP using combination therapy (taken permission from Elsevier, Jafari and Rezvanpour [29], license no:5778851004722)

into the shell. As illustrated in Fig. 10, PEG was also utilized as a ligand to achieve a customized distribution to the tumor.

Additionally, a magnetic liposomal upconversion nanoparticle (UCNP) can be utilized in chemo-phototherapy to monitor the impact of the drug through UCL (Upconversion Luminescence) and MR (Magnetic Resonance) imaging, as illustrated in Fig. 11 [96].

3.3 UCNPs for PDT

For cancer therapy, photodynamic therapy (PDT) has attracted a lot of interest. In this case, what we have is a photosensitizer that accumulates ions in the tumor site and is then activated with light and creates singlet oxygen $^1O_2$, which kills the tumor cells.

Because the majority of the photosensitizers are excited with visible light, therefore the penetration is not that deep. So, when we are treating the tumor deeply, it actually cannot reach them. Therefore, we could use upconversion nanoparticles because they are excited in near-infrared (NIR). They can also emit visible light to reach the deeper site of the tumor and excite the photosensitizer at the same time. Unlike other forms of treatment, such as surgery or chemotherapy, PDT enables the

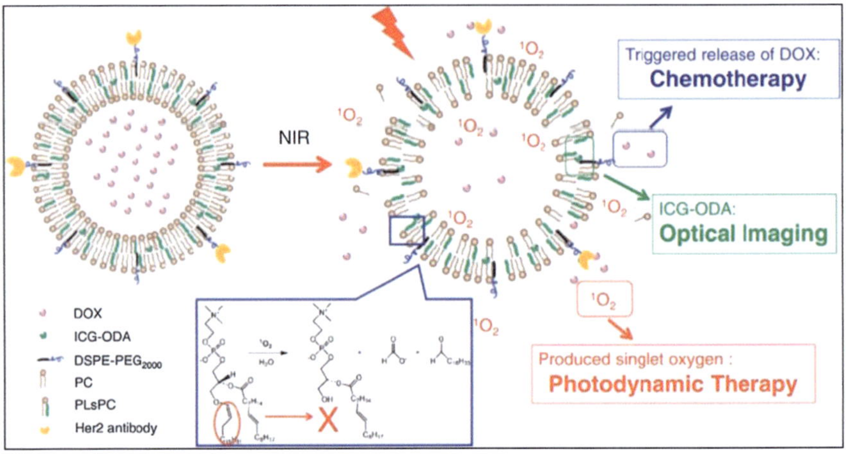

Fig. 11 A magnetic liposomal upconversion nanoparticle (UCNP) can be utilized in chemophototherapy to monitor the impact of the drug through UCL (upconversion luminescence) and MR (magnetic resonance) imaging (taken permission from Elsevier, Jafari and Rezvanpour [29], license no:5798870322011).

targeted application to certain areas where the tumor is located. A further benefit of PDT over other therapy methods is its quick recovery time; in contrast, surgery, for example, requires a longer time for wound healing [97, 98].

4 Conclusion

This study provides an overview of the basic structure of the UCNPs, different synthesis methods, synthesis approaches for drug delivery systems based on upconversion nanoparticles, and uses of UCNP-enriched composite nanomaterials in cancer treatment and drug administration. More specifically, the explanation highlighted the possibility of regulating the depth of penetration into the target tissue by understanding the operation of the upconversion (UC) technique. It was emphasized that upconversion nanoparticles (UCNPs) could be beneficial in Photodynamic Therapy. The activation of the photosensitizer (PS) is influenced by the type of light source used, impacting its effectiveness. Targeted drug delivery to the site where the tumor is located is an additional factor influencing cellular absorption. Key aspects of an ideal photoresponsive drug delivery system include precise external control, photolysis without side products, responsiveness to near-infrared light, prevention of premature release, and clean photolysis. It has been demonstrated that combining PDT with other traditional treatments is an effective approach for treating and eliminating malignant cells.

Nanoparticles known as upconversion nanoparticles (UCNPs) are useful tools in cancer treatment and drug delivery systems, demonstrating high effectiveness and minimal adverse effects. Recent studies have placed particular focus on pH-sensitive complexes in comparison to other drug types. Research suggests that it is feasible to incorporate various drug combinations to explore opportunities for more than two types of treatments. For instance, combining a pH-sensitive medication with a sensitizer possessing photosensitizer (PS) capabilities can create a unique drug compound suitable for a combination of chemotherapy and photodynamic therapy. This combination aims to increase the likelihood of inducing cell death in cancerous cells. In conclusion, there are many prospects for creative research in this fascinating subject since upconversion nanoparticle (UCNP)-based drug delivery systems are still in their early phases of study. Shortly, further ground-breaking breakthroughs in the detection and treatment of disease are expected to come from this extremely active field of study.

References

1. J.C. Boyer, F. Vetrone, L.A. Cuccia, J.A. Capobianco, Synthesis of colloidal upconverting $NaYF_4$ nanocrystals doped with Er^{3+}, Yb^{3+} and Tm^{3+}, Yb^{3+} via thermal decomposition of lanthanide trifluoroacetate precursors. J. Am. Chem. Soc. **128**(23), 7444–7445 (2006)
2. J. Zhao, D. Jin, E.P. Schartner, Y. Lu, Y. Liu, A.V. Zvyagin, et al., Nat. Nanotechnol. **8**, 729–734 (2013)
3. F. Auzel, Upconversion and anti-stokes processes with f and d ions in solids. Chem. Rev. **104**(1), 139–174 (2004)
4. V.V. Ovsyakin, P.P. Feofilov, Cooperative sensitization of luminescence in crystals activated with rare earth ions. SPIE Milest. Ser. **161**, 9–10 (2000)
5. J.V. Frangioni, In vivo near-infrared fluorescence imaging. Curr. Opin. Chem. Biol. **7**(5), 626–634 (2003)
6. S. Wu, G. Han, D.J. Milliron, S. Aloni, V. Altoe, D.V. Talapin, et al., Non-blinking and photostable upconverted luminescence from single lanthanide-doped nanocrystals. Proc. Natl. Acad. Sci. **106**(27), 10917–10921 (2009)
7. F. Vetrone, R. Naccache, A.J. de la Fuente, F. Sanz-Rodríguez, A. Blazquez-Castro, E.M. Rodriguez, et al., Intracellular imaging of HeLa cells by non-functionalized $NaYF_4$: Er^{3+}, Yb^{3+} upconverting nanoparticles. Nanoscale **2**(4), 495–498 (2010)
8. G. Chen, T.Y. Ohulchanskyy, S. Liu, W.C. Law, F. Wu, M.T. Swihart, et al., Core/shell $NaGdF_4$: $Nd^{3+}/NaGdF_4$ nanocrystals with efficient near-infrared to near-infrared downconversion photoluminescence for bioimaging applications. ACS Nano **6**(4), 2969–2977 (2012)
9. Q. Zhan, S. He, J. Qian, H. Cheng, F. Cai, Optimization of optical excitation of upconversion nanoparticles for rapid microscopy and deeper tissue imaging with higher quantum yield. Theranostics **3**(5), 306 (2013)
10. Z. Song, Y.G. Anissimov, J. Zhao, A.V. Nechaev, A. Nadort, D. Jin, et al., Background free imaging of upconversion nanoparticle distribution in human skin. J. Biomed. Opt. **18**(6), 061215–061215 (2013)
11. G.D.C.E.O. AD, S. Aloni, M.V.P. Altoe, E.S. Barnard, B. Sanii, J.J. Urban, D.J. Milliron, B.E. Cohen, P.J. Schuck, Nat. Nano **9**, 300–305 (2014)
12. S.H. Nam, Y.M. Bae, Y.I. Park, J.H. Kim, H.M. Kim, J.S. Choi, et al., Long-term real-time tracking of lanthanide ion doped upconverting nanoparticles in living cells. Angew. Chem. Int. Ed. **50**(27), 6093–6097 (2011)
13. J. Zhou, Q. Liu, W. Feng, Y. Sun, F. Li, Upconversion luminescent materials: Advances and applications. Chem. Rev. **115**(1), 395–465 (2015)

14. J. Shen, G. Chen, A.M. Vu, W. Fan, O.S. Bilsel, C.C. Chang, G. Han, Engineering the upconversion nanoparticle excitation wavelength: Cascade sensitization of tri-doped upconversion colloidal nanoparticles at 800 nm. Adv. Opt. Mater. **1**(9), 644–650 (2013)
15. X. Xie, N. Gao, R. Deng, Q. Sun, Q.H. Xu, X. Liu, Mechanistic investigation of photon upconversion in Nd^{3+}-sensitized core–shell nanoparticles. J. Am. Chem. Soc. **135**(34), 12608–12611 (2013)
16. Y.F. Wang, G.Y. Liu, L.D. Sun, J.W. Xiao, J.C. Zhou, C.H. Yan, Nd^{3+}-sensitized upconversion nanophosphors: Efficient in vivo bioimaging probes with minimized heating effect. ACS Nano **7**(8), 7200–7206 (2013)
17. H. Wen, H. Zhu, X. Chen, T.F. Hung, B. Wang, G. Zhu, et al., Upconverting near-infrared light through energy management in core-shell-shell nanoparticles. Angew. Chem. **125**(50) (2013)
18. U. Rocha, K.U. Kumar, C. Jacinto, I. Villa, F. Sanz-Rodríguez, del Carmen Iglesias de la Cruz, M., … & Jaque, D., Neodymium-doped LaF_3 nanoparticles for fluorescence bioimaging in the second biological window. Small **10**(6), 1141–1154 (2014)
19. L.D. Sun, Y.F. Wang, C.H. Yan, Paradigms and challenges for bioapplication of rare earth upconversion luminescent nanoparticles: Small size and tunable emission/excitation spectra. Acc. Chem. Res. **47**(4), 1001–1009 (2014)
20. Y. Liu, K. Ai, J. Liu, Q. Yuan, Y. He, L. Lu, A high-performance ytterbium-based nanoparticulate contrast agent for in vivo X-ray computed tomography imaging. Angew. Chem. **124**(6), 1466–1471 (2012)
21. S. Zeng, M.K. Tsang, C.F. Chan, K.L. Wong, J. Hao, PEG modified $BaGdF_5$: Yb/Er nanoprobes for multi-modal upconversion fluorescent, in vivo X-ray computed tomography and biomagnetic imaging. Biomaterials **33**(36), 9232–9238 (2012)
22. H. Liu, W. Lu, H. Wang, L. Rao, Z. Yi, S. Zeng, J. Hao, Simultaneous synthesis and amine-functionalization of single-phase $BaYF_5$: Yb/Er nanoprobe for dual-modal in vivo upconversion fluorescence and long-lasting X-ray computed tomography imaging. Nanoscale **5**(13), 6023–6029 (2013)
23. W. Zheng, P. Huang, D. Tu, E. Ma, H. Zhu, X. Chen, Lanthanide-doped upconversion nanobioprobes: Electronic structures, optical properties, and biodetection. Chem. Soc. Rev. **44**(6), 1379–1415 (2015)
24. F. Li, C. Li, J. Liu, X. Liu, L. Zhao, T. Bai, et al., Aqueous phase synthesis of upconversion nanocrystals through layer-by-layer epitaxial growth for in vivo X-ray computed tomography. Nanoscale **5**(15), 6950–6959 (2013)
25. S. Zeng, H. Wang, W. Lu, Z. Yi, L. Rao, H. Liu, J. Hao, Dual-modal upconversion fluorescent/X-ray imaging using ligand-free hexagonal phase $NaLuF_4$: Gd/Yb/Er nanorods for blood vessel visualization. Biomaterials **35**(9), 2934–2941 (2014)
26. Z. Yi, S. Zeng, W. Lu, H. Wang, L. Rao, H. Liu, J. Hao, Synergistic dual-modality in vivo upconversion luminescence/X-ray imaging and tracking of amine-functionalized $NaYbF_4$: Er nanoprobes. ACS Appl. Mater. Interfaces **6**(6), 3839–3846 (2014)
27. Y. Min, J. Li, F. Liu, P. Padmanabhan, E.K. Yeow, B. Xing, Recent advance of biological molecular imaging based on lanthanide-doped upconversion-luminescent nanomaterials. Nanomaterials **4**(1), 129–154 (2014)
28. R. Wang, F. Zhang, NIR luminescent nanomaterials for biomedical imaging. J. Mater. Chem. B **2**(17), 2422–2443 (2014)
29. M. Jafari, A. Rezvanpour, Upconversion nano-particles from synthesis to cancer treatment: A review. Adv. Powder Technol. **30**(9), 1731–1753 (2019)
30. Y. Wu, X. Xu, X. Chen, R. Yang, Q. Xiao, Y. Li, Mesoporous silica coated $Gd_2(CO_3)_3$: Eu hollow nanospheres for simultaneous cell imaging and drug delivery. RSC Adv. **6**(67), 62320–62326 (2016)
31. D. Yang, P.A. Ma, Z. Hou, Z. Cheng, C. Li, J. Lin, Current advances in lanthanide ion (ln^{3+})-based upconversion nanomaterials for drug delivery. Chem. Soc. Rev. **44**(6), 1416–1448 (2015)
32. N.M. Idris, M.K.G. Jayakumar, A. Bansal, Y. Zhang, Upconversion nanoparticles as versatile light nanotransducers for photoactivation applications. Chem. Soc. Rev. **44**(6), 1449–1478 (2015)

33. F. Wang, X. Liu, Recent advances in the chemistry of lanthanide-doped upconversion nanocrystals. Chem. Soc. Rev. **38**(4), 976–989 (2009)
34. S.V. Eliseeva, J.C.G. Bünzli, Lanthanide luminescence for functional materials and biosciences. Chem. Soc. Rev. **39**(1), 189–227 (2010)
35. S. Mahlik, F. Diaz, P. Boutinaud, Luminescence quenching in $KYb(WO_4)_2$-Tb^{3+}: An example of temperature-pressure equivalence. J. Lumin. **191**, 18–21 (2017)
36. Z. Yahiaoui, M.A. Hassairi, M. Dammak, E. Cavalli, F. Mezzadri, Tunable luminescence and energy transfer properties in YPO_4: Tb^{3+}, Eu^{3+}/Tb^{3+} phosphors. J. Lumin. **194**, 96–101 (2018)
37. X.X. Ma, W. Zhou, Z. Zhang, X.R. Wang, B. Zhang, Y.C. Guo, Preparation and luminescence properties of $Ba_3LaNa\ (PO_4)_3F$: Ce^{3+}, Tb^{3+} phosphors. J. Lumin. **199**, 82–86 (2018)
38. P. Dai, Enhanced red emission induced by Tb^{3+} doping in europium-based molybdate phosphors. Mater. Res. Bull. **94**, 64–69 (2017)
39. C. Liang, H. You, Y. Fu, X. Teng, K. Liu, J. He, Luminescence properties of a tunable blue-green-emitting $Ca_{10}(PO_4)_6S$: Ce^{3+}, Tb^{3+} phosphors for UV-excited white LEDs. Optik **131**, 335–342 (2017)
40. X.Y. Sun, T.T. Han, D.L. Wu, F. Xiao, S.L. Zhou, Q.M. Yang, J.P. Zhong, Investigation on luminescence properties of Dy^{3+}-, Eu^{3+}-doped, and Eu^{3+}/Dy^{3+}-codoped $SrGd_2O_4$ phosphors. J. Lumin. **204**, 89–94 (2018)
41. Y. Zhang, A. Abraha, R. Zhang, T. Shahbazyan, M. Fadavi, E. Heydari, Q. Dai, Luminescence properties of $CaWO_4$ and $CaWO_4$: Eu^{3+} nanostructures prepared at low temperature. Opt. Mater. **84**, 115–122 (2018)
42. C. Gheorghe, S. Hau, L. Gheorghe, F. Voicu, M. Greculeasa, M. Enculescu, et al., Yellow laser potential of cubic $Ca_3(Nb,Ga)_5O_{12}$: Dy^{3+} and $Ca_3(Li,Nb,Ga)_5O_{12}$: Dy^{3+} single crystals. J. Alloys Compd. **739**, 806–816 (2018)
43. G. Tian, X. Zhang, Z. Gu, Y. Zhao, Recent advances in upconversion nanoparticles-based multifunctional nanocomposites for combined cancer therapy. Adv. Mater. **27**(47), 7692–7712 (2015)
44. N. Erathodiyil, J.Y. Ying, Functionalization of inorganic nanoparticles for bioimaging applications. Acc. Chem. Res. **44**(10), 925–935 (2011)
45. Li, W., & Zhao, D. (2013). Extension of the Stöber method to construct mesoporous SiO_2 and TiO_2 shells for uniform multifunctional core–shell structures
46. G. Chen, H. Qiu, P.N. Prasad, X. Chen, Upconversion nanoparticles: Design, nanochemistry, and applications in theranostics. Chem. Rev. **114**(10), 5161–5214 (2014)
47. W. Stöber, A. Fink, E. Bohn, Controlled growth of monodisperse silica spheres in the micron size range. J. Colloid Interface Sci. **26**(1), 62–69 (1968)
48. F.J. Arriagada, K. Osseo-Asare, Synthesis of nanosize silica in a nonionic water-in-oil microemulsion: Effects of the water/surfactant molar ratio and ammonia concentration. J. Colloid Interface Sci. **211**(2), 210–220 (1999)
49. F. Li, C. Li, X. Liu, Y. Chen, T. Bai, L. Wang, et al., Hydrophilic, upconverting, multicolor, lanthanide-doped $NaGdF_4$ nanocrystals as potential multifunctional bioprobes. Chemistry–a Eur. J. **18**(37), 11641–11646 (2012)
50. B. Liu, C. Li, D. Yang, Z. Hou, P.A. Ma, Z. Cheng, et al., Upconversion-luminescent Core/mesoporous-silica-Shell-structured β-$NaYF_4$: Yb^{3+}, Er^{3+}@ SiO_2@ $mSiO_2$ composite Nanospheres: Fabrication and drug-storage/release properties. Eur. J. Inorg. Chem. **2014**(11), 1906–1913 (2014)
51. Q. Xiao, X. Zheng, W. Bu, W. Ge, S. Zhang, F. Chen, et al., A core/satellite multifunctional nanotheranostic for in vivo imaging and tumor eradication by radiation/photothermal synergistic therapy. J. Am. Chem. Soc. **135**(35), 13041–13048 (2013)
52. H. Xing, W. Bu, S. Zhang, X. Zheng, M. Li, F. Chen, et al., Multifunctional nanoprobes for upconversion fluorescence. MR and CT trimodal imaging. Biomaterials **33**(4), 1079–1089 (2012)
53. Y. Liu, Y. Liu, W. Bu, C. Cheng, C. Zuo, Q. Xiao, et al., Hypoxia induced by upconversion-based photodynamic therapy: Towards highly effective synergistic bioreductive therapy in tumors. Angew. Chem. **127**(28), 8223–8227 (2015)

54. Y. Yuan, Y. Min, Q. Hu, B. Xing, B. Liu, NIR photoregulated chemo-and photodynamic cancer therapy based on conjugated polyelectrolyte–drug conjugate encapsulated upconversion nanoparticles. Nanoscale **6**(19), 11259–11272 (2014)
55. H. Yuan, H. Chong, B. Wang, C. Zhu, L. Liu, Q. Yang, et al., Chemical molecule-induced light-activated system for anticancer and antifungal activities. J. Am. Chem. Soc. **134**(32), 13184–13187 (2012)
56. S. Yang, N. Li, Z. Liu, W. Sha, D. Chen, Q. Xu, J. Lu, Amphiphilic copolymer coated upconversion nanoparticles for near-infrared light-triggered dual anticancer treatment. Nanoscale **6**(24), 14903–14910 (2014)
57. J.E. Lee, N. Lee, T. Kim, J. Kim, T. Hyeon, Multifunctional mesoporous silica nanocomposite nanoparticles for theranostic applications. Acc. Chem. Res. **44**(10), 893–902 (2011)
58. C. Li, J. Liu, S. Alonso, F. Li, Y. Zhang, Upconversion nanoparticles for sensitive and in-depth detection of cu^{2+} ions. Nanoscale **4**(19), 6065–6071 (2012)
59. H. Park, G.Y. Yoo, M.S. Kim, K. Kim, C. Lee, S. Park, W. Kim, Thin film fabrication of upconversion lanthanide-doped $NaYF_4$ by a sol-gel method and soft lithographical nanopatterning. J. Alloys Compd. **728**, 927–935 (2017)
60. A. Patra, C.S. Friend, R. Kapoor, P.N. Prasad, Upconversion in Er^{3+}: ZrO_2 nanocrystals. J. Phys. Chem. B **106**(8), 1909–1912 (2002)
61. V. Venkatramu, D. Falcomer, A. Speghini, M. Bettinelli, C.K. Jayasankar, Synthesis and luminescence properties of Er^{3+}-doped $Lu_3Ga_5O_{12}$ nanocrystals. J. Lumin. **128**(5–6), 811–813 (2008)
62. C. Li, Z. Hou, Y. Dai, D. Yang, Z. Cheng, P.A. Ma, J. Lin, A facile fabrication of upconversion luminescent and mesoporous core–shell structured β-$NaYF_4$: Yb^{3+}, Er^{3+}@ $mSiO_2$ nanocomposite spheres for anti-cancer drug delivery and cell imaging. Biomater. Sci. **1**(2), 213–223 (2013)
63. S. Gai, P. Yang, C. Li, W. Wang, Y. Dai, N. Niu, J. Lin, Synthesis of magnetic, up-conversion luminescent, and mesoporous core–shell-structured nanocomposites as drug carriers. Adv. Funct. Mater. **20**(7), 1166–1172 (2010)
64. L. Li, C. Liu, L. Zhang, T. Wang, H. Yu, C. Wang, Z. Su, Multifunctional magnetic–fluorescent eccentric-(concentric-Fe_3O_4@SiO_2)@ polyacrylic acid core–shell nanocomposites for cell imaging and pH-responsive drug delivery. Nanoscale **5**(6), 2249–2253 (2013)
65. L. Chen, L. Li, L. Zhang, S. Xing, T. Wang, Y.A. Wang, et al., Designed fabrication of unique eccentric mesoporous silica nanocluster-based core–shell nanostructures for pH-responsive drug delivery. ACS Appl. Mater. Interfaces **5**(15), 7282–7290 (2013)
66. J. Liu, F. Liu, K. Gao, J. Wu, D. Xue, Recent developments in the chemical synthesis of inorganic porous capsules. J. Mater. Chem. **19**(34), 6073–6084 (2009)
67. Y. Wang, P. Yang, P.A. Ma, F. Qu, S. Gai, N. Niu, et al., Hollow structured $SrMoO_4$: Yb^{3+}, ln^{3+}(ln= tm, Ho, tm/Ho) microspheres: Tunable up-conversion emissions and application as drug carriers. J. Mater. Chem. B **1**(15), 2056–2065 (2013)
68. H. Ren, L. Zhang, T. Wang, L. Li, Z. Su, C. Wang, Universal and facile synthesis of multicolored upconversion hollow nanospheres using novel poly (acrylic acid sodium salt) microspheres as templates. Chem. Commun. **49**(54), 6036–6038 (2013)
69. X. Kang, D. Yang, Y. Dai, M. Shang, Z. Cheng, X. Zhang, et al., Poly (acrylic acid) modified lanthanide-doped $GdVO_4$ hollow spheres for up-conversion cell imaging. MRI and pH-dependent drug release. Nanoscale **5**(1), 253–261 (2013)
70. W. Feng, L.D. Sun, Y.W. Zhang, C.H. Yan, Solid-to-hollow single-particle manipulation of a self-assembled luminescent $NaYF_4$: Yb, Er nanocrystal monolayer by electron-beam lithography. Small (Weinheim an der Bergstrasse, Germany) **5**(18), 2057–2060 (2009)
71. D. Yang, Y. Dai, P. Ma, X. Kang, Z. Cheng, C. Li, J. Lin, One-step synthesis of small-sized and water-soluble $NaREF_4$ upconversion nanoparticles for in vitro cell imaging and drug delivery. Chemistry–a Eur. J. **19**(8), 2685–2694 (2013)
72. B. Ungun, R.K. Prud'homme, S.J. Budijono, J. Shan, S.F. Lim, Y. Ju, R. Austin, Nanofabricated upconversion nanoparticles for photodynamic therapy. Opt. Express **17**(1), 80–86 (2008)

73. P.A. Ma, H. Xiao, X. Li, C. Li, Y. Dai, Z. Cheng, et al., Rational design of multifunctional upconversion nanocrystals/polymer nanocomposites for cisplatin (IV) delivery and biomedical imaging. Adv. Mater. **25**(35), 4898–4905 (2013)
74. G. Tian, Z. Gu, L. Zhou, W. Yin, X. Liu, L. Yan, et al., Mn^{2+} dopant-controlled synthesis of $NaYF_4$: Yb/Er upconversion nanoparticles for in vivo imaging and drug delivery. Adv. Mater. (Deerfield Beach, Fla.) **24**(9), 1226–1231 (2012)
75. G. Tian, W. Yin, J. Jin, X. Zhang, G. Xing, S. Li, et al., Engineered design of theranostic upconversion nanoparticles for tri-modal upconversion luminescence/magnetic resonance/X-ray computed tomography imaging and targeted delivery of combined anticancer drugs. J. Mater. Chem. B **2**(10), 1379–1389 (2014)
76. L. Cheng, K. Yang, Y. Li, J. Chen, C. Wang, M. Shao, et al., Facile preparation of multifunctional upconversion nanoprobes for multimodal imaging and dual-targeted photothermal therapy. Angew. Chem. Int. Ed. **50**(32), 7385–7390 (2011)
77. L. Cheng, K. Yang, Y. Li, X. Zeng, M. Shao, S.T. Lee, Z. Liu, Multifunctional nanoparticles for upconversion luminescence/MR multimodal imaging and magnetically targeted photothermal therapy. Biomaterials **33**(7), 2215–2222 (2012)
78. Y. Tang, H. Liu, J. Gao, X. Liu, X. Gao, X. Lu, et al., Upconversion particle@Fe_3O_4@molecularly imprinted polymer with controllable shell thickness as high-performance fluorescent probe for sensing quinolones. Talanta **181**, 95–103 (2018)
79. B. Dong, H. Song, H. Yu, H. Zhang, R. Qin, X. Bai, et al., Upconversion properties of Ln^{3+} doped $NaYF_4$/polymer composite fibers prepared by electrospinning. J. Phys. Chem. C **112**(5), 1435–1440 (2008)
80. G. Li, Z. Hou, C. Peng, W. Wang, Z. Cheng, C. Li, et al., Electrospinning derived one-dimensional LaOCl: Ln^{3+} (ln= Eu/Sm, Tb, tm) nanofibers, nanotubes and microbelts with multicolor-tunable emission properties. Adv. Funct. Mater. **20**(20), 3446–3456 (2010)
81. Z. Hou, P. Yang, C. Li, L. Wang, H. Lian, Z. Quan, J. Lin, Preparation and luminescence properties of YVO_4: Ln and Y (V, P) O_4: Ln (ln= Eu^{3+}, Sm^{3+}, Dy^{3+}) nanofibers and microbelts by sol-gel/electrospinning process. Chem. Mater. **20**(21), 6686–6696 (2008)
82. Z. Hou, X. Li, C. Li, Y. Dai, P.A. Ma, X. Zhang, et al., Electrospun upconversion composite fibers as dual drugs delivery system with individual release properties. Langmuir **29**(30), 9473–9482 (2013)
83. T.T.T. Nguyen, C. Ghosh, S.G. Hwang, L.D. Tran, J.S. Park, Characteristics of curcumin-loaded poly (lactic acid) nanofibers for wound healing. J. Mater. Sci. **48**, 7125–7133 (2013)
84. A. Schneider, X.Y. Wang, D.L. Kaplan, J.A. Garlick, C. Egles, Biofunctionalized electrospun silk mats as a topical bioactive dressing for accelerated wound healing. Acta Biomater. **5**(7), 2570–2578 (2009)
85. P.S. Low, W.A. Henne, D.D. Doorneweerd, Discovery and development of folic-acid-based receptor targeting for imaging and therapy of cancer and inflammatory diseases. Acc. Chem. Res. **41**(1), 120–129 (2008)
86. T. Cao, Y. Yang, Y. Gao, J. Zhou, Z. Li, F. Li, High-quality water-soluble and surface-functionalized upconversion nanocrystals as luminescent probes for bioimaging. Biomaterials **32**(11), 2959–2968 (2011)
87. D. Yang, X. Kang, P.A. Ma, Y. Dai, Z. Hou, Z. Cheng, et al., Hollow structured upconversion luminescent $NaYF_4$: Yb^{3+}, Er^{3+} nanospheres for cell imaging and targeted anti-cancer drug delivery. Biomaterials **34**(5), 1601–1612 (2013)
88. C. Wang, L. Cheng, Z. Liu, Drug delivery with upconversion nanoparticles for multi-functional targeted cancer cell imaging and therapy. Biomaterials **32**(4), 1110–1120 (2011)
89. S. Jiang, Y. Zhang, K.M. Lim, E.K. Sim, L. Ye, NIR-to-visible upconversion nanoparticles for fluorescent labeling and targeted delivery ofsiRNA. Nanotechnology **20**(15), 155101 (2009)
90. X. Yang, Q. Xiao, C. Niu, N. Jin, J. Ouyang, X. Xiao, D. He, Multifunctional core–shell upconversion nanoparticles for targeted tumor cells induced by near-infrared light. J. Mater. Chem. B **1**(21), 2757–2763 (2013)
91. R.R. Allison, K. Moghissi, Photodynamic therapy (PDT): PDT mechanisms. Clin. Endosc. **46**(1), 24–29 (2013)

92. X. Wang, C.X. Yang, J.T. Chen, X.P. Yan, A dual-targeting upconversion nanoplatform for two-color fluorescence imaging-guided photodynamic therapy. Anal. Chem. **86**(7), 3263–3267 (2014)
93. L. Liang, Y. Lu, R. Zhang, A. Care, T.A. Ortega, S.M. Deyev, et al., Deep-penetrating photodynamic therapy with KillerRed mediated by upconversion nanoparticles. Acta Biomater. **51**, 461–470 (2017)
94. W. Zhao, H. Chen, Y. Li, L. Li, M. Lang, J. Shi, Uniform rattle-type hollow magnetic mesoporous spheres as drug delivery carriers and their sustained-release property. Adv. Funct. Mater. **18**(18), 2780–2788 (2008)
95. V.V. Mody, A. Cox, S. Shah, A. Singh, W. Bevins, H. Parihar, Magnetic nanoparticle drug delivery systems for targeting tumor. Appl. Nanosci. **4**, 385–392 (2014)
96. Q. Li, W. Li, H. Di, L. Luo, C. Zhu, J. Yang, et al., A photosensitive liposome with NIR light triggered doxorubicin release as a combined photodynamic-chemo therapy system. J. Control. Release **277**, 114–125 (2018)
97. H. Qiu, M. Tan, T.Y. Ohulchanskyy, J.F. Lovell, G. Chen, Recent progress in upconversion photodynamic therapy. Nanomaterials **8**(5), 344 (2018)
98. M.R. Hamblin, Upconversion in photodynamic therapy: Plumbing the depths. Dalton Trans. **47**(26), 8571–8580 (2018)

Future Prospects of Up-Conversion Materials: Advancements, Applications, and Challenges

Vineet Sharma and Jyoti Singh

Abstract Over the last few decades, a significantly important category of nanoparticles made of "up-conversion" materials has emerged as a promising candidate for various scientific applications. Up-conversion phosphors (UCPs) are well-known for their ability to convert multiple longer-wavelength photons to shorter-wavelength fluorescence. Because of their relatively bright and narrow-band fluorescence, they have been useful and utilized in a variety of applications. This chapter explores the potential applications of up-conversion materials in commercial energy-efficient lighting, safe bioimaging, high-efficiency solar cells, and high-density data storage applications. Recent advances highlight the capacity of UCPs to revolutionize technologies and contribute to sustainable energy solutions. UCPs provide a distinct advantage in lighting applications by converting low-energy light into high-energy emissions. This process can be used to improve the overall efficiency and light yield of lighting systems, thereby making them more energy-efficient and environmentally friendly. Furthermore, the versatility of UCPs enables the generation of specific colors by controlling the material composition. This capability expands the possibilities for dynamic and customizable lighting solutions. Furthermore, the chapter investigates the use of UCPs in emerging technologies like quantum dots and nanophotonics, which broadens the possibilities for advanced lighting systems. It investigates how combining up-conversion processes with these cutting-edge technologies can result in novel solutions for next-generation displays and illumination. These phosphors have distinct properties in bioimaging that make them useful tools for improving imaging resolution and sensitivity. This chapter explores recent developments, including how UCPs can be tailored for specific biological markers, allowing for more precise and efficient imaging techniques in medical diagnostics and research. Furthermore, the incorporation of UCPs into solar cells represents an

V. Sharma · J. Singh (✉)
Department of Applied Physics (SOVSAS), Gautam Buddha University, Greater Noida, Uttar Pradesh, India
e-mail: jyoti.singh.phy@gmail.com

V. Sharma
Department of Applied Physics, Delhi Technological University, Bawana Road, Delhi 110042, India

V. Dubey et al. (eds.), *Upconversion Luminescence of Phosphors*, Progress in Optical Science and Photonics 39, https://doi.org/10.1007/978-981-95-1762-6_10

exciting opportunity for advancements in renewable energy. These materials' significant ability to convert low-energy sunlight into higher-energy photons demonstrates their potential to improve solar energy conversion efficiency. This chapter investigates strategies for optimizing up-conversion processes within solar cells, addresses challenges, and proposes future research directions to maximize the impact of this technology on sustainable energy solutions. Data storage is another area where UCPs have exciting applications. Their optical properties could transform data encoding and retrieval methods. This chapter discusses recent experiments and theoretical frameworks that investigate the viability of using UCPs for high-density data storage and pave the way for novel approaches to information technology. Despite the optimistic outlook, challenges remain, and the chapter critically examines current limitations and potential barriers. Addressing efficiency, cost-effectiveness, and scalability issues, the discussion seeks to guide future research efforts and foster collaboration within the scientific community.

1 Introduction

The rapid evolution of photonic technologies has created a substantial need for materials that can effectively convert light into different wavelengths, supporting various applications from biomedical imaging to energy harvesting [1–5]. Among these materials, up-conversion substances have received notable attention for their exceptional ability to transform low-energy photons into higher-energy photons via a process called up-conversion [6–10]. This rare occurrence, which challenges the standard energy conversion processes, unlocks a wide array of possibilities in sectors such as biomedicine, sensing, lighting, and photovoltaics [10–15]. As advancements in technology continue to expand the realm of possibilities, the research and application of up-conversion materials are becoming increasingly essential for future innovations. Up-conversion materials, which mainly consist of rare-earth ions doped into host matrices, possess extraordinary optical properties. They can absorb two or more low-energy photons, typically from the infrared (IR) region, and emit a single photon of higher energy, often in the visible or ultraviolet (UV) range [7, 8, 16–20]. This unique capability has attracted considerable interest, as it allows for the effective use of light in spectral regions that are usually underutilized in traditional technologies. For instance, these materials have the potential to enhance solar cell efficiency by converting IR radiation into visible light, which can be absorbed more effectively by photovoltaic materials. The promising applications of up-conversion materials make them a significant avenue for the development of sustainable energy solutions.

Up-conversion fundamentally operates through the interaction of dopant ions, which are usually rare-earth elements like erbium (Er), ytterbium (Yb), and thulium (Tm), situated within a crystalline host matrix. These ions have multiple energy levels that enable the sequential absorption of low-energy photons, followed by the emission of higher-energy photons [8, 21–25]. The efficiency of this up-conversion process is significantly influenced by various factors, including the choice of host matrix, the concentration of dopant ions, the type of excitation source, and the environmental conditions [8, 26–29]. Recent advancements in nanotechnology and material science have led to notable improvements in the synthesis and tuning of up-conversion materials, enhancing their performance and broadening their potential applications. Up-conversion materials are emerging as a highly promising technology in the biomedical sector, providing distinct advantages over traditional fluorescent probes. Up-conversion nanoparticles (UCNPs) are characterized by their exceptional photostability, low toxicity, and reduced background autofluorescence, making them particularly suitable for bioimaging and biosensing applications [15, 18–20]. Their unique capability to transform near-infrared (NIR) light into visible light facilitates deeper tissue penetration, thereby minimizing damage to biological tissues and enhancing imaging clarity. This has led to extensive investigations into the use of UCNPs for cancer cell imaging, drug delivery tracking, and real-time monitoring of cellular activities. Additionally, up-conversion materials have demonstrated potential in photodynamic therapy (PDT), where they activate photosensitizers using NIR light, offering a targeted and less invasive treatment alternative for various medical conditions [30–34]. In addition to their applications in the biomedical field, up-conversion materials are making significant advancements in energy and environmental sustainability. In the realm of photovoltaic technology, these materials have the capability to exceed the efficiency limitations of conventional solar cells by transforming sub-bandgap infrared light into wavelengths that silicon-based photovoltaic cells can effectively utilize. This enhancement in light harvesting and utilization paves the way for more efficient solar energy conversion, which is essential for meeting the increasing demand for renewable energy. Furthermore, up-conversion materials are being employed in solid-state lighting, where they enhance the efficiency and color quality of light-emitting diodes (LEDs), as well as in anti-counterfeiting measures, leveraging their distinctive optical properties for secure labeling and authentication.

The advancements and potential applications of up-conversion materials are indeed promising; however, significant challenges must be addressed to fully realize their capabilities. A primary concern is the relatively low quantum efficiency of the up-conversion processes, which limits their practical implementation in various applications. High up-conversion efficiency typically requires careful control over the synthesis of the materials, optimization of dopant concentrations, and thoughtful selection of host matrices, making the process complex and resource-intensive [7, 8, 35–37]. Furthermore, the development of cost-effective and scalable manufacturing techniques is a critical challenge that must be overcome for the widespread adoption of up-conversion materials, particularly in commercial applications. One of the primary challenges is the integration of up-conversion materials into established technologies. In photovoltaic applications, for example, integrating these materials into solar cells necessitates meticulous engineering to guarantee compatibility with existing designs and to optimize energy conversion efficiency [38–40]. Similarly, in biomedical fields, it is essential to thoroughly assess the biocompatibility and potential toxicity of up-conversion nanoparticles to ensure their safe application in clinical environments. Tackling these issues demands a multidisciplinary strategy that merges knowledge from materials science, chemistry, physics, and engineering to create solutions that improve the performance, efficiency, and usability of up-conversion materials.

In Conclusion, the field of up-conversion materials is an exciting area of research with vast potential for future technological progress. Their unique ability to convert low-energy photons into higher-energy emissions opens up remarkable opportunities in sectors such as healthcare, energy, security, and environmental monitoring. While challenges remain, continuous research and development are effectively addressing the obstacles associated with these materials, setting the stage for their integration into advanced technologies. These present chapter aims to explore the latest advancements, a variety of applications, and the future outlook of up-conversion materials, providing a detailed overview of the current state of this vibrant field and highlighting the challenges that must be overcome to fully harness their capabilities.

2 Advance Applications and Challenges of Up-Conversion Materials

2.1 Biomedical Application of Up-Conversion Materials

The utilization of up-conversion materials holds significant potential for the enhancement of imaging techniques and applications in the fields of clinical practice and biological research. An in-depth analysis of the potential applications of up-conversion materials in the field of biomedicine is presented here [13–18].

2.1.1 Theranostic Applications

Up-conversion materials have emerged as promising platforms for theranostic applications that combine diagnostic imaging and targeted therapy into a single device. Researchers can selectively transport medicines or therapeutic agents to specific biological targets by functionalizing up-conversion nanoparticles with targeting ligands and therapeutic payloads, while using imaging tools to monitor treatment responses in real time. Theranostic platforms based on up-conversion materials have the potential to transform personalized medicine by enabling precise and targeted treatment techniques tailored to individual patient characteristics, thereby increasing therapeutic success and reducing side effects.

2.1.2 Deep Tissue Imaging

Up-conversion materials in bioimaging have a significant advantage in overcoming common limitations of fluorescence imaging, such as tissue autofluorescence and light scattering, which hinder the ability to penetrate deep into tissues. Up-conversion materials produce light with longer wavelengths when stimulated by near-infrared (NIR) radiation, enabling enhanced tissue penetration and imaging at greater depths in biological samples. This feature is particularly valuable for investigating intricate biological systems such as dense tissue specimens, whole organisms, and in vivo imaging applications. These scenarios often pose challenges for conventional fluorescent imaging techniques due to limitations imposed by tissue depth and scattering effects.

2.1.3 High-Resolution Imaging

Up-conversion materials have high photostability and narrow emission bandwidths, enabling high-resolution imaging with minimal background noise and photobleaching. Using the unique emission properties of up-conversion materials, researchers can obtain sharper and more detailed images of subcellular structures, organelles, and molecular interactions, leading to a better understanding of cellular processes and disease mechanisms. High-resolution imaging with up-conversion materials is required to study dynamic biological activities such as protein trafficking, membrane dynamics, and synaptic activity at unprecedented spatial and temporal resolution.

2.1.4 In Vivo Imaging and Biosensing

Up-conversion materials are ideal for in vivo imaging and biosensing applications because they are biocompatible, have low cytotoxicity, and cause less photodamage to living tissues. Researchers can image physiological processes in living creatures in real-time by enclosing up-conversion nanoparticles in biocompatible matrices or addressing them to specific cellular or tissue targets. In addition to imaging, up-conversion materials can be developed as biosensors capable of detecting biomolecules, analytes, and physiological factors in vivo, with potential applications in illness diagnostics, drug screening, and biomedical research.

2.1.5 Multimodal Imaging

Up-conversion materials provide distinct advantages for multimodal imaging by permitting the simultaneous viewing of numerous biological targets with great spatial and temporal resolution. Researchers can create probes that emit diverse wavelengths of light in response to varied excitation wavelengths by adjusting the composition and properties of up-conversion materials, allowing for multiplexed imaging of various biomolecules and cellular structures. Multimodal imaging using up-conversion materials can provide detailed insights into complicated biological processes such as cell signaling networks, gene expression, and disease progression, allowing for more precise diagnosis and therapy monitoring.

Overall, the possibilities for up-conversion materials in bioimaging are extremely exciting, with revolutionary capabilities for improving our understanding of biological systems, diagnosing diseases, and designing tailored therapeutics. Continued research and innovation in this field are likely to improve the performance, adaptability, and clinical translation of up-conversion-based imaging systems, hence creating new prospects in biomedical imaging and personalized medicine.

2.2 Challenges

While up-conversion materials have massive potential for bioimaging applications, various difficulties must be overcome to exploit their full capabilities in the bioimaging sector.

2.2.1 Targeting and Specificity

To achieve targeted imaging of certain biomolecules, cells, or tissues, up-conversion materials must be functionalized with targeting ligands or biomolecular recognition elements. Ensuring the specificity and selectivity of targeting strategies, especially in complex biological contexts, presents difficulties due to off-target binding, nonspecific interactions, and biological variability. Developing effective targeting and delivery tactics, including surface modification techniques, molecular engineering methodologies, and bioconjugation procedures, is crucial for enabling precise and selective imaging of biological targets with up-conversion materials.

2.2.2 Tissue Penetration and Imaging Depth

While up-conversion materials are better for deep tissue imaging than classic fluorescent imaging techniques, their penetration depth is still limited, especially in thick tissue specimens and in vivo imaging applications. Addressing tissue scattering and absorption effects to increase imaging depth and resolution is still a serious challenge. Developing sophisticated imaging techniques like adaptive optics, multi-photon excitation, and optical clearing methods can help overcome these obstacles and improve the depth and resolution of up-conversion-based bioimaging.

2.2.3 Clinical Translation and Regulatory Approval

Up-conversion-based bioimaging systems must be rigorously validated, standardized, and regulated before they can be used in clinical practice. Demonstrating the safety, efficacy, and clinical utility of up-conversion materials and imaging systems in preclinical and clinical trials is critical for regulatory approval and widespread clinical use. Additional obstacles that must be addressed to aid the clinical translation and commercialization of up-conversion-based bioimaging technologies include overcoming regulatory barriers, managing intellectual property issues, and dealing with cost considerations.

2.2.4 Biocompatibility and Toxicity

Ensuring the biocompatibility of up-conversion materials is critical for their safe usage in biological imaging and therapy. Some materials, especially those containing heavy metals or complex chemical molecules, may cause cytotoxicity or immunogenicity issues. Developing biocompatible surface coatings and changing material compositions to reduce cytotoxicity while retaining imaging capability is an important research objective. Comprehensive biocompatibility studies, including in vitro and in vivo testing, are required to determine the long-term safety and potential immunogenicity of up-conversion materials for biomedical applications.

2.2.5 Signal-to-Noise Ratio and Sensitivity

In imaging applications, a high signal-to-noise ratio (SNR) and sensitivity are required to identify weak signals from biological targets while minimizing background noise. Up-conversion materials often have lower photon conversion efficiencies and signal intensities than conventional fluorophores, which limits their ability to detect low-abundance biomolecules and delicate biological processes. To improve up-conversion materials' performance in bioimaging applications, material engineering, and optimization of excitation and detection systems are required.

To address these problems, academics, engineers, physicians, and regulatory agencies will need to work together to design and execute up-conversion materials for bioimaging applications. Despite these challenges, ongoing research and investment in this field have the potential to transform biomedical imaging, diagnosis, and therapy, resulting in better patient outcomes and advances in customized medicine.

2.3 Data Storage Application of Up-Conversion Materials

Up-conversion materials have gained popularity in a variety of industries due to their capacity to convert lower-energy photons into higher-energy ones. Up-conversion materials have potential in data storage for several reasons [8, 19–21], including:

2.3.1 Increased Storage Density

Higher storage densities can be achieved by using up-conversion materials, which allow data to be stored using shorter-wavelength light. This could lead to better use of storage space. Traditional optical storage systems employ longer wavelengths, which limits the quantity of data that can be stored in a given region. Data can be encoded using shorter wavelengths by using the capacity of up-conversion materials to transform lower-energy photons into higher-energy ones, resulting in more compact storage and increased data density. This improved storage density may lead to breakthroughs in a variety of applications, such as high-capacity optical discs, holographic storage, and three-dimensional (3D) optical memories.

2.3.2 Non-Destructive Readout

Up-conversion materials provide a non-destructive readout, in which data can be read without affecting or deteriorating the stored information. This is especially useful for applications where maintaining data integrity is critical. Unlike some traditional storage systems, where reading data might damage or deteriorate the stored information, up-conversion materials allow data to be retrieved without affecting the stored state. The non-destructive readout is especially critical for applications that need data integrity, such as archival storage, where the original information must be preserved without degradation over lengthy periods.

2.3.3 Compatibility with Existing Technologies

Up-conversion materials can be designed to emit light at certain wavelengths, allowing them to work with current optical data storage systems. This interoperability allows for easy integration into existing data storage systems without requiring large infrastructure changes. One advantage of up-conversion materials is that they work with existing optical data storage methods. These materials can be designed to emit light at precise wavelengths, making them suitable for incorporation into existing data storage systems with minimal modification. This compatibility enables the easy integration of up-conversion materials into existing optical storage platforms, such as CD, DVD, and Blu-ray discs, as well as upcoming technologies like holographic storage and photonic integrated circuits. Up-conversion materials can speed adoption and commercialization for a wide range of applications, from consumer electronics to enterprise-level data centers, by using the existing infrastructure of optical data storage systems.

2.3.4 Multilevel Data Storage

The multi-level structure of up-conversion materials enables the encoding of many bits of data per storage unit. By utilizing distinct energy levels or excitation routes, these materials can store information more compactly and efficiently. This is accomplished by taking advantage of the material's multiple energy states or excitation pathways. By storing information across various energy levels or excitation states, up-conversion materials can attain larger storage capacities within a given volume, resulting in more effective storage space utilization. Multilevel data storage is critical for satisfying the growing demand for larger storage capacities in current data storage systems, including applications like high-density optical discs and solid-state optical memory.

2.3.5 Security

Up-conversion materials may improve data security by enabling encryption techniques that manipulate photon energy levels. This could lead to the development of stronger encryption technologies for data storage. Up-conversion materials have prospects to improve data security through novel encryption approaches based on photon energy manipulation. Up-conversion materials' unique energy levels and excitation routes can be used to create encryption systems that are intrinsically more resistant to hacking and data breaches. These encryption methods may include encoding data with specific energy levels or excitation pathways inside the material, making it difficult for unauthorized parties to access or understand the encoded information. Enhanced data security is critical in a variety of applications, including sensitive data storage, financial transactions, and secure communications, where data must be protected from unauthorized access.

2.4 Challenges

Despite several encouraging potentials, some drawbacks must be overcome before up-conversion materials may be extensively used for data storage.

2.4.1 Materials Stability

Many up-conversion materials are susceptible to environmental conditions like temperature, humidity, and light exposure. Creating stable materials that retain their qualities throughout time is critical for ensuring accurate data storage. Many up-conversion materials degrade over time owing to environmental conditions such as temperature changes, humidity, and light exposure. Ensuring the long-term stability of up-conversion materials is critical for dependable data storage applications, especially in cases where data must be kept over time. Researchers are working to develop stable up-conversion materials by integrating protective coatings, altering material compositions, and investigating novel material synthesis processes that improve resistance to environmental deterioration.

2.4.2 Cost-Effectiveness

The cost of developing up-conversion materials, especially those based on rare earth elements, can be prohibitively expensive. Finding cost-effective synthesis methods and alternative materials with equivalent qualities is critical to commercial viability. The expense of manufacturing up-conversion materials, particularly those based on rare-earth elements, can be a substantial impediment to their broad use in data storage applications. Finding cost-effective synthesis methods and alternative materials with comparable up-conversion capabilities is critical to making these compounds economically viable for commercial use. To minimize production costs, researchers are looking into approaches such as scalable synthesis, rare-earth recycling, and the development of alternative up-conversion materials based on abundant and inexpensive elements.

2.4.3 Integration with Read/Write Systems

Up-conversion materials must be integrated into read/write systems that work with current data storage technologies. This necessitates the development of efficient readout mechanisms and writing techniques specific to the properties of up-conversion materials. Integrating up-conversion materials into existing read/write systems presents technical challenges, particularly in developing efficient readout mechanisms and writing techniques that are tailored to the materials' specific properties. Up-conversion materials necessitate specialized detection systems capable of precisely measuring emitted photons at various energy levels, which may necessitate the development of custom optical components and signal-processing algorithms. Furthermore, achieving reliable and precise data writing onto up-conversion materials necessitates novel techniques for controlling the excitation and emission processes within the material, as well as optimizing spatial resolution and data transfer speeds.

2.4.4 Efficiency

Improving the efficiency of up-conversion processes is crucial in practical applications. Improving quantum yield and decreasing energy losses during conversion are continuing research objectives. One of the key problems for up-conversion materials is increasing their efficiency in converting lower-energy photons to higher-energy ones. Many up-conversion processes have poor quantum yields, which means that only a small fraction of the incident photons are converted, resulting in energy losses and decreased overall efficiency. Improving the efficiency of up-conversion processes necessitates a deeper knowledge of the fundamental principles driving photon absorption, energy transfer, and emission in these materials. Researchers are investigating a variety of tactics to improve efficiency, such as improving material composition, creating innovative nanostructures, and developing better synthesis and fabrication procedures.

Addressing these issues is critical to maximizing the promise of up-conversion materials in practical data storage applications. By overcoming these challenges, researchers can reap the multiple benefits of up-conversion materials, such as greater storage density, non-destructive readout, layered data encoding, improved security, and compatibility with existing optical storage systems.

2.5 *Photovoltaic Application of Up-Conversion Materials*

A photovoltaic cell is an electrical device that directly transforms light energy into electricity through the photovoltaic effect. When the light source is sunlight, it is referred to as a solar cell. Solar cells have garnered significant interest from researchers due to the fact that sunlight is one of the most affordable and cleanest energy sources available, and it is more readily accessible compared to other renewable energy options such as wind, geothermal, tidal, and nuclear energy. Nevertheless, effectively harnessing sunlight for electricity generation via the photovoltaic effect remains a challenge.

Materials that absorb light play a crucial role in the functionality of solar cells. These materials need to possess properties that align with the spectrum of light they encounter to effectively absorb and convert it. Solar cells can be categorized into various subtypes depending on the light-absorbing materials utilized, including silicon solar cells, dye-sensitized solar cells (DSSCs), organic or polymer solar cells, and quantum dot solar cells (QDSCs) [22–27]. Quantum dot solar cells (QDSCs) utilize the architecture of Gratzel cells or dye-sensitized solar cells (DSSCs), but they incorporate low band gap semiconductor nanoparticles with sufficiently small crystallite sizes to create quantum dots, such as CdS, CdSe, Sb_2S_3, and PbS. The size quantization of these quantum dots enables the tuning of the band gap by varying the particle size [22, 23]. Additionally, they exhibit high extinction coefficients and demonstrate the potential for multiple exciton generation. The actual efficiency of a solar cell is generally lower than the expected efficiency. This discrepancy can be attributed to two main loss mechanisms inherent to the semiconductor material, which impose a fundamental limit on the energy conversion efficiency of solar cells. The first mechanism is sub-band gap energy photon loss, where photons lack sufficient energy to excite the semiconductor and create electron–hole pairs. For instance, crystalline silicon solar cells experience approximately 20% energy loss from the sun's spectrum (AM1.5) due to this issue. The second mechanism involves the thermalization of charge carriers, which occurs when high-energy photons, possessing energy greater than the solar cell's band gap, are absorbed. Although these photons can generate one electron–hole pair, the surplus energy is dissipated as heat. In addition to these primary losses related to photon absorption, several other loss mechanisms are associated with the electronic characteristics of solar cells, such as contact voltage loss, recombination loss due to subpar interface or material quality, junction loss, and reflection loss at the module's surface interfaces. Collectively, these fundamental losses result in a theoretical efficiency limit of around 30% for all single-junction cells under non-concentrated AM1.5 illumination.

To minimize thermalization losses, the implementation of rare earth-based down-conversion (REDC) and downshift, often known as photoluminescence (PL) conversion, has been proposed to optimize the performance of solar cells. REDC represents a distinctive mechanism associated with rare earth ions, similar to up-conversion (UC), and is believed to offer greater potential for performance improvement compared to PL conversion [24, 25]. In the down-conversion process, research has shown that RE-ion-doped phosphors can absorb a single high-energy photon and effectively transform it into two or more lower-energy photons, such as converting ultraviolet light to visible light [26]. Typically, these down-conversion layers are positioned in front of solar cells. However, this arrangement can lead to reflection losses at the air-downconverter interface. While theoretical models suggest that combining down-conversion with silicon solar cells could yield a conversion efficiency of up to 38.6%, various loss mechanisms, including the aforementioned reflection losses, have hindered the achievement of new record efficiencies using down-conversion materials [24]. UC materials are defined by their inability to transmit photons that do not resonate with the band gap. This characteristic allows for their integration on the backside of solar cells, eliminating the reflection losses that typically limit the efficiency of DC materials. It is noteworthy that approximately 52% of the total solar energy flux exists in the infrared range ($\lambda > 700$ nm), indicating a significant opportunity for performance enhancement through efficient UC [27–30]. RE-ion-doped phosphors are known for their broad emission capabilities, covering near-infrared (NIR), visible, and ultraviolet (UV) wavelengths, and have increasingly attracted attention in the realm of solar cell research in recent years. Apart from up-conversion (UC), several device strategies have been devised to mitigate sub-band gap losses in solar cells. One prominent strategy is spectral beam splitting technology, which was first introduced in the early 1960s [31]. This technique involves partitioning the solar spectrum into two segments: the high-energy segment is utilized by a photovoltaic cell, while the low-energy segment is either converted into heat or re-converted into electricity through thermophotovoltaic conversion. Another approach is the tandem or multi-band gap cells, a concept initially proposed by Wolf [31] in 1960 and further examined by Barnham and Duggan [32], as well as Keevers and Green [33]. These systems feature multiple cells with distinct band gaps that are connected in series. The top cell captures UV and visible photons, while the remaining infrared photons are absorbed by the subsequent cells with lower band gaps. In contrast to these strategies, UC is unique in that it does not require specific engineering of solar cells.

This characteristic allows UC materials to be incorporated into any solar cell type, including crystalline, thin-film, and organic/hybrid cells. Typically, a layer of UC material is positioned on the backside (or both sides) of a solar cell, where it absorbs the sub-band gap portion of the solar spectrum and re-emits visible light that aligns with the solar cell's band gap. To enhance photon collection, a reflective layer is often incorporated alongside the UC layer [33, 34].

In comparison to UC bulk materials, UCNPs offer several advantages, including optical properties that are dependent on size and shape. They are poised to play significant roles in the development of new optoelectronic devices in the near future. Notably, UCNPs have been integrated with various types of solar cells, such as crystalline silicon, amorphous silicon thin films, organic cells, and dye-sensitized solar cells (DSSCs) [34–39]. Furthermore, UCNPs can be seamlessly incorporated into nanocomposite materials. These innovative structures, created by combining two or more materials, exhibit complementary properties that are crucial for solar cell applications. The interactions at the nanoscale surface among different compositions enhance energy transfer within the nanostructure, leading to improved solar cell efficiency.

2.5.1 Recent Advancement in Up-Conversion-Assisted Photovoltaic Application

The concept of photon up-conversion (UC) using lanthanide-doped materials is gaining traction as a means to enhance the efficiency of solar energy conversion. By employing UC materials, it is possible to address the issue of photon loss due to sub-band gap energy by transforming two sub-band gap photons into a single above-band gap photon. In 1996, Zahraman et al. [36] introduced a device that utilized the principle of two-photon UC to capture infrared photons that would otherwise be lost due to transparency in a GaAs solar cell. They fabricated an ultrathin GaAs cell through the epitaxial lift-off (ELO) technique and positioned it atop a 100-μm-thick vitroceramic doped with Yb^{3+} and Er^{3+}. The cell was illuminated with a Ti–sapphire IR laser operating at 1.391 V, which is below the GaAs band gap, with an input power reaching approximately 1 W. Despite the excitation wavelength of 891 nm not aligning with the Yb3 + absorption peak at 980 nm, an efficiency of 2.5% was achieved. The authors described the mechanism as a two-photon UC process involving sequential absorption and energy transfer from Yb^{3+} to Er^{3+}, resulting in the emission of a green photon that generated a photoresponse in the GaAs cell. It is important to note that the excitation density employed was significantly higher than the standard terrestrial solar irradiation (AM1.5). Richards and Shalav [38] demonstrated that under an excitation density of 2.4 W/cm^2, equivalent to a solar concentration of 1000 ×), a crystalline silicon solar cell incorporating Er^{3+} doped $NaYF_4$ as an upconverter achieved external quantum efficiency (EQE) of 3.4% at a wavelength of 1,523 nm. This finding aligns with the results reported by Shalav et al. [38]. Additionally, Shan and Demopoulos [34] introduced the innovative use of an Yb^{3+}/Er^{3+} codoped LaF_3–TiO_2 layer to create a "triple-layer" working electrode on fluorine-doped tin oxide (FTO) glass for dye-sensitized solar cells (DSSCs). When illuminated with a 980-nm wavelength fiber laser at a power of 2.5 W, the system produced an open-circuit voltage (VOC) of 0.40 V and a short-circuit current (ISC) of 0.036 mA. This distinctive FTO/transparent TiO_2/UC-TiO_2/scattering TiO_2 configuration presents both advantages and challenges. For instance, the in situ conversion of NIR light to dye-absorbable visible light minimizes losses from the converted light. However, the red light generated by the UC–TiO_2 nanocomposite at approximately 655 nm is not effectively absorbed by the dye N719. Another significant challenge lies in preventing charge recombination to maximize the utilization of the upconverted light. Further enhancements to the UC–TiO_2 photoanode design can be pursued by optimizing the morphological characteristics, including the structure and composition of the nanocomposites.

Wang et al. [22] illustrated the potential of up-conversion nanoparticles (UCNPs) in organic solar cells composed of P3HT: PCBM, achieving a photocurrent density of $\sim$16.5 μA/cm^2 under a laser illumination intensity of approximately 25 mW/cm^2. More recently, Su et al. [35] introduced an innovative approach to incorporate UCNPs into a hetero-nanostructured photoanode, aimed at improving near-infrared (NIR) light absorption. Their research detailed the strategic design and construction of a three-dimensional interconnected porous network of hetero-nanostructured photoanodes, which significantly enhances NIR light harvesting. The photoelectrode features a titanium oxide (TiO_2) inverse opal structure that efficiently conducts electrons while minimizing carrier losses through its continuous electron pathways. This three-dimensional hierarchical porous network not only offers a substantial interfacial surface area for effective UCNP and quantum dot sensitizer loading but also facilitates excellent interaction between the electrolyte and quantum dots. Furthermore, the inverse opal structure enhances light-matter interaction due to strong light scattering and an increased light path length within the electrode. Utilizing the developed photoanode, they successfully demonstrated hydrogen generation through a photoelectrochemical process powered by NIR irradiation, confirming that up-conversion was responsible for the photoresponse under NIR exposure. This marks the first validation of photoelectrochemical electrodes featuring UCNPs embedded in porous photonic crystals. The interconnected pores of these photonic crystals position the nanoparticles in close proximity to CdSe quantum dots, thereby enhancing energy transfer and light harvesting.

Wang et al. [22] successfully developed monodispersed and uniform $NaGdF_4$:Er^{3+}@$NaGdF_4$:Ho^{3+}@NaGdF4 nanocrystals featuring an active-core/active-shell/inert-shell architecture. In these nanocrystals, the intermediate shells serve a dual purpose: they enhance the up-conversion (UC) luminescence of Er^{3+} in the core through passivation and provide a hosting environment for the UC luminescence of Ho^{3+}. Notably, the strategic doping of Er^{3+} in the core and Ho^{3+} in the shell effectively mitigates detrimental energy transfers between the two ions, thereby reducing UC emission quenching for both dopants. These innovative core/shell nanocrystals can be excited by ultrabroadband near-infrared (NIR) photons that are transparent to crystalline silicon, resulting in the emission of visible to NIR photons that exceed the band gap of crystalline silicon. This occurs through two primary UC pathways: for Er^{3+}, transitions include $^2H_{11/2}$ and $^4S_{3/2}$ to $^4I_{15/2}$ (539 nm), $^4F_{9/2}$ to $^4I_{15/2}$ (655 nm), $^4I_{9/2}$ to $^4I_{15/2}$ (809 nm), and $^4I_{11/2}$ to $^4I_{15/2}$ (980 nm) under 1480–1600-nm excitation; for Ho^{3+}, transitions include $^5F5$ to $^5I8$ (645 nm) and $^5I5$ to $^5I8$ (900 nm) under 1140–1250-nm excitation. All emitted light can be absorbed by crystalline silicon, suggesting that this novel nanomaterial holds significant potential for enhancing energy efficiency in crystalline silicon solar cells.

In a recent study, Li et al. [23] presented an efficient technique for the fabrication of UC $NaYF_4$: Yb/Er/Gd (18/2/30 mol%) nanorods, which are coated with gold nanoparticles or feature gold shell nanostructures. This nanocomposite was integrated onto the front side of a flexible standard p–i–n hydrogenated amorphous silicon (a-Si:H) solar cell on a steel substrate. It was capable of capturing the sub-band gap segment of the solar spectrum and subsequently reemitting visible light that matched the band gap of the solar cell. When compared to a reference cell that did not incorporate UC nanomaterials, the solar cells utilizing $NaYF_4$: Yb/Er/Gd nanorods with different gold nanostructures exhibited a substantial increase in short-circuit current, with enhancements ranging from 16 to 72 times. The device configuration with $NaYF_4$:Yb/Er/Gd nanorods with gold nanoparticles exhibited the highest near-infrared photovoltaic performance, recording a short-circuit current of 1.16 mA and an external quantum efficiency (EQE) of 0.14% when exposed to a 980-nm laser at an excitation power of 1100 mW. Notably, this method did not require any alteration or destruction of the intrinsic structure of the solar cells, nor did it necessitate changes to the solar cell fabrication processes [23, 24, 40].

2.5.2 Challenges and Future Potential for Up-Conversion-Assisted Solar Cells

In conclusion, the up-conversion (UC) process represents a sophisticated approach to enhance the efficiency of various types of solar cells. Recent years have seen notable advancements, both in theoretical studies and experimental results, within the realm of solar UC. Nevertheless, there has yet to be a substantial improvement in performance under natural solar irradiation conditions (AM1.5). This chapter outlines several factors contributing to this situation:

i. The effectiveness of existing up-conversion (UC) materials is still inadequate. The reported quantum yield for UC nanomaterials is persistently under 1%. Even bulk UC materials do not exceed a quantum yield of 3.5%. In comparison, quantum dots and organic dyes achieve much greater quantum yields, typically between 10 and 90%.
ii. Lanthanide ions demonstrate a very low absorption cross section, with Nd^{3+} measuring 1.2×10^{-19} cm^2 at wavelengths between 790 and 810 nm, and Yb^{3+} at 1.2×10^{-20} cm^2 within the 900 to 1000 nm range. Other lanthanide ions, including Er^{3+}, Tm^{3+}, and Ho^{3+}, exhibit even lower absorption cross sections. In contrast, quantum dots and organic dyes typically have absorption cross sections

that are several orders of magnitude higher. Consequently, many studies on UC-assisted solar cells utilize NIR lasers as the irradiation source instead of relying on natural solar light.

iii. In addition to the challenges posed by low quantum yield and limited absorption cross section, the range of the absorption spectrum is also quite restricted. The primary absorption wavelengths for various rare earth ions are as follows: Yb^{3+} spans 900–1000 nm ($^2F_{7/2} \rightarrow {}^2F_{5/2}$); Er^{3+} covers 1480–1600 nm ($^4I_{15/2} \rightarrow {}^4I_{13/2}$); and Ho^{3+} is effective in the 1140–1250 nm range ($^5I_8 \rightarrow {}^5I_6$), among others. Furthermore, the absorption cross sections vary significantly, as previously mentioned. It is evident that relying solely on one or two types of lanthanide ions is insufficient for absorbing the entirety of NIR light, as each ion type can only capture a limited portion of the NIR spectrum. Despite ongoing efforts to design and synthesize multi-doped nanocomposites for broader NIR absorption, the quenching effects, also known as cross-relaxation, between lanthanide ions (such as Er^{3+} with Ho^{3+}, Er^{3+} with Nd^{3+}, and Tm^{3+} with Tm^{3+}) must be taken into account. Consequently, a complex nanostructure is necessary to accommodate multi-doped lanthanide-based nanocomposites.

iv. The host matrices of the most effective up-conversion nanoparticles (UCNPs), such as $NaYF_4$ and $NaGdF_4$, are insulating. This characteristic is not ideal for solar cells, as increased electrical resistance can lead to reduced efficiency.

In response to the aforementioned limitations, the following suggestions are proposed as a perspective for the UC assisted solar cell:

i. Up-conversion nanoparticles (UCNPs) exhibit a notably low quantum yield, suggesting substantial potential for improvement. To enhance their performance, it is crucial to design advanced nanostructures that effectively reduce crystal defects and modify the host's crystal field. Moreover, increasing the doping concentration could be a viable approach to enhance the efficiency of UCNPs, provided that the quantum yield remains stable, since a higher doping concentration can facilitate greater light absorption.

ii. To address the inadequate electrical conductivity of conventional UCNPs, it is essential to develop a more conductive host matrix that exhibits high-efficiency up-conversion properties. Semiconductors commonly utilized in quantum dots present a promising option. Recent studies have suggested semiconductor nanocrystals, such as ZnS, as potential UC host materials. However, there remains ongoing debate regarding the positioning of lanthanide ions within these nanocrystals, specifically whether they predominantly reside on the outermost layer due to significant ionic size discrepancies between the host and dopant ions, or if they are uniformly integrated within the host lattice. Consequently, additional research is necessary to enhance the electrical conductivity of UCNPs, thereby improving solar cell performance.

iii. The absorption wavelength of up-conversion nanoparticles (UCNPs) is influenced by the lanthanide ions that are doped into them. It is important to note that altering the absorption wavelength of a specific lanthanide ion is not feasible.

However, if the lanthanide ions do not directly contribute to the absorption characteristics but instead facilitate the up-conversion of absorbed light, the situation changes. The material MS: Eu, Sm (where M = Ca, Sr) has been recognized as an electron trapping medium since the 1980s. The host matrix MS is capable of absorbing UV and blue light, leading to a down-conversion process that benefits the efficiency of solar cells by storing absorbed energy in the electron trapping levels created by the doped lanthanide ions. The electrons that are trapped in these deep energy levels (approximately 1 eV) cannot be thermally excited to higher states at room temperature. However, when exposed to infrared light, these electrons can return to their original state and recombine with holes, resulting in the emission of visible light and the generation of near-infrared up-conversion luminescence. Notably, the NIR excitation light can span a wide spectral range (800–1600 nm) at low power densities (mW), aligning well with natural sunlight.

Considerable advancements have been made in the area of up-conversion (UC) for photovoltaic applications, encompassing both theoretical studies and experimental successes. Nevertheless, the efficiency of the UC materials currently under investigation remains insufficient to enhance the practical performance of solar cells under standard solar irradiation conditions. There is a pressing need for further research to create and refine UC materials with higher efficiency. Additionally, it is essential to develop more advanced engineering strategies regarding the architecture of solar cells to fully leverage the potential of UC. Given the recent enhancements in the UC efficiency of lanthanide-doped phosphors, the future of this technology appears promising.

References

1. Z. Shi, Y. Duan, X. Zhu, Q. Wang, D. Li, K. Hu, W. Feng, F. Li, C. Xu, Dual functional NaYF4:Yb^{3+}, Er^{3+}@$NaYF_4$:Yb^{3+}, Nd^{3+} core-shell nanoparticles for cell temperature sensing and imaging. Nanotechnology **29**, 940019–940023 (2018)
2. A. Pilch, C. Wuerth, M. Kaiser, D. Wawrzynczyk, M. Kurnatowska, S. Arabasz, K. Prorok, M. Samoc, W. Strek, U. Resch-Genger, A. Bednarkiewicz, Shaping luminescent properties of Yb^{3+} and Ho^{3+} co-doped upconverting core-shell beta- $NaYF_4$ nanoparticles by dopant distribution and dpacing. Small **13**, 1701635–1701647 (2017)
3. Q. Wu, B. Huang, X. Peng, S. He, Q. Zhan, Non-bleaching fluorescence emission difference microscopy using single 808-nm laser excited red upconversion emission. Opt. Express **25**, 30885–30894 (2017)
4. Q. Wu, H. Chen, A. Fang, X. Wu, M. Liu, H. Li, Y. Zhang, S. Yao, Universal multifunctional nanoplatform sased on target-induced in situ promoting Au seeds growth to quench fluorescence of upconversion nanoparticles. ACS Sens. **2**, 1805–1813 (2017)
5. B. Ling, Y. Ma, H. Chen, L. Wang, A SPR aptamer sensor for mercury based on AuNPs@$NaYF_4$:Yb, Tm, Gd upconversion luminescent nanoparticles. Anal. Method. UK **9**, 6032–6037 (2017)
6. Z. Farka, M.J. Mickert, A. Hlavacek, P. Skladal, H.H. Gorris, Single molecule upconversion-linked immunosorbent assay with extended dynamic range for the sensitive detection of diagnostic biomarkers. Anal. Chem. **89**, 11825–11830 (2017)

7. J. Yao, M. Yang, Y. Duan, Chemistry, biology, and medicine of fluorescent nanomaterials and related systems: new insights into biosensing, bioimaging, genomics, diagnostics, and therapy. Chem. Rev. **114**, 6130–6178 (2014)
8. F. Auzel, Upconversion and anti-Stokes processes with f and d ions in solids. Chem. Rev. **104**, 139–173 (2004)
9. F. Wang, X. Liu, Recent advances in the chemistry of lanthanide-doped upconversion nanocrystals. Chem. Soc. Rev. **38**, 976–989 (2009)
10. D.R. Gamelin, H.U. Gudel, *Upconversion Processes in Transition Metal and Rare Earth Metal Systems*, vol. 214 (Springer, Berlin, Heidelberg, 2001), pp.1–56
11. M. Oheim, D.J. Michael, M. Geisbauer, D. Madsen, R.H. Chow, Principles of two photon excitation fluorescence microscopy and other nonlinear imaging approaches. Adv. Drug Deliv. Rev. **58**, 788–808 (2006)
12. L.R. Dalton, A.W. Harper, R. Ghosn, W.H. Steier, M. Ziari, H. Fetterman, Y. Shi, R.V. Mustacich, A.K.Y. Jen, K.J. Shea, Synthesis and processing of improved organic second-order nonlinear optical materials for applications in photonics. Chem. Mater. **7**, 1060–1081 (1995)
13. L. Cheng, C. Wang, Z. Liu, Upconversion nanoparticles and their composite nanostructures for biomedical imaging and cancer therapy. Nanoscale **5**, 23–37 (2013)
14. H. Wang, W. Li, K. Dong, P. Zhang, Y. Sang, J. Ren, X. Qu, Remote and reversible control of in vivo bacteria clustering by NIR-driven multivalent upconverting nanosystems. Biomaterials **217**, 119310–119316 (2019)
15. Z. Yi, X. Li, Z. Xue, X. Liang, W. Lu, H. Peng, H. Liu, S. Zeng, J. Hao, Remarkable NIR enhancement of multifunctional nanoprobes for in vivo trimodal bioimaging and upconversion optical/T-2-weighted MRI-guided small tumor diagnosis. Adv. Funct. Mater. **25**, 7119–7129 (2015)
16. D. Wang, L. Zhu, Y. Pu, J. Wang, J. Chen, L. Dai, Transferrin-coated magnetic upconversion nanoparticles for efficient photodynamic therapy with near-infrared irradiation and luminescence bioimaging. Nanoscale **9**, 11214–11221 (2017)
17. X. Ge, L. Dong, L. Sun, Z. Song, R. Wei, L. Shi, H. Chen, New nanoplatforms based on UCNPs linking with polyhedral oligomeric silsesquioxane (POSS) for multimodal bioimaging. Nanoscale **7**, 7206–7215 (2015)
18. M. Chan, S. Chen, C. Chen, Y. Chan, R.J. Lin, D.P. Tsai, M. Hsiao, R. Chung, X. Chen, R. Liu, Single 808 nm laser treatment comprising photothermal and photodynamic therapies by using gold nanorods hybrid upconversion particles. J. Phys. Chem. C **122**, 2402–2412 (2018)
19. Q. Zhang, S. Yue, H. Sun, X. Wang, X. Hao, S. An, Nondestructive readout in Er/Yb co-doped $Na_{0.5}Bi_{2.5}Nb_2O_9$ for optical data storage. J. Appl. Phys. **123**, 245102 (2017)
20. X. Ai, J. Aw, B. Xing, Upconversion nanoparticles for bioimaging and data storage. Chem. Soc. Rev. **39**, 229–240 (2016)
21. M. Wuttig, N. Yamada, Phase-change materials for rewriteable data storage. Nat. Mater. **6**, 824–832 (2007)
22. H.Q. Wang, M. Batentschuk, A. Osvet, L. Pinna, C.J. Brabec, Rare-earth ion doped upconversion materials for photovoltaic applications. Adv. Mater. **23**, 2675–2680 (2011)
23. X.M. Li, F. Zhang, D.Y. Zhao, Highly efficient lanthanide upconverting nanomaterials: progresses and challenges. Nano Today **8**, 643–676 (2013)
24. B.S. Richards, Luminescent layers for enhanced silicon solar cell performance: downconversion. Sol. Energy Mater. Sol. Cells **90**, 1189–1207 (2006)
25. T. Trupke, M.A. Green, P. Wurfel, Improving solar cell efficiencies by down-conversion of high-energy photons. J. Appl. Phys. **92**, 1668–1674 (2002)
26. E. Klampaftis, D. Ross, K.R. McIntosh, B.S. Richards, Enhancing the performance of solar cells via luminescent down-shifting of the incident spectrum: a review. Sol. Energy Mater. Sol. Cells **93**, 1182–1194 (2009)
27. R.T. Wegh, H. Donker, K.D. Oskam, A. Meijerink, Visible quantum cutting in $LiGdF_4$: Eu^{3+} through downconversion. Science **283**, 663–666 (1999)
28. C. Struempel, M. McCann, G. Beaucarne, V. Arkhipov, A. Slaoui, V. Svrcek, C. del Canizo, I. Tobias, Modifying the solar spectrum to enhance silicon solar cell efficiency–an overview of available materials. Sol. Energy Mater. Sol. Cells **91**, 238–249 (2007)

29. A. Shalav, B.S. Richards, M.A. Green, Luminescent layers for enhanced silicon solar cell performance: up-conversion. Sol. Energy Mater. Sol. Cells **91**, 829–842 (2007)
30. A.G. Imenes, D.R. Mills, Spectral beam splitting technology for increased conversion efficiency in solar concentrating systems: a review. Sol. Energy Mater. Sol. Cells **84**, 19–69 (2004)
31. M. Wolf, Limitations and possibilities for improvement of photovoltaic solar energy converters.1. considerations for Earth's surface operation. in *Proceedings of the Institute of Radio Engineers,* vol. 48 (1960), 1246–1263
32. K.W.J. Barnham, G. Duggan, A new approach to high-efficiency multi-band-gap solar-cells. J. Appl. Phys. **67**, 3490–3493 (1990)
33. M.J. Keevers, M.A. Green, Efficiency improvements of silicon solar-cells by the impurity photovoltaic effect. J. Appl. Phys. **75**, 4022–4031 (1994)
34. G.B. Shan, G.P. Demopoulos, Near-infrared sunlight harvesting in dye-sensitized solar cells via the insertion of an upconverter-TiO2 nanocomposite layer. Adv. Mater. **22**, 4373–4377 (2010)
35. L.T. Su, S.K. Karuturi, J.S. Luo, L.J. Liu, X.F. Liu, J. Guo, T.C. Sum, R.R. Deng, H.J. Fan, X.G. Liu, A.I.Y. Tok, Photon upconversion in hetero-nanostructured photoanodes for enhanced near-infrared light harvesting. Adv. Mater. **25**, 1603–1607 (2013)
36. P. Gibart, F. Auzel, J.C.G.J.C. Guillaume, K.Z.K. Zahraman, Below band-gap IR response of substrate-free GaAs solar cells using two-photon up-conversion. Jpn. J. Appl. Phys. **35**(8R), 4401 (1996)
37. B.S. Richards, A. Shalav, Enhancing the near-infrared spectral response of silicon optoelectronic devices via up-conversion. IEEE. T. Electron Dev. **54**, 2679–2684 (2007)
38. A. Shalav, B.S. Richards, T. Trupke, K.W. Kramer, H.U. Gudel, Application of $NaYF_4$: Er^{3+} up-converting phosphors for enhanced near-infrared silicon solar cell response. Appl. Phys. Lett. **86**, 013505 (2005)
39. D.Q. Chen, L. Lei, A.P. Yang, Z.X. Wang, Y.S. Wang, Ultra-broadband near-infrared excitable upconversion core/shell nanocrystals. Chem. Commun. **48**, 5898–5900 (2012)
40. Z.Q. Li, X.D. Li, Q.Q. Liu, X.H. Chen, Z. Sun, C. Liu, X.J. Ye, S.M. Huang, Core/shell structured $NaYF_4$:Yb^{3+}/Er^{3+}/Gd^{+3} nanorods with Au nanoparticles or shells for flexible amorphous silicon solar cells. Nanotechnology **23**, 025402 (2012)